New Perspectives on Seed Germination

Edited by Ertan Yildirim, Sıtkı Ermiş and Eren Özden

Published in London, United Kingdom

New Perspectives on Seed Germination
http://dx.doi.org/10.5772/intechopen.1003321
Edited by Ertan Yildirim, Sıtkı Ermiş and Eren Özden

Contributors
Andrea Mastinu, Anirban Roy, Bahman Fazeli-Nasab, Beata Dedicova, Berna Baş, Boris Nemzer, Debashis Paul, Eren Özden, Ertan Yildirim, Faquir C. Garg, Gabriel Guillén Solís, Gipumi Torres-Abe, Gugu Cynthia Maseko, Hanieh Sadeghi, Hao Feng, İbrahim Demir, Juliana de Souza Rodrigues, Junzhou Ding, Karina Alejandra Balderas-Ruíz, Kingsley K. Kanu, M. Albertus Jan-Willem Vos, Masoumeh Asadi Aghbolaghi, Ramin Piri, Riyaz Sayyed, Samantha J. Bowen, Sourav Garai, Sudhir Kumar, Suman Dutta, Sıtkı Ermiş, Thobile Perseverance Mkhwanazi, Timothy L. Grey, Vincent C. Kanu, Zahra Shakeri Chaleshtori, Zakheleni Palane Dube, Ángel Arturo Guevara-García

First published in London, United Kingdom, 2025 by IntechOpen
IntechOpen is the global imprint of INTECHOPEN LIMITED, registered in England and Wales, registration number: 11086078, 167-169 Great Portland Street, London, W1W 5PF, United Kingdom

For EU product safety concerns: IN TECH d.o.o., Prolaz Marije Krucifikse Kozulić 3, 51000 Rijeka, Croatia, info@intechopen.com or visit our website at intechopen.com.

British Library Cataloguing-in-Publication Data
A catalogue record for this book is available from the British Library

New Perspectives on Seed Germination
Edited by Ertan Yildirim, Sıtkı Ermiş and Eren Özden
p. cm.

This title is part of the Physiology Book Series, Volume 29
Topic: Plant Physiology
Series Editor: Tomasz Brzozowski
Topic Editor: Jen-Tsung Chen

Print ISBN 978-0-85466-119-0
Online ISBN 978-0-85466-118-3
eBook (PDF) ISBN 978-0-85466-120-6
ISSN 2631-8261

IntechOpen Book Series
Physiology
Volume 29

Aims and Scope of the Series

Modern physiology requires a comprehensive understanding of the integration of tissues and organs throughout the mammalian body, including the cooperation between structure and function at the cellular and molecular levels governed by gene and protein expression. While a daunting task, learning is facilitated by identifying common and effective signaling pathways mediated by a variety of factors employed by nature to preserve and sustain homeostatic life. As a leading example, the cellular interaction between intracellular concentration of Ca+2 increases, and changes in plasma membrane potential is integral for coordinating blood flow, governing the exocytosis of neurotransmitters, and modulating gene expression and cell effector secretory functions. Furthermore, in this manner, understanding the systemic interaction between the cardiovascular and nervous systems has become more important than ever as human populations' life prolongation, aging and mechanisms of cellular oxidative signaling are utilised for sustaining life. Altogether, physiological research enables our identification of distinct and precise points of transition from health to the development of multimorbidity throughout the inevitable aging disorders (e.g., diabetes, hypertension, chronic kidney disease, heart failure, peptic ulcer, inflammatory bowel disease, age-related macular degeneration, cancer). With consideration of all organ systems (e.g., brain, heart, lung, gut, skeletal and smooth muscle, liver, pancreas, kidney, eye) and the interactions thereof, this Physiology Series will address the goals of resolving (1) Aging physiology and chronic disease progression (2) Examination of key cellular pathways as they relate to calcium, oxidative stress, and electrical signaling, and (3) How changes in plasma membrane produced by lipid peroxidation products can affect aging physiology, covering new research in the area of cell, human, plant and animal physiology.

Meet the Series Editor

Prof. Dr. Thomas Brzozowski works as a professor of Human Physiology and is currently a Chairman at the Department of Physiology and is V-Dean of the Medical Faculty at Jagiellonian University Medical College, Cracow, Poland. His primary area of interest is physiology and pathophysiology of the gastrointestinal (GI) tract, with a major focus on the mechanism of GI mucosal defense, protection, and ulcer healing. He was a postdoctoral NIH fellow at the University of California and the Gastroenterology VA Medical Center, Irvine, Long Beach, CA, USA, and at the Gastroenterology Clinics Erlangen-Nuremberg and Munster in Germany. He has published 290 original articles in some of the most prestigious scientific journals and seven book chapters on the pathophysiology of the GI tract, gastroprotection, ulcer healing, drug therapy of peptic ulcers, hormonal regulation of the gut, and inflammatory bowel disease.

Meet the Volume Editors

Ertan Yildirim is a Full Professor in the Department of Horticulture, Agriculture Faculty at Atatürk University in Turkey. His research focuses on teaching and researching topics such as vegetable growing, breeding, greenhouse management, seed germination and physiology, organic agriculture, and stress physiology. He is involved in many projects. He has been a visiting scientist at Cornell University. He served as Director of the Vocational Training School and Director of the Graduate School of Natural and Applied Sciences. Member of national and international social and governmental organizations. He has a registered garden cress cultivar and has applied to register for two bean and two pinto bean cultivars. Professor Yildirim has published over 350 articles. He has served as a reviewer for high-impact-factor journals. He has attended many international courses, congresses and symposiums.

Sıtkı Ermis is an Associate Professor in the Department of Horticulture at the Agriculture Faculty at Eskişehir Osmangazi University in Türkiye. His research focuses on teaching and researching vegetable growing, DUS testing on vegetables, seed technology, and seed storage physiology. He is involved in many projects. He has conducted research on plant molecular genetics at the University of Ciheam Crete. Before joining the University, he had been working as a DUS Expert for 16 years, registering and conducting more than 4,000 experiments with various vegetable varieties. Dr. Ermiş has published over 50 articles. He has performed review work for publications with high impact factors. He has attended several training sessions, international courses, congresses and symposiums.

Eren Özden is an Associate Professor in the Department of Horticulture, Agriculture Faculty at Iğdır University in Türkiye, and the Department of Horticulture and Agronomy, Faculty of Agriculture, Kyrgyz-Turkish Manas University in Kyrgyzstan. His research is focused on teaching and researching seed germination and physiology, seed technology, and seed stress physiology. He is involved in many projects. He served as Director of the Agricultural Application and Research Centre at Iğdır University. Member of national and international social and governmental organizations. Dr. Özden has published over 60 articles. He has served as a reviewer for high-impact-factor journals. He has attended many international courses, congresses and symposiums.

Contents

Preface

Germinable seeds are needed for the continuity of viability and reproduction in plants. The seed germination mechanism remains shrouded in physiological and molecular mysteries, concealing its secrets. Germination is a complex event that yields different results across species, and the same seed lot can exhibit varying viability responses when exposed to different ecological conditions. In fact, in parallel with the efforts of scientists and the advancement of technology, new methods, models, and applications are being developed to increase germination rates in seed lots. The effects of these developed applications on the viability of plant species vary. Because the internal structure of each seed can produce variable results due to its unique physiological and molecular composition. Seed germination is so simple that it can be put into a tiny observation, but at the same time, it is so complex that it cannot be explained.

This book was prepared to reveal the morphological, physiological, and molecular aspects of seed germination in the light of current information. Seed technology applications directly related to germination include research on the physiological and molecular basis of germination and adaptation. It covers research on seed technology treatments, the physiological basis of germination, molecular mechanisms, and adaptation directly related to the germination process.

We believe this book presents a new level of material that will interest researchers, as well as advanced undergraduate students and others seeking a more comprehensive understanding of seed germination and its mechanisms.

Ertan Yildirim
Faculty of Agriculture,
Atatürk University,
Erzurum, Türkiye

Sıtkı Ermiş
Faculty of Agriculture,
Osmangazi University,
Eskişehir, Türkiye

Eren Özden
Faculty of Agriculture,
Kyrgyz-Turkish Manas University,
Bishkek, Kyrgyzstan

Chapter 1

Perspective Chapter: Insights into Seed Germination – Physiological and Environmental Mechanisms

Eren Özden, Sıtkı Ermiş, Ertan Yildirim and İbrahim Demir

Abstract

Seed germination represents a pivotal phase in the plant life cycle, marking the transition from dormancy to active growth. This complex process is governed by a network of physiological, biochemical, and molecular mechanisms that respond dynamically to both internal and external prompts. In this chapter, we provide a comprehensive overview of the regulatory pathways underlying seed germination, beginning with the critical role of water uptake and metabolic reactivation. We examine the intricate interplay of phytohormones like abscisic acid, gibberellins, auxins, cytokinins, and ethylene discussing their influence on gene expression, enzyme activity, and cellular development. Special emphasis is placed on the cross talk between hormonal signals and the roles of reactive oxygen species (ROS) and nitric oxide (NO) as signaling molecules. Environmental factors such as temperature, light, moisture, and nutrient availability are discussed for their modulatory effects on germination timing and success. The agricultural and ecological significance of understanding seed germination is also highlighted, particularly in the context of improving crop resilience and sustainability. Finally, we outline emerging research directions in molecular regulation and stress physiology that are poised to advance the field. This synthesis aims to bridge foundational knowledge with modern advancements to inform future strategies in seed technology and crop improvement.

Keywords: seed germination, water uptake, phytohormonal crosstalk, environmental regulation, seed physiology, stress adaptation

1. Introduction

Seed germination is one of the most critical stages of the plant life cycle and plays a fundamental role in ecosystem sustainability. This process involves a complex sequence of biochemical and physiological events during which the seed becomes metabolically active, and embryonic development resumes. In the early stages of plant development, seed germination not only ensures the regeneration of vegetation in natural environments but also directly impacts agricultural productivity. Therefore, understanding the fundamental mechanisms of seed germination physiology and the influence of environmental factors is an important research topic in agricultural sciences and biology.

 IntechOpen

Seed germination encompasses various biological processes that occur in succession, including water uptake (imbibition), enzymatic activation, cellular respiration, and embryonic cell division, as well as the molecular stages underlying these events. Water uptake depends on factors such as seed coat permeability, the water-holding capacity of seed structure, and environmental humidity levels [1]. During this stage, seed cells swell by absorbing water, leading to increased mechanical pressure on the cell walls. After imbibition, stored enzymes within the seed become active, metabolic processes accelerate, and embryonic cells rapidly divide. Each of these steps triggers dynamic biochemical changes essential for the progression of germination [2–5].

Environmental factors also significantly affect the germination process. Temperature, oxygen levels, water availability, and light are key determinants of the timing and rate of germination. For example, temperature fluctuations influence metabolic rates, while oxygen deficiency can trigger anaerobic processes, limiting embryonic development [6]. Light can act as a trigger for germination in some plants, while inhibiting it in others [7]. Understanding the impact of these factors on seed physiology provides profound insights into plant adaptation capabilities and enables the development of strategies to cope with environmental stresses such as climate change.

1.1 Bibliometric data of seed germination physiology

In the Web of Science system, a search using the keyword "Seed Germination Physiology" in the fields of Biology, Plant Science, Agronomy, Forestry, and Horticulture reveals 1217 sources from 1995 to nowadays. When a bibliometric analysis was conducted using the Bibliometrix package in R-studio, the areas studied are presented in **Figure 1**. As seen from bibliometric data, seed germination physiology and the factors affecting this process are a subject of intense interest. It was observed that germination physiology is closely related to temperature, abscisic acid, and plant cultivation. At the species level, Arabidopsis and tomato are the most frequently studied, while seed dormancy and plant metabolism are the most focused research groups. Prominent studies have highlighted determining the level of cellular degradation, water uptake, and the application of enzymes and hormones. Here, it is seen that especially among the ecological factors, temperature and the germination inhibitor ABA are taken into consideration.

Figure 1.
Bibliometric analysis of seed germination physiology.

Figure 2.
Heat map of bibliometric data seed germination physiology.

When a heat map of the Web of Science bibliometric data obtained with the keyword "Seed Germination Physiology" was created using the R-Studio program, it was noted that 6% of the 1217 studies focused on the themes of physiology, temperature, and germination; 5% on abscisic acid and growth; and 3% on Arabidopsis, dormancy, and stress (**Figure 2**).

When examining the Co-occurrence Network relationship, it was found that the interactions between plant growth-stress-abscisic acid, physiology-temperature-germination, dormancy-development dynamics, mean germination time-seed repair, and cellular degradation mechanisms-antioxidants stand out (**Figure 3**). Of course,

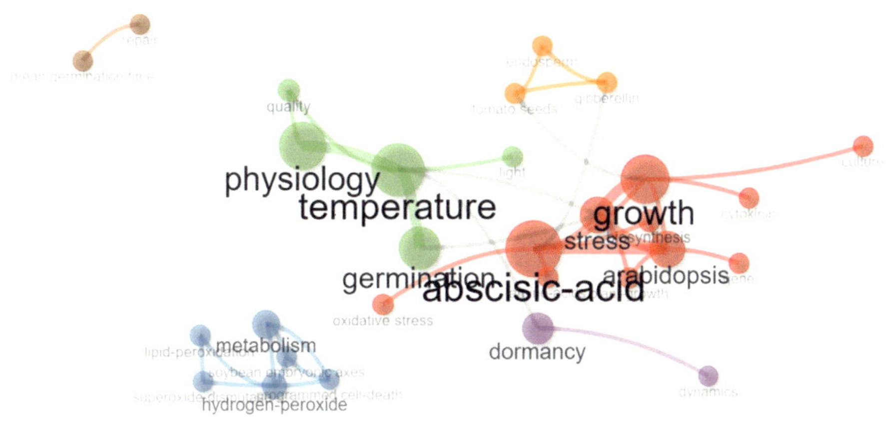

Figure 3.
Co-occurrence network of seed germination physiology.

studies have been carried out on the germination indicators mentioned above for many years, sometimes addressing the germination physiology of a single parameter, while others reveal multiple and complex relationships [6].

This study aims to analyze the mechanisms governing seed germination physiology and the effects of environmental factors on this process. It also seeks to evaluate the implications of germination physiology for plant production, shedding light on applications aimed at achieving agricultural sustainability and developing stress-tolerant plants. Understanding the mechanisms of seed germination in detail is of great importance for modern agricultural practices and genetic improvement studies. In this context, investigating the biochemical pathways regulating germination lays a strategic foundation for producing high-quality and efficient seeds. This chapter synthesizes current knowledge on the physiological, biochemical, and molecular mechanisms of seed germination, with emphasis on hormonal regulation, environmental interactions, and implications for crop improvement.

2. Seed germination process

Seed germination is one of the most critical stages of the plant life cycle and is influenced by various biological processes and environmental factors [8]. The seed germination process consists of three main stages:

1. Imbibition (Water Uptake) Phase

2. Lag (Waiting) Phase

3. Radicle (Root) Emergence

Specific external conditions must be met for seed germination to occur. These conditions include environmental factors such as water, temperature, oxygen, and light. Additionally, molecular and physiological changes, such as enzyme activation, energy production, protein synthesis, and hormone production, occur during the germination process.

Understanding the seed germination process is essential for agriculture and plant cultivation. Providing appropriate germination conditions plays a critical role in ensuring that seedlings grow strong and healthy. For instance, presoaking seeds in water before sowing can expedite the imbibition phase and increase germination rates [9].

2.1 Relationship between seed germination and water uptake

The first and most critical stage of germination is the uptake of water by the seed. When there is not enough water in the soil, imbibition does not occur, which prevents the metabolism from starting [2]. This is a very limiting situation for seed production in areas with drought conditions or for successful germination conditions after planting existing seeds. With global warming, serious problems may occur in seed germination in the future and varieties suitable for this situation should be developed or coating technologies that can meet the moisture requirement of the seed during the germination process should be developed.

Understanding the relationship between seed germination and water uptake requires considering both biological mechanisms and environmental effects:

2.1.1 Role of water uptake during germination phases

During the imbibition phase, water is absorbed by the seed, initiating the process known as germination (**Figure 4**). At this stage, water is rapidly taken up, causing the seed to swell and activating metabolic processes. This initial water uptake is critical for reviving cellular metabolism and triggering the action of growth hormones. Ermiş et al. [11] reported a high correlation between water uptake and germination/emergence in white coated french bean seeds.

In the second stage, known as the activation phase, water continues to be absorbed at a slower rate compared to the first stage. However, the amount of water uptake is sufficient to facilitate the biochemical reactions and cellular activities required for growth.

The third stage, radicle emergence, marks the final phase of germination, characterized by the breaking of the seed coat by the root tip (radicle). This process occurs through cell expansion and takes place when water uptake reaches a specific threshold [12, 13].

Water primarily enters the seed through specialized structures such as the hilum, micropyle, or lens. The micropyle is a region consisting solely of parenchyma cells and lacks the macrosclereid layer. The hilum contains well-developed tracheid bars and astrosclereids, along with both palisade and counter-palisade layers. It is chemically distinct and particularly rich in hydrophilic compounds like pectins. The lens, located near the hilum and opposite the micropyle, lacks an osteosclerotic layer [14].

The efficiency of water uptake through these entry points varies depending on the plant species. For instance, in seeds of bean [15], mungbean [16], cress [4], and soybean [12, 17], water primarily enters through the hilum, which is enriched with hydrophilic compounds that enhance water absorption. In other examples, water entry occurs *via* the micropyle in canola species [18] and through the lens in *Delonix regia* [19].

2.1.2 Seed coat properties and water permeability

The outermost covering layer of the seed, known as the seed coat (testa), develops from the ovule's outer integuments [20]. The seed coat protects the embryo and the endosperm from mechanical or environmental damage while maintaining seed

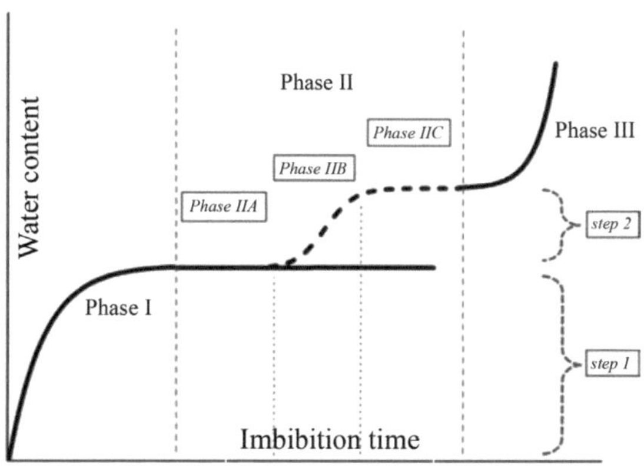

Figure 4.
Seed germination phases [10].

integrity [21]. Although it typically constitutes 10–20% of the seed's weight, the seed coat plays a crucial role in regulating water uptake and influencing the germination process [12].

Seeds with thick coats absorb water more slowly. This can protect against damage caused by rapid water uptake as well as may delay germination [22]. Studies on beans and soybeans have shown that seeds with dark-colored and thick coats exhibit more controlled and slower water absorption. In contrast, seeds with thinner or lighter-colored coats absorb water more quickly, which can speed up germination but increases the risk of imbibition damage [13, 23–25].

2.1.3 Seed size and water uptake dynamics

Smaller seeds (thousand of weight 10≤), due to their higher surface area-to-volume ratio, absorb water more rapidly. This increased contact area leads to faster germination as well as makes smaller seeds more susceptible to structural damage. A study on *Copaifera langsdorffi* reported that smaller seeds exhibited approximately twice the water permeability of larger seeds and had higher germination rates [26]. Larger seeds (thousand of weight 100≥); on the other hand, absorb water at a steadier rate due to their lower surface area-to-volume ratio, which can be advantageous in environments with fluctuating water availability [13, 23, 27].

Environmental conditions such as soil moisture significantly influence germination rates. In soils with low water potential, seeds absorb water more slowly, delaying germination but reducing the risk of water stress. Conversely, high soil moisture levels promote rapid water absorption and faster germination; however, uncontrolled water uptake can lead to imbibition damage [23].

Understanding factors affecting water uptake aids in selecting seeds with optimal germination traits. Breeding programs focus on seed coat properties and water absorption rates to develop seeds resistant to changing environmental conditions, particularly in regions with variable water availability [12, 25].

2.2 Relationship between seed germination and nutrient uptake

The relationship between seed germination and nutrient uptake is a fundamental dynamic of plant development. These processes are interconnected through a complex array of biochemical and physiological mechanisms that significantly affect plant growth performance and productivity [28].

When germination begins, the seed's internal reserves serve as the initial nutrient sources for the embryo. The diversity and quantity of these nutrients play a crucial role in embryo development. Specifically, macronutrients such as nitrogen, phosphorus, potassium, and magnesium are essential for cell division, DNA synthesis, protein production, and energy transfer [29, 30]. Micronutrients, including iron, zinc, and copper, support enzyme activation and hormone synthesis [31, 32]. Advanced technologies like X-ray photoemission spectroscopy (XPS) or scanning electron microscopy have been used to analyze ion distribution in seed coats, revealing new mechanisms involved in osmopriming that facilitate the diffusion of beneficial substances between the inner and outer seed coat layers during germination [33].

Several factors influence a seed's nutrient uptake capacity, including soil conditions and pH. Soil pH can either enhance or hinder the absorption of certain nutrients. During germination, favorable pH conditions can significantly improve the roots' ability to absorb nutrients [34]. Additionally, soil moisture is critical, as germination is initiated by water availability. Water not only drives the germination process but also aids in the transport and dissolution of nutrients [2].

Plant hormones play an essential role in both germination and nutrient uptake. For instance, gibberellins accelerate the conversion of starch and other reserve materials into energy during germination. Abscisic acid (ABA) suppresses germination, maintaining seed dormancy until the embryo matures. Auxins and cytokinins enhance root growth and nutrient absorption surface areas, promoting nutrient uptake [6, 35].

Certain seeds have hard outer coats or chemical inhibitors that make nutrient uptake challenging. Such seeds require mechanical or biochemical treatments to become germination-ready and begin nutrient absorption [36]. The germination capacity and root development of seeds vary among plant species. Some species are genetically adapted to low-nutrient environments, while others thrive in nutrient-rich conditions [37]. For example, some wild plants successfully germinate in low-nutrient soils [38].

The availability of nutrients directly influences germination rates. In nutrient-rich environments, seed germination occurs more rapidly, while in nutrient-deficient conditions, embryo development may slow or cease entirely.

3. Seed germination and environmental factors

Seed germination is the process of the seed becoming ready for growth, culminating in the emergence of the radicle and shoot. Germination is a highly sensitive and complex process controlled by environmental factors such as temperature, light, water, and oxygen. These conditions directly influence germination rate and duration, potentially affecting the plant's successful growth or causing delays.

3.1 Temperature

Temperature has a decisive impact on seed germination. Each seed species has an optimal temperature range; germination slows or ceases outside this range. For instance, temperate plants typically germinate within a 20–30°C range, while tropical plants prefer higher temperatures. Temperature affects metabolic processes by influencing enzyme activity. At very low temperatures, enzymes are insufficiently active, inhibiting germination, whereas high temperatures may cause protein denaturation and seed damage [36].

3.2 Water (moisture)

Recalcitrant seeds remain metabolically and physiologically active and are sensitive to desiccation, as they either lack or fail to activate the protective mechanisms required for acquiring desiccation tolerance, which allows orthodox seeds to survive at extremely low moisture levels [39].

Water is critical in the initial phase of germination, known as imbibition. When seeds come into contact with water, their cell walls swell, activating biochemical processes. A seed's water absorption capacity depends on the seed coat structure, with some seeds cracking their coat to facilitate water uptake. However, excessive water can create anaerobic conditions, leading to seed decay or reduced germination rates [40].

3.3 Oxygen

Energy required for embryo development during germination is supplied *via* oxygen-dependent cellular respiration. The availability of oxygen is crucial; oxygen-deficient environments can hinder embryo growth and halt germination. Water-saturated soils can reduce oxygen levels, suffocating the seed [41].

3.4 Light

The effect of light on germination varies between plant species. Some plants, such as certain forest species, germinate in the dark, while others require light to germinate. In light-sensitive plants, light acts as a signal to start or halt germination, ensuring seeds germinate at the right time and place in natural settings [42]. Lettuce (*Lactuca sativa* L.) seeds are phototropic and often necessitate light for germination. Light stimulates the phytochrome system, which is essential for initiating the germination process. This light dependency guarantees that lettuce seeds germinate in optimal environmental circumstances [2].

3.5 Soil

Soil structure directly influences water absorption and oxygen availability to seeds. Clay soils retain water better, whereas sandy soils provide more aeration. Most seeds prefer neutral or slightly acidic soils for germination, though some plant species thrive in alkaline soils. Soil structure is particularly important for preventing water-logging after rain, which could suffocate seeds [2].

3.6 Interaction of environmental factors

During germination, these factors interact with one another. For instance, maintaining a balance between water and oxygen is critical for healthy seed growth. Similarly, temperature fluctuations can alter the effects of water and oxygen within the seed. In agricultural production, achieving optimal environmental conditions is essential for successful germination rates [36, 40].

4. Seed germination and antioxidants

During seed germination, processes such as water uptake, cell expansion, and embryo activation may lead to oxidative stress. The uncontrolled increase in reactive oxygen species (ROS) levels can cause cellular damage, though moderate ROS levels soften endosperm cell walls and promote embryo growth [43].

In the germination process, both enzymatic (e.g., catalase, superoxide dismutase) and nonenzymatic (e.g., vitamins C and E) antioxidants play roles. These antioxidants neutralize oxidative stress and protect cellular structures through defense mechanisms [44]. They also prevent harmful lipid peroxidation and maintain cellular membrane integrity [45, 46].

4.1 Antioxidant activity during germination

Temperature, water availability, oxygen levels, and light significantly influence antioxidant activities during germination. Additionally, plant species and genetic differences result in varied antioxidant defense mechanisms against oxidative stress during germination. The seed's maturity and storage conditions also determine antioxidant activity effectiveness during germination.

To measure antioxidant activity, markers such as hydrogen peroxide (H_2O_2) and malondialdehyde (MDA) levels indicate cellular breakdown, while total antioxidant activity (DPPH, FRAP, ABTS) or specific antioxidant activities (e.g., SOD, CAT, APX, GPX, PPO) are evaluated [47–49].

Nitric oxide (NO) works alongside ROS to regulate oxidative balance and promote cell growth [50, 51]. NO also acts as a signaling molecule in breaking seed dormancy, especially by reducing abscisic acid (ABA) levels and initiating germination. Studies have shown that NO regulates hormone levels such as gibberellin (GA) and ABA, suppressing ABA's inhibitory effect while enhancing GA's growth-promoting role [52, 53].

Drought slows down seed germination and also leads to the accumulation of reactive oxygen species (ROS), which damage cell membranes. Drought-tolerant seeds can minimize this damage through their antioxidant enzyme systems [43].

Agricultural practices, such as seed coating and pre-treatment techniques, aim to enhance the antioxidant content of seeds. However, these interventions are not always effective, as stress formation and suppression involve many physiological and molecular parameters. Factors such as hormones, amino acids, sugars, and alkaloids influence antioxidant activity directly or indirectly.

5. Seed germination and hormones

The main plant hormones regulating the seed germination process include abscisic acid (ABA), gibberellins (GA), ethylene, cytokinins, and auxins. Each hormone plays a specific role and has distinct effects during germination. According to various studies and theories, ABA and GA are directly influential in the physiology of germination, while other hormones either support or inhibit germination through interactions with ABA or GA. These hormones regulate the complexity of seed germination, water uptake, and radicle protrusion *via* transcription factors involved in stress-response gene networks [6, 7, 32].

5.1 Abscisic acid (ABA)

ABA plays an inhibitory role in seed germination, helping seeds remain dormant. When ABA levels are high, seeds do not initiate germination and remain in a dormant state [54]. By acting as a delaying agent, ABA enables seeds to maintain dormancy for varying periods, thus ensuring their storability. ABA may also minimize at the end of the seed development stage since seeds get ready to germinate.

Under stress conditions such as salinity and drought, ABA levels increase to reduce water loss and slow metabolic activities, creating a survival mechanism for seeds under unfavorable conditions.

The balance between ABA and GA is critical during protrusion or germination. For germination to begin, ABA levels must decrease, while GA levels must rise. The ABA/GA ratio contributes to determining seed germination potential and rates [42, 54].

5.2 Gibberellins (GA)

Gibberellins are key hormones in seed germination. GA is synthesized in the seed embryo and initiates the germination process by loosening cell walls and promoting cell growth [55]. Increased GA levels are considered an indicator of germination initiation [56].

GA also stimulates the synthesis of enzymes like amylase, which hydrolyze stored nutrients such as starch in the seed's storage tissues into simple sugars, providing energy for the embryo [57].

5.3 Ethylene

Ethylene facilitates water uptake and supports root development during germination. Under stress conditions, ethylene production increases, triggering germination [58].

Ethylene suppresses the inhibitory effects of ABA, helping seeds overcome dormancy. It is believed to work alongside GA to promote germination. Ethylene also softens hard seed coats and enables embryo development, particularly in seeds under environmental stress [59].

Ethylene accelerates water uptake and cell division through cell wall expansion, aiding in radicle emergence and germination progression [60]. Under stress conditions such as salinity or drought, ethylene supports germination by maintaining ion balance and promoting embryo growth [61].

5.4 Cytokinins

Cytokinins promote cell division and support metabolic processes necessary for cellular growth during germination. They are assumed to play a role in both the initiation of germination and the formation of plant parts [62].

Cytokinins aid in the hydrolysis of proteins in seed storage tissues, providing amino acids essential for embryo development [63]. They also facilitate the mobilization of stored nutrients, enhancing the availability of carbohydrates and proteins for embryo growth [64].

Research suggests that cytokinins can sometimes reduce ABA's inhibitory effects, initiating germination [65, 66].

5.5 Auxins

Auxins promote root cell elongation and differentiation, which is critical for the healthy development of a seedling's root system [67, 68].

Auxins influence germination in response to environmental signals such as light and temperature. Increased auxin levels under favorable environmental conditions promote germination [69]. They can also support germination under stressful conditions like salinity and drought [70].

While low levels of auxins may not directly initiate germination, they create a favorable environment for early embryonic development, aiding in breaking dormancy [71].

5.6 Interactions among hormones

ABA-GA Balance: For germination to commence, ABA levels must decrease, and GA levels must increase. This balance depends on environmental factors and the seed's genetic makeup [42, 56].

Ethylene-GA Collaboration: Ethylene enhances GA's germination-promoting effects by suppressing ABA's influence [42, 59, 72].

Effects of Other Phytohormones: Cytokinins, auxins, salicylic acid, and brassinosteroids have been reported to indirectly promote germination, but their mechanisms—whether by stimulating GA or suppressing ABA—are not fully understood. Jasmonic acid, on the other hand, is suggested to inhibit germination, particularly under stress conditions [73].

Environmental Impact: Environmental conditions (temperature, humidity, and light) affect hormone levels, thereby either initiating or inhibiting germination. For example, low temperatures increase ABA levels, while high temperatures stimulate GA synthesis. Variable temperature conditions have been reported to correlate positively with auxin and salicylic acid levels, reducing ABA's effects and increasing germination rates [74].

6. Molecular mechanisms of seed germination

The molecular mechanisms of seed germination require the coordinated interaction of genetic, biochemical, and signaling pathways to ensure seeds germinate at the appropriate time.

6.1 Embryo activation

The seed germination process involves the regulation by multiple hormones, initiated through various gene expressions. The balance between gibberellins (GA) and abscisic acid (ABA) determines the breaking of seed dormancy and germination. GA promotes cell expansion and enzyme activation, whereas ABA inhibits this process [75].

Gibberellic acid binds to GA receptors in embryo cells, triggering the degradation of DELLA proteins. DELLA proteins act as suppressors of germination, and their breakdown initiates the expression of genes required for germination [76]. While GA induces DELLA degradation, transcription factors like SPATULA, associated with germination, become active. These factors regulate the expression of germination-initiating genes [77]. GA interacts with the GID1 (GA Insensitive Dwarf 1) receptor, facilitating DELLA degradation through the SCF (SKP-Cullin-F-box) complex [78].

Genes involved in GA biosynthesis and signaling, such as GA_3ox and $GA_{20}ox$, play critical roles in controlling germination [79]. The effects of GA occur through receptors in plant cells, activating cellular responses required for germination.

ABA, on the other hand, maintains dormancy and prevents germination. ABA signaling pathways also regulate seed responses to environmental stress. High ABA levels persist under conditions like drought and cold to maintain dormancy by suppressing metabolic activities [80].

Genes responsible for cytokinin biosynthesis (IPT, LOG) and their signaling pathways (AHK/AHP/ARR) are active during germination. These genes regulate cytokinin synthesis and effects [81].

Auxins regulate gene expression linked to growth responses in cells. This process occurs *via* AUX/IAA and ARF proteins [82]. Environmental interactions are also necessary for the expression of genes required for germination. For instance, photoreceptors (like phytochromes) detect light signals and initiate molecular responses essential for germination [83]. Germination requires optimal temperature and humidity. Environmental stresses alter gene expression and hormonal balances, affecting germination [54].

Changes in ABA and GA levels regulate dormancy release at the epigenetic level. Activation of certain genes during germination is controlled by epigenetic mechanisms such as histone methylation and DNA methylation. Changes in histone acetylation and methylation influence whether genes associated with germination remain active or inactive. Histone acetyltransferases (HATs) and deacetylases (HDACs) can affect the transcriptional activity of genes during germination [84].

6.2 Storage protein and starch degradation

Genes responsible for mobilizing storage proteins and starches in seeds become active during germination. For example, LEA (Late Embryogenesis Abundant) genes and α-amylase genes play crucial roles during this process [3].

The most common storage proteins in seeds—globulins (e.g., vicilin, legumin), albumins, prolamins, and glutelins—are converted into amino acids during germination. These amino acids are then used for new protein synthesis, cell growth, and energy metabolism [85]. GA induces the degradation of storage proteins by activating the aleurone layer in endosperm cells to increase protease synthesis [86]. ABA, acting antagonistically to GA, suppresses protease production and slows the degradation of storage proteins [87]. Grudkowska and Zagdanska ([88] reported that genes encoding Cathepsin-like proteases and Cys proteases exhibit increased expression during germination.

The breakdown of storage starch during germination is also critical. Typically, α-amylase synthesized in the aleurone layer of endosperm cells converts starch molecules into sugars like maltose and glucose, providing energy for embryo development [57]. Genes encoding α-amylase, such as Amy1, Amy2, and Amy3, are part of multigene families and respond to environmental and hormonal signals [89]. Furthermore, Özden et al. [6] reported that endo-β-mannanase enzyme activity and MAN2 gene expression increase with seed maturation and germination rates, likely hydrolyzing galactomannans.

7. Conclusion

Seed germination is a highly orchestrated developmental event involving hormonal regulation, enzymatic activation, molecular signaling, and environmental responsiveness. The dynamic balance between growth-promoting and inhibitory signals. These includes particularly the antagonism between gibberellins and abscisic acid. It plays a central role in determining germination success. External factors such as temperature, moisture, light, and nutrient availability further modulate these internal processes, ensuring that germination occurs under favorable conditions.

Recent advances in molecular biology, including transcriptomics, hormone signaling networks, and epigenetic modifications, have greatly enhanced our understanding of how seeds perceive and respond to both intrinsic and extrinsic cues. Moreover, the role of antioxidants, reactive oxygen species, and nitric oxide in fine-tuning germination has opened new frontiers in stress physiology.

Understanding these complex germination mechanisms has far-reaching implications in agriculture and ecosystem sustainability. It informs breeding programs for stress-tolerant crops, seed priming technologies, and conservation strategies for native and endangered species. Future research should focus on integrating omics tools, gene editing technologies, and systems biology approaches to unravel the remaining mysteries of seed germination and to translate basic insights into applied innovations.

Both now and in the future, scientists will continue to devote considerable effort to understanding the mechanisms of seed germination, and this process is expected to become increasingly well understood. In this regard, fully elucidating the roles of hormones, enzymes, sugars, and amino acids in seed physiology, and supporting these findings with molecular-level studies, is of critical importance. This knowledge is anticipated to contribute to the development of stronger, more productive, and climate-resilient cultivars in the coming years.

Author details

Eren Özden[1,2]*, Sıtkı Ermiş[3], Ertan Yildirim[4] and İbrahim Demir[5]

1 Faculty of Agriculture, Department of Horticulture and Agronomy, Kyrgyz-Turkish Manas University, Bishkek, Kyrgyz Republic

2 Faculty of Agriculture, Department of Horticulture, Iğdır University, Iğdır, Türkiye

3 Faculty of Agriculture, Department of Horticulture, Eskişehir Osmangazi University, Eskişehir, Türkiye

4 Faculty of Agriculture, Department of Horticulture, Atatürk University, Erzurum, Türkiye

5 Faculty of Agriculture, Department of Horticulture, Ankara University, Ankara, Türkiye

*Address all correspondence to: eren.ozden@manas.edu.kg

References

[1] Copeland LO, McDonald MF. Principles of Seed Science and Technology. Berlin/Heidelberg, Germany: Springer Science & Business Media; 2012

[2] Bewley JD, Black M. Seeds: Physiology of Development and Germination. Berlin/Heidelberg, Germany: Springer Science and Business Media; 1994

[3] Nonogaki H, Bassel G, Bewley J. Germination—Still a mystery. Plant Science. 2010;**179**:574-581

[4] Özden E. Imbibition-induced changes in cell membrane on germination and some physiological parameters in aged cress (*Lepidium sativum* L.) seeds. Turkish Journal of Agriculture and Forestry. 2022;**46**(4):453-465. DOI: 10.55730/1300-011X.3017

[5] Ranal MA, Santana DG. Como e Por Que Medir o Processo de Germinação? Brazilian Journal of Botany. 2006;**29**:1-11

[6] Özden E, Türkhan A, Gözen V, Aydın A. Physiological and molecular responses of immature cucumber seed under after-ripening treatments. Scientia Horticulturae. 2024;**337**:113487. DOI: 10.1016/j.scienta.2024.113487

[7] Khan AA, Samimy C. Hormones in Relation to Primary and Secondary Seed Dormancy. The Phsiology and Biochemistry of Seed Development, Doprmancy and Germination. USA: Elseiver Bioemedical Press; 1982. pp. 216-228

[8] Demir I, Kadioglu N. The effects of oxygen availability in the seed container during storage on seed germination in tomato, onion, cabbage and marrow seeds. Horticultural Studies. 2023;**41**(1):1-5. DOI: 10.16882/hortis.1403403

[9] Farooq M, Basra SMA, Afzal I, Khaliq A. Optimization of hydropriming techniques for rice seed invigoration. Seed Science and Technology. 2006;**34**(2):507-512

[10] Toorop PE. Nitrate controls testa rupture and water content during release of physiological dormancy in seeds of *Sisymbrium officinale* (L.) Scop. Seed Science Research. 2015;**25**:138-146

[11] Ermiş S, Öktem G, Demir İ. Seed water uptake relates germination and emergence potential of white coated French bean cultivars. Black Sea Journal of Agriculture. 2022;**5**(2):100-103

[12] Upretee P, Bandara MS, Tanino KK. The role of seed characteristics on water uptake preceding germination. Seeds. 2024;**3**:559-574. DOI: 10.3390/seeds3040038

[13] Vidak M, Lazarević B, Javornik T, Šatović Z, Carović-Stanko K. Seed water absorption, germination, emergence and seedling phenotypic characterization of the common bean landraces differing in seed size and color. Seeds. 2022;**1**:324-339. DOI: 10.3390/seeds1040027

[14] Zablatzká L, Balarynová J, Klˇcová B, Kopecký P, Smýkal P. Anatomy and Histochemistry of seed coat development of wild *Pisum sativum* Subsp. elatius (M. Bieb.) Asch. Et Graebn. And domesticated pea (*Pisum sativum* Subsp. sativum L.). International Journal of Molecular Sciences. 2021;**22**:4602

[15] Korban SS, Coyne DP, John L. Weihing. Rate of water uptake and sites of water entry in seeds of different

cultivars of dry bean. Horticultural Science. 1981;**16**:545-546

[16] Miano AC, Pereira JDC, Castanha N, Júnior MDDM, Augusto PED. Enhancing mung bean hydration using the ultrasound technology: Description of mechanisms and impact on its germination and Main components. Scientific Reports. 2016;**6**:38996

[17] Pietrzak LN, Frégeau-Reid J, Chatson B, Blackwell B. Observations on water distribution in soybean seed during hydration processes using nuclear magnetic resonance imaging. Canadian Journal of Plant Science. 2011;**82**:513-519

[18] De Paula AS, Delgado CML, Paulilo MTS, Santos M. Breaking physical dormancy of Cassia leptophylla and Senna macranthera (Fabaceae: Caesalpinioideae) seeds: Water absorption and alternating temperatures. Seed Science Research. 2012;**22**:259-267

[19] Jaganathan GK, Wu GR, Han YY, Liu BL. Role of the lens in controlling physical dormancy break and germination of *Delonix regia* (Fabaceae: Caesalpinioideae). Plant Biology. 2017;**19**:53-60

[20] Lemes EM, Rodrigues HC, Catão M. Soybean seed coat cracks and green seeds—Predisposing conditions, identification and management. Seeds. 2024;**3**:133-148

[21] Moïse JA, Han S, Gudynaitę-Savitch L, et al. Seed coats: Structure, development, composition, and biotechnology. In Vitro Cellular & Developmental Biology - Plant. 2005;**41**:620-644. DOI: 10.1079/IVP2005686

[22] Nakayama N, Komatsu S. Water uptake by seeds in yellow-seeded soybean (Glycine max (L.) Merrill) cultivars with contrasting imbibition behaviors. Plant Production Science. 2008;**11**:415-422

[23] Bradford KJ. A water relations analysis of seed germination rates. Plant Physiology. 1990;**94**(2):840-849. DOI: 10.1104/pp.94.2.840

[24] Chachalis D, Smith ML. Seed coat regulation of water uptake during imbibition in soybeans (*Glycine max* (L.) Merr.). Seed Science and Technology. 2001;**29**:401-412

[25] Ermis S, Özden E, Yıldırım E. Seeds of resilience: Physiology and mechanisms of Hardseededness. In: Seed Biology - New Advances. London, UK: IntechOpen; 2024. DOI: 10.5772/intechopen.1003847

[26] Souza ML, Fagundes M. Seed size as key factor in germination and seedling development of *Copaifera langsdorffii* (Fabaceae). American Journal of Plant Sciences. 2014;**2014**:2566-2573

[27] Taylor AG. Seed storage, germination, quality, and enhancements. In: The Physiology of Vegetable Crops. Wallingford, UK: CABI; 2020. pp. 1-30

[28] Martínez-Ballesta MC, Egea-Gilabert C, Conesa E, Ochoa J, Vicente MJ, Franco JA, et al. The importance of ion homeostasis and nutrient status in seed development and germination. Agronomy. 2020;**10**:504. DOI: 10.3390/agronomy10040504

[29] Eckert C, Offenborn JN, Heinz T, Armarego-Marriott T, Schultke S, Zhang C, et al. The vacuolar calcium sensors CBL2 and CBL3 affect seed size and embryonic development in Arabidopsis thaliana. The Plant Journal. 2014;**78**:146-156

[30] Wen D, Xu H, Xie L, He H, Hou H, Zhang C. A loose endosperm structure of

wheat seed produced under low nitrogen level promotes early germination by accelerating water uptake. Scientific Reports. 2017;7:1-11

[31] Guo J, Wang J, Xi L, Huang WD, Liang J, Chen JG. RACK1 is a negative regulator of ABA responses in Arabidopsis. Journal of Experimental Botany. 2009;**60**:3819-3833

[32] Kim SG, Lee AK, Yoon HK, Park CM. A membrane-bound NAC transcription factor NTL8 regulates gibberellic acid-mediated salt signaling in Arabidopsis seed germination. The Plant Journal. 2009;**55**:77-88

[33] Gómez-Ramírez A, López-Santos C, Cantos M, García JL, Molina R, Cortino J, et al. Surface chemistry and germination improvement of quinoa seeds subjected to plasma activation. Scientific Reports. 2017;**7**:5924

[34] Havlin JL, Beaton JD, Tisdale SL, Nelson WL. Soil Fertility and Fertilizers; an Introduction to Nutrient Management. 6th ed. Upper Saddle River, NJ: Prentice Hall; 2014

[35] Finkelstein R, Reeves W, Ariizumi T, Steber C. Molecular aspects of seed dormancy. Annual Review of Plant Biology. 2008;**59**:387-415

[36] Baskin CC, Baskin JM. Seeds: Ecology, Biogeography, and Evolution of Dormancy and Germination. USA: Cambridge University Press; 2014

[37] Grime JP. Plant Strategies, Vegetation Processes, and Ecosystem Properties. USA: Wiley; 2001

[38] Smith SE, Read DJ. Mycorrhizal Symbiosis. New York: Academic Press; 2008. pp. 637-768. DOI: 10.1016/B978-0-12-370526-6.X5001-6

[39] Berjak P, Pammenter NW. Recalcitrance. In: Encyclopedia of Applied Plant Sciences. 2nd ed. Netherlands: Academic Press; 2017. pp. 532-539. DOI: 10.1016/B978-0-12-394807-6.00048-4

[40] Bewley JD, Bradford K, Hilhorst H, Nonogaki H. Seeds: Physiology of Development, Germination and Dormancy. 3rd ed. New York: Springer Science & Business Media; 2013. p. 392. DOI: 10.1007/978-1-4614-4693-4

[41] Bradford KJ. Water relations in seed germination. Seed Science Research. 1995;**5**(1):1-17

[42] Kucera B, Cohn MA, Leubner-Metzger G. Plant hormone interactions during seed dormancy release and germination. Seed Science Research. 2005;**15**(4):281-307

[43] Bailly C. Active oxygen species and antioxidants in seed biology. Seed Science Research. 2004;**14**(2):93-107

[44] Gill SS, Tuteja N. Reactive oxygen species and antioxidant machinery in abiotic stress tolerance in crop plants. Plant Physiology and Biochemistry. 2010;**48**(12):909-930

[45] Gokdas Z, Yildirim E, Gupta S, Demir I. A Karrikinolide stimulated seed germination of artificially aged marrow, cabbage and pepper seeds through repair of cell structure and enzyme activity. South African Journal of Botany. 2022;**151**(Part A):208-213. DOI: 10.1016/j.sajb.2022.09.049

[46] Shams M, Yildirim E, Khadivi A, Ekinci M, Muhie SH. Differences in seed germination, plant growth, and Proline content of some pepper (*Capsicum annuum* L.) genotypes under salinity stress. Journal of Crop

Health. 2024;**76**(1):1-8. DOI: 10.1007/s10343-023-00962-w

[47] Ekinci M, Yildirim E, Dursun A, Turan M. Mitigation of salt stress in lettuce (*Lactuca sativa* L. var. Crispa) by seed and foliar 24-epibrassinolide treatments. HortScience Horts. 2012;**47**(5):631-636. DOI: 10.21273/HORTSCI.47.5.631

[48] Oracz K et al. Hydrogen peroxide as a signal controlling seed dormancy and germination. New Phytologist. 2007;**175**(3):501-509

[49] Özden E, Light ME, Demir İ. Alternating temperatures increase germination and emergence in relation to endogenous hormones and enzyme activities in aubergine seeds. South African Journal of Botany. 2021;**139**:130-139. DOI: 10.1016/j.sajb.2021.02.015

[50] Ekinci M, Ors S, Yildirim E, et al. Determination of physiological indices and some antioxidant enzymes of chard exposed to nitric oxide under drought stress. Russian Journal of Plant Physiology. 2020;**67**:740-749. DOI: 10.1134/S1021443720040056

[51] Sarath G, Hou G, Baird LM, Mitchell RB. Reactive oxygen species, ABA and NO in seed germination. Plant Physiology and Biochemistry. 2007;**45**(3-4):209-214

[52] Bethke PC, Libourel IGL, Jones RL. Nitric oxide in seed dormancy and germination. Annual Review of Plant Biology. 2006;**57**:451-482

[53] Liu Y, Ye N, Liu R, Chen M, Zhang J. H2O2 mediates gibberellin-induced activation of the mitogen-activated protein kinase Cascade in wheat Aleurone cells. Plant Physiology. 2009;**149**(3):1499-1510

[54] Finch-Savage WE, Leubner-Metzger G. Seed dormancy and the control of germination. New Phytologist. 2006;**171**(3):501-523

[55] Yamaguchi S. Gibberellin metabolism and its regulation. Annual Review of Plant Biology. 2008;**59**:225-251

[56] Miransari M, Smith DL. Plant hormones and seed germination. Environmental and Experimental Botany. 2014;**99**:110-121

[57] Sun T, Gubler F. Molecular mechanisms of gibberellin signaling in plants. Annual Review of Plant Biology. 2004;**55**:197-223

[58] Kepczynski J, Kepczynska E. Ethylene in seeddormancy and germination. Physiologia Plantarum. 1997;**101**:720-726

[59] Matilla J, Matilla-Vázquez MA. Involvement of ethylene in seed physiology. Plant Science. 2008;**175**(1-2):87-97. DOI: 10.1016/j.plantsci.2008.01.014

[60] Linkies A, Müller K, Morris K, et al. Ethylene interacts with ABA, GA and ROS Signalling pathways during germination. Plant Journal. 2009;**58**(2):322-334

[61] Zhang H, Li W, Chen J, Yang Y, Zhang Z, Zhang H, et al. Transcriptional activator TSRF1 reversely regulates pathogen resistance and osmotic stress tolerance in tobacco. Plant Molecular Biology. 2007;**63**:63-71

[62] Kaminek M. Regulation of Cytokinin biosynthesis and degradation in higher plants. Physiologia Plantarum. 1992;**86**(4):603-608

[63] Mok DWS, Mok MC. Cytokinin metabolism and action. Annual Review

of Plant Physiology and Plant Molecular Biology. 2001;**52**:89-118

[64] Werner T et al. Role of Cytokinin in plant development. Journal of Experimental Botany. 2003;**54**(393):1-11

[65] Davies PJ. Plant Hormones: Biosynthesis, Signal Transduction, Action! Dordrecht, The Netherlands: Kluwer Academic Publishers; 2004

[66] Nishimura C et al. Cytokinins and Abscisic acid act antagonistically in the regulation of germination and post-germination growth. Plant & Cell Physiology. 2004;**45**(11):1647-1654

[67] Overvoorde P, Fukaki H, Beeckman T. Auxin control of root development. Cold Spring Harbor Perspectives in Biology. 2010;**2**(6):a001537

[68] Teale WD, Paponov IA, Palme K. Auxin in action: Signalling, transport and the control of plant growth and development. Nature Reviews Molecular Cell Biology. 2006;**7**(11):847-859

[69] Vandenbussche F, Pierik R, Millenaar FF, Voesenek LA, Straeten VD. Reaching out of the shade. Current Opinion in Plant Biology. 2005;**8**:462-468

[70] Zhao Y. Auxin biosynthesis and its role in plant development. Annual Review of Plant Biology. 2010;**61**:49-64

[71] Woodward AW, Bartel B. Auxin: Regulation, action, and interaction. Annals of Botany. 2005;**95**(5):707-735. DOI: 10.1093/aob/mci083

[72] Corbineau F, Xia Q, Bailly C, El-Maarouf-Bouteau H. Ethylene, a key factor in the regulation of seed dormancy and germination. Frontiers in Plant Science. 2014;**5**:539

[73] Pan J, Wang H, You Q, Cao R, Sun G, Yu D. Jasmonate-regulated seed germination and crosstalk with other phytohormones. Journal of Experimental Botany. 2023;**74**(4):1162-1175

[74] Çayan S, Sarıkamış G, Özmen CY, Kibar U, Özden E, Ergül A. The influence of exogenous gibberellic acid (GA_3) and 24-epibrassinolide (24-EpiBL) on seed germination and the expression of genes involved in GA and BR synthesis/signalling in pepper (*Capsicum annuum* L.). Acta Scientiarum Polonorum Hortorum Cultus. 2021;**20**(5):15-23. DOI: 10.24326/asphc.2021.5.2

[75] Holdsworth MJ, Bentsink L, Soppe WJJ. Molecular networks regulating Arabidopsis seed maturation, after-ripening, dormancy and germination. New Phytologist. 2008;**179**(1):33-54

[76] Penfield S et al. Temperature sensing is mediated by the Abscisic acid-dependent regulation of flowering locus C expression. Nature. 2006;**432**(7013):1034-1037

[77] Feng S et al. Gibberellin-regulated DELLA proteins are the master growth regulators in plants. Plant Journal. 2008;**54**(6):885-901

[78] Achard P, Genschik P. Releasing the brakes of plant growth: How GAs shutdown DELLA proteins. Trends in Plant Science. 2009;**14**(8):463-471

[79] Hedden P, Thomas SG. Gibberellin biosynthesis and its regulation. Biochemical Journal. 2012;**444**(1):11-25

[80] Finkelstein R et al. Abscisic acid signaling in seeds and seedlings. The Plant Cell. 2002;**14**(S1):S15-S45

[81] Sakakibara H. Cytokinins: Activity, biosynthesis, and translocation.

Annual Review of Plant Biology. 2006;**57**:431-449. DOI: 10.1146/annurev.arplant.57.032905.105231

[82] Guilfoyle TJ, Hagen G. Auxin response factors. Current Opinion in Plant Biology. 2007;**10**(5):453-460. DOI: 10.1016/j.pbi.2007.08.014. Epub 2007 Sep 27

[83] Oh E et al. Light and ABA signaling in seed germination. Trends in Plant Science. 2009;**14**(10):597-603

[84] Shu K et al. Abscisic acid and gibberellins antagonistically mediate plant development and abiotic stress responses. Frontiers in Plant Science. 2013;**4**:447

[85] Shewry PR et al. Plant Storage Proteins. The Plant Cell. 1995;7(7):945-956

[86] Bethke PC et al. Role of gibberellins in regulating protease gene expression in barley Aleurone layers. Journal of Experimental Botany. 2007;**58**(3):425-434

[87] Nambara E et al. Abscisic acid and the control of seed dormancy and germination. Nature Reviews Molecular Cell Biology. 2010;**11**(8):611-622

[88] Grudkowska M, Zagdanska B. Hydrolytic enzymes in the programmed cell death of plants. Acta Biochimica Polonica. 2004;**51**(3):621-635

[89] Loreti E et al. Glucose and disaccharide-specific induction of α-amylase genes in barley embryos. Plant Physiology. 2000;**123**(3):939-948

Chapter 2

Germination of Lesser-Known Species of Acacia and Other Caesalpinioideae

M. Albertus Jan-Willem Vos

Abstract

The subfamily Caesalpinioideae contains 5096 accepted species and 519 unplaced species according to the World Flora Online (WFO) Plant List. For the past 10 years, we have experimented with 585 taxa of Acacias and other Caesalpinioideae. These seeds have very similar germination profiles. While there has been extensive research, especially in Australia, we have only found 173 research papers for 319 taxa. Upon the time of writing, we have done 1,754 experiments. This paper proposes a return to our experiments especially for species without any known research. Our methods are based on nursery work in view of planting in the botanical garden of Château Pérouse. For this to work, we have developed a range of tools in order to create microclimates as well as substrates for the nursery and the garden. These also give the possibility of testing the response to climate change in the garden.

Keywords: Acacia, Caesalpinioideae, seed, germination protocol, experiences

1. Introduction

The current Botanical Garden Château Pérouse is a proving ground for the much bigger Botanical Garden Château Pérouse to be created in the future. In this context, the current Botanical Garden Château Pérouse is a proof of concept and a research facility combined. In its Living Plant Collection Policy, the garden specifically stresses the need to know how to reproduce each taxon from seeds. Although other reproduction techniques are important, reproduction by seeds is by far the way the garden adds taxa to its Living Plant Collection. We have sown an average of 4000 taxa each year in 5000 tests for the last 6 years.

In 2019, the Garden obtained the French label "National collection of Acacias of temperate regions,". This occasion was used to improve the information tools for the garden by adding extensive botanical information and a nature photo to our information-page per taxon, testing Acacias from climate zones more dissimilar to our own and testing other germination techniques for better results. This also created the need to test different microclimates in the garden mostly based on the elements: more or less shade, more or less air-humidity trough different ways of watering, and/or using indigenous poplar trees.

At the same time, we started working on tree trials for climate change research for our region. Our targets are foremost trees from continental and subtropical climate zones with a dry summer and a cold winter. We also tested the species adaptation for a clay soil instead of our garden substrates, which are much more draining. From 2022 onwards, we have also been working in the coastal areas of the Netherlands on the same subject.

1.1 Facts about the Château Pérouse botanical gardens

The botanical garden is situated at 50 km from the coast in the South of France; it receives 550 mm of rainwater annually with at least 4 months of no rain or dew at all during summer. It's, however, connected to an irrigation system fed by the Rhône river, which provides water all year round.

The natural soil of the botanical garden consists of a river deposit starting with a layer of red clay from 0.5 to 2.5 m thick, under which layers of permeable grit of limestone and impermeable red clay alternate. While testing plants, we discovered that the topsoil keeps too much water in the winter, and this makes plants much more sensitive to frost. So, we started to create substrates, giving us much more drainage.

The temperatures have gradually moved up. The annual minimal temperatures at the beginning of the 2000s were – 5°C (23°F) for a few nights, but for the last 8 years, this has moved to –2°C (28°F) from 05:30-07:30 for two or three nights measured in the coldest parts of the garden. This is followed by a sunny day with temperatures in excess of 10°C (50°F). On overcast days, there is no frost.

The annual maximum temperatures at the beginning of the 2000s were 35°C (95°F) in the shade during daytime for 1 or 2 weeks, but for the last 8 years, we have seen already three summers where this has moved to 42°C (108°F). The minimum night temperatures in these warm weeks moved from 23°C (74°F) to 28°C (82°F).

The botanical garden is influenced by the Mistral winds but is outside the strongest wind path. It experiences gusts of over 120 km/h each year.

1.2 Basic principles for our propagation

Our foremost guideline for propagation has always been the successful introduction of the propagated plants into the garden. We found out that while working with our climate chamber, the germination results were better than the results in our greenhouses, but the resulting plants were more fragile and had losses in the greenhouses getting adapted to a less protected environment. Only 5% of all our germination tests occur nowadays in the climate chamber. These are mostly succulent taxa that are less difficult to adapt to a greenhouse environment.

We try to make sure that there is as less as possible a selection of the germinating plants by adapting our germination and repotting strategies to this. We try to sow early enough to give early starters their chance and keep the potting trays for 10-12 months in the greenhouses. We repot each plant when it's ready, not the seedlot, so the germinated plants are normally in more than one nursery area depending on their pot size. We store no information per plant of this, but it ensures a little bit more adaptability of each plant species by letting the climate being the biggest deciding factor.

Following earlier tests, in 2023, we tested a clay-based substrate (20% clay +10% compost+20% Coco fiber+50% sand) versus a sand-based substrate for 32 Acacia (27 have germinated) and 39 Eucalyptus species. We do these tests for climate-change research where the problem of these species is the frost-tenderness in clay soils due to the storage of too much water in these soils in wintertime, which prepares them badly for frosts. In the first stage, the seeds are sown in our normal mixture in which 20% of the fine sand is replaced by red clay. At the second stage, we repot with a mixture of 30% clay in 3 L forest containers. The fourth stage will have 50% clay in a 6 L container. For the last stage, we will evaluate per species the follow-up mixture in 55 L containers, with a protocol where only the substrate is changed.

We use small greenhouses (5 × 16 m = 80 m^2) that we can adapt by changing the polyethylene covering used in wintertime by pulling a shade cloth, with 50% or 70% obstruction, over it in spring. At the beginning of spring, the polyethylene covering gets pulled out from under the shade cloth. The front and rear facades are changed from polyethylene to light shade cloth, which gives the wind the opportunity to strengthen the seedlings. In many of the greenhouses, a fog system and/or ventilation system is installed. Two are equipped with antifreeze heating and one with a water basin of 70m^3 under the surface. These equipment give us the opportunity to create different microclimates in the greenhouses.

1.3 Seed germination at Château Pérouse

We started sowing in 2014, first using vendor information. When this left us with many questions, we formed a team of gardeners/botanists who little by little improved their knowledge through books and articles.

From the start of 2021 onward, we systematically searched for research papers for each taxon sown as well as taking into account our previous results and their germination protocols.

Reasons for sowing are the introduction of new species (or lineages of species = different collecting sites) and thus testing their protocols, tests for improving protocols, tests for different substrates (clay versus sand), and tests for the viability of seeds harvested in our garden as well as their viability over the years by dry storage at 18°C (64°F).

New germinations are recorded each week. For a full overview of our tests, we invite you to visit our website at https://www.chateau-perouse.com/en/about/germinations-en where you will find overviews and downloadable PDF's with all the yearly updated germination tests for the Fabaceae/Caesalpinioideae subfamily and others [1–173].

Equally of interest might be our online-database https://www.chateau-perouse.com/tuinen/index.php?route=product/category&path=1 where you can access our species information by clicking on the link-icon. In these information pages, on a per-species level, all the botanical-, germination, and horticultural-information and external references are made accessible.

1.4 Microclimates at Château Pérouse

The default microclimate in the garden is similar to a Californian dry Chaparral, Chilean dry Matorral, South-African Little Karoo, and Australian Kwongan. With more or less shade from indigenous pine trees and more or less watering (drip systems), we can make microclimates for most Mediterranean climates.

As for forests, we can recreate many types by planting young trees (height 2 m) at an irregular pattern of on average 2-3 m distance and accepting to lose the less fast-growing trees. Of course, these are all planted on new substrates and have watering by bubblers (1-2 l/min) either close to the tree trunk or in a drainage pipe laid around the root ball.

If the forest needs to stay humid, we have tested sprinkler systems at different heights in the canopy. This works well provided that there is enough coverage to maintain a humidity bubble by watering at the beginning and the end of the day. This also lowers the summer temperatures by 5-10°C (3-6°F) depending on the foliage coverage and the quantity of watering. Some examples of these are *Acacia alata, A. applanata, A. huegelii,* and *A. menzelii,* which seem to benefit from this microclimate. For others, we still need time to reach a definite conclusion.

For now, we have not succeeded to recreate Mediterranean alpine climates and colder-climate forests (like Nothofagus), but for this, we have a test garden in the Netherlands where work started 3 years ago.

2. Materials and methods

We use a shorthand notation system on our labels, which I will use to explain the different materials and methods. The long version of the germination protocol can be found in each of our html pages per species with the inclusion of research papers studied for this protocol. The question marks indicate places which are filled with the relevant numbers according to the specific information.

Spring, Autumn: Sow in spring or autumn.

??-??°C or > 25°C: The day-temperature range in which the sowing has to occur.

Freezer: Dry storage of seeds in a commercial freezer at -15°C (5°F) enveloped in a zip-bag.

Soak 20°C??h Damp Strat. ??wk: Humid stratification, for a given number of weeks, in a commercial refrigerator at 4°C (39°F) enveloped in a zip-bag with an upfront immersion in a small glass containers for a given duration.

Dry Strat. ??wk: Dry stratification, for a given number of weeks, in a commercial refrigerator at 4°C (39°F) enveloped in a zip-bag.

Soak 20°C ??h: Immersion in a small glass containers for a given duration.

Oven ??°C ??min: Heating the seeds, enveloped in aluminum sheet, in a commercial oven pre-heated at the indicated temperature. Scar: Manual scarification with sandpaper (no.80), leaving the tegument intact.

Scar: Manual scarification with sandpaper (no.80), leaving the tegument intact.

Soak ??°C ??sec, ??h SmokeDisk: Immersion in a small glass container at a given temperature for a given duration followed by an immersion in, smoke infused, ambient water in a glass container for a given duration. The smoke infusion comes from a smoke impregnated paper disk from which a pie-part is cut and put at the bottom of the glass container.

Soak ??°C ??sec, ??h: Immersion in a small glass container at a given temperature for a given duration followed by an immersion in ambient water in a glass container for a given duration.

Soak ??°C ??h: Immersion in a small glass container at a given temperature for a given duration, letting the water cool-down by its self.

??%GA ??min: Immersion in a small glass container ambient water with a percentage of gibberellic acid GA3 for a given duration.

Sowing mix+Sand: Well drained seed sowing mix.

20Clay: 20% clay+10% compost+20% Coco fibers+50% sand.

Sowing mix: 20% compost+40% clay+40% sand.

Damp substrate+Light: Keep the substrate moist and in daylight.

Marsh: The seed-plate with the seeds is placed in a saucer of water until germination occurs. As a result, the moisture reaches the seeds by capillary action and ensures that the seeds do not dry out.

No direct sun: During germination, keep the substrate moist and protected from direct sunlight.

Seedling damping off: During germination, there is a risk of damping off, so water sparingly and keep in daylight.

Damp substrate+Salt+Light: Sprinkle 1gr of table salt on the substrate. Keep the substrate moist and in daylight during germination.

3. Germination tests on Acacias and other Caesalpinioideae

This year, we tested the storage of *Acacia harpophylla* [96] and *Libidibia ferrea* [112, 152] in the freezer and then sowed them at the right time with no results whatsoever. In 2020, we had a six out of 12 result without freezer storage.

We tested dry stratification of *Chamaecrista viscosa* with no results whatsoever. When we switched to humid stratification, we had 50% results [132, 169, 171].

1588 of the 1741 tests were done with a protocol including dipping in water of 99°C (210°F) for a small duration and then soaked in ambient water for the duration of 12 or 24 hours. Some seeds can be soaked in hot water, which then cools down for the duration of 12 or 24 hours. For some seeds, 99°C (210°F) is too hot; the seed-coating is not hard enough, and the seeds get boiled, so then we have to find a lower water temperature and/or a shorter duration of immersion.

Other protocols we tested were based on: heating seeds in the oven, manual scarification with sandpaper on bigger seeds, and scarification with gibberellic acid. One has to keep in mind that we are working with seed-lots of 12 seeds, so some mechanical scarification procedures are impossible to implement, while other seeds are too small.

We tested with success *Acacia elata* impregnation with smoke through soaking for 12 hours in ambient water with a slice of a smoke disk, but we have to wait for the end results in order to compare them with the results of just a hot-water treatment followed by a 12-hour immersion in ambient water in 2021 with 7 out of 20 result [11, 96, 111, 151].

We tested oven-heating [5, 9, 23, 26, 60, 64, 66, 81, 88, 92, 96, 108, 111, 120, 133, 145, 147, 151, 153] with *Acacia dealbata*, *A. falciformis*, *A. leiocalyx*, *A. myrtifolia*, *A. pulchella*, *A. pulchella var goadbyi*, *A. sophorae*, and *Albizia kalkora* for one test each. Almost all gave similar results with hot-water treatment. Only *Acacia myrtifolia* seemed to give a better result compared to hot-water treatment. We are waiting on the results for some others in the next months.

We tested manual scarification with sandpaper (no. 80), leaving the tegument intact, followed by 24 hours of watering with better results than hot-water treatment for *Acacia auriculiformis, A. baileyana, A. pendula, Cassia abbreviata ssp beareana, Delonix regia* and *Parkia biglandulosa* of 60 species (with some 76 research papers studied).

For a few species we use different substrates than a well-drained seed-sowing mix. A sand substrate is used for *Anadenanthera colubrina, A.colubrina var cebil* [35, 100, 101, 131], *Entada burkei* [114], and *Hoffmannseggia repens*. The species of the genus

Anadenanthera and *Hoffmannseggia repens* are difficult for us, and we have no results whatsoever. *Entada burkei* gave good results in 2018. Species that like the humidity-holding capacity of a clay substrate (20% compost+40% clay +40% sand) are *Cassia abbreviata, C. abbreviata ssp beareana, C. brewsteri, Senna artemisioides, S. barclayana, S. didymobotrya, S. ferraria, S. glutinosa, and S. x floribunda*. Most of the more recent tests prescribe sowing in spring, and they work very well. The older tests had sowing in autumn, and this did not work.

For 35 species, we tested an improved humidification system where we kept watering as usual and added the placement of the seed-plate with the seeds in a saucer of water until germination occurred. As a result, the moisture reached the seeds by capillary action. *Acacia cincinnata, A. dimidiata, A. gladiiformis, A. lateriticola, A. mearnsii, A. pruinocarpa, Albizia adianthifolia, A. anthelmintica, A. forbesii, A. harveyi, A. petersiana, Chamaecrista viscosa, Erythrophleum africanum, Erythrostemon gilliesii, Neltuma juliflora, Paraserianthes lophantha, Parkinsonia praecox, Peltophorum dubium, Pseudalbizzia niopoides, Senna artemisioides, S. artemisioides ssp helmsii, S.artemisioides ssp oligophylla, S. artemisioides ssp sturtii, S. costata, S. ferraria, S. glaucifolia, S. glutinosa, S. hamersleyensis, S. notabilis, S. occidentalis, S. pleurocarpa, S.polyphylla, S.reticulata,* and *S.venusta* were tested in this way.

10 Species: Acacia cincinnata, A.gladiiformis, A.pruinocarpa, Chamaecrista viscosa, Erythrophleum africanum, Senna artemisioides, S.glaucifolia, S.hamersleyensis, S.pleurocarpa, and S.reticulata appear to benefit substantially from this treatment. For this treatment, there was no research paper directly explaining this method, but from environmental descriptions, we tried this method.

Anadenanthera colubrina, A.colubrina var cebil [35, 100, 101, 131], *Cenostigma eriostachys* [104], and *Libidibia ferrea* [112, 152] were tested with protection from direct sunlight. *Cenostigma eriostachys* had very good results, and for *Libidibia ferrea*, it's too early to say.

Figure 1.
Seedling trays with Acacias and repotted Acacias in 1.5 and 2.5 L forest pots (high pots).

Parkinsonia florida and *P. microphylla* [104] can have the risk of damping off, so water sparingly and keep in daylight. We usually have 30% results with our protocols.

Another special protocol is for *Acacia ampliceps* [113, 162, 164] where we sow 1gm of table salt on the substrate with 50% results. We will test a new in 2025 to see the results without the salt.

For 38 species, we could not continue our research because of sourcing problems, and we could do much more with access to all the species not tested this far (**Figure 1**).

4. Growing Acacias and other Caesalpinioideae in the nursery

We started out sowing 20 seeds in small containers (8 × 16 × 5 cm). In 2020, we moved on to forest-tree-sowing plates with 60 holes and 18.5 cm height. We are now sowing six holes per test with two seeds in opposite sides. With this move, we lost most root interactions between the small seedlings as well as faced losses of disentangling the roots of the seedlings when repotting in 1.5 L forest containers.

We moved at the same time from a peat/coarse sand-based substrate to a cooked wood/coco-fibers/fine sand-based substrate. With the forest plates we can move either to a 2.5L forest container or a 1.5L forest container and from these to a 6L container or a 12L Airpot®. For slow growing species we repot from a 1.5L forest container to a 2.5L forest container before moving on to bigger pots.

After repotting to 6 L containers or 12 L Airpot®'s, the plants go to the shade house (23 × 75 m = 1725 m^2 or 76 × 246 ft. = 18,696 ft^2) where they will stay, or after 6 months, they will be moved to a place in the sun. Plants needing a lot of humidity get a place in the test gardens with sprinkler systems.

Normally, we repot the Acacias from 6 L containers to 55 L containers or 12 L Airpot®'s to 45 L Airpot®'s. This gives the shrubs/trees 3 to 5 years in the nursery. We plant using smaller sized pots when the shrubs/trees go to a microclimate, which offers more humidity or shade. We use Airpot®'s for dry-climate species and normal containers for more-humidity-loving species.

Most South-African and other subtropical Acacias have to stay, during the winter, in a frost-free greenhouse until they reach a height of 2 m. After this, they are able to withstand some hours of frost. Anyway, the average yearly temperature in the garden is to low so with us they are slow growers.

5. Acacias and other Caesalpinioideae in the botanical garden

In the beginning, we tested the possibilities of the site through just buying plants in nurseries and then planting them in connection to the watering system. Some suffered from frost. We then created a new test garden with an improved soil by adding 1.5 m of very well-drained substrate including silicate sand, pozzolan pebbles, and an amount of coco fiber. This substrate is quite heavy, which helps to anchor the trees against the Mistral winds.

6. Conclusion

Most Acacias and other Caesalpinioideae can be sown with good results using hot-water treatment and soaking for 12-24 hours. Only a few need different treatments to have results or substantially improved results.

All the research has taught us to use less and less seeds as well as to predict the most probable germination protocol just by studying the seeds. Twelve seeds can be enough to check the protocol, or a multiple thereof in case of problems, more vendors, or different collection sites.

By studying Fabaceae and other tree species, we get an idea of the microclimate that a plant community needs and the translation we need to get them to thrive in the botanical garden (sun/shade, watering regimes).

It's difficult to get seeds of other species in the same subfamily, and this hinders us to substantially expand our research. Still, we were able to do some 260 germination tests each year for the last 6 years.

Author note

Document written by M. A.J.W. Vos on behalf of "Les Jardins de Pérouse s.a.r.l." having the mission of directing the "Botanical Garden Château Pérouse".

Author details

M. Albertus Jan-Willem Vos
Les Jardins de Pérouse s.a.r.l., Château Pérouse, route de Nîmes, Saint-Gilles, France

*Address all correspondence to: ajwvos@gmail.com

References

[1] Baskin J, Nan X, Baskin C. A comparative study of seed dormancy and germination in an annual and perennial species of *Senna (Fabaceae)*. Seed Science Research. 1998;**8**(4):501-512. DOI: 10.1017/S0960258500004475

[2] Merritt DJ, Martyn AJ, Ainsley P, Young RE, Seed LU, Thorpe M, et al. A continental-scale study of seed lifespan in experimental storage examining seed, plant, and environmental traits associated with longevity. Biodiversity and Conservation. 2014;**23**:1081-1104. DOI: 10.1007/s10531-014-0641-6

[3] Went FW, Munz PA. A long term test of seed longevity. Aliso: A Journal of Systematic and Floristic Botany. 1949;**2**(1):63-75. Available from: https://scholarship.claremont.edu/cgi/viewcontent.cgi?article=1874&context=aliso

[4] Went FW. A long term test of seed longevity. II. Aliso. A Journal of Systematic and Floristic Botany. 1969;7(1):1-12. Available from: https://scholarship.claremont.edu/cgi/viewcontent.cgi?article=1659&context=aliso

[5] Bell DT. A review of ecological factors affecting seed germination of species used in the reclamation of Western Australian bauxite mining. In: Proceedings of the 12th Annual National Meeting of the American Society for Surface Mining and Reclamation. Gillette, Wyoming. 1995;**1**:125-135. Available from: https://www.asmr.us/Portals/0/Documents/Conference-Proceedings/1995/0125-Bell.pdf

[6] Negash L. A selection of African native trees: Biology, uses, propagation and restoration techniques. Addis Ababa University Press; 2021. ISBN: 9789994430864

[7] Negash L. A Selection of Ethiopia's Indigenous Trees: Biology, Uses and Propagation Techniques. Addis Ababa University Press; 2022. pp. 302-320. ISBN: 9789994452279

[8] Thirupathi M, Kumar MK, Srinivas D, Rajendar K, Reddy KJ. A study on seed germination of *Cassia Alata Linn* an antiallergenic plant. Journal of Phytology. 2012;**4**(5):13-15. Available from: https://www.academia.edu/2548063/A_study_on_Seed_germination_of_Cassia_alata_Linn_an_antiallergenic_plant

[9] Horlock F, Faragher J. Acacia cut flower and foliage production manual. 2000. Available from: https://www.agrifutures.com.au/wp-content/uploads/publications/00-057.pdf

[10] Bell S, Driscoll C. Acacia pendula (weeping Myall) in the Hunter Valley of New South Wales: Early explorers' journals, database records and habitat assessments raise doubts over naturally occurring populations. Cunninghamia. 2014;**14**:179-200. DOI: 10.7751/cunninghamia.2014.14.009

[11] Ryan PA, Podberscek M, Raddatz CG, Taylor DW. Acacia species trials in Southeast Queensland, Australia. ACIAR Proceedings. 1987;**16**:81-85. ISBN: 9780949511263

[12] Jin M, Wang Z, He Z, Jiang L, Liu J, Lan Y, et al. Allelopathic effect of *Castanopsis kawakamii* forest litter on seed germination of small Philippine Acacia (Acacia confusa). Applied Ecology and Environmental Research.

2019;**17**:15103-15116. DOI: 10.15666/aeer/1706_1510315116

[13] Hanna PJ. Anatomical features of the seed coat of *Acacia kempeana* (Mueller) which relate to increased germination rate induced by heat treatment. New Phytologist. 1984;**96**:23-29. DOI: 10.1111/j.1469-8137.1984.tb03539.x

[14] Neri CJC, Collazos SR, Manuel O, Huamá HE, Garcì JV. Aplicación de la escarificació fìica y mecáica en la emergencia y crecimiento de semillas de tara (Caesalpinia espinosa). Revista de Investigació de Agroproducció Sustentable. 2018;**2(2)**:45-53. DOI: 10.25127/aps.20182.392

[15] Kheloufi A, Mansouri LM, Boukhatem FZ. Application and use of sulphuric acid pretreatment to improve seed germination of three acacia species. Reforesta. 2017;**3**:1-10. DOI: 10.21750/REFOR.3.01.25

[16] Netshiluvhi T. Aspects of Seed Propagation of Commonly Utilised Medicinal Trees of KwaZulu-Natal. University of Natal, Durban. 1996

[17] Kaplan, H. Assessing the invasiveness of *Acacia stricta* and *Acacia implexa*: Is eradication an option? 2012. Available from: https://core.ac.uk/download/pdf/37348766.pdf

[18] Howes J. Australian plants society. 2020. Available from: https://resources.austplants.com.au/plant-database/

[19] Dunphy M, McAlpin S, Nelson P, Chapman M, Nicholson H. Australian rainforest seeds: A guide to collecting, processing and propagation. Csiro publishing; 2020. p. 43. ISBN: 9781486311507

[20] Letnic M. Bet-hedging and germination in the Australian arid zone shrub *Acacia* Ligulata. Austral Ecology. 2000;**25**(4):368-374. DOI: 10.1111/j.1442-9993.2000.tb00042.x

[21] Ferreira VM. Biometric characterization and seed germination of giant mimosa [Mimosa bimucronata (DC) O. Kuntze]. Australian Journal of Crop Science. 2018;**12**(1):108-115. Available from: http://www.cropj.com/melo_12_1_2018_108_115.pdf

[22] Paula A, Delgado CML, Maria SP, Santos M. Breaking physical dormancy of *Cassia leptophylla* and *Senna macranthera* (*Fabaceae: Caesalpinioideae*) seeds: Water absorption and alternating temperatures. Seed Science Research. 2012;**22**:259-267. DOI: 10.1017/S096025851200013X

[23] Kheloufi A, Mansouri LM, Aziz N, Sahnoune M, Boukemiche S, Ababsa B. Breaking seed coat dormancy of six tree species. Reforesta. 2018;**5**:4-14. DOI: 10.21750/REFOR.5.02.48

[24] Ottewell KM, Bickerton D, Lowe AJ. Can a seed bank provide demographic and genetic rescue in a declining population of the endangered shrub *Acacia pinguifolia*? Conservation Genetics. 2011;**12**:669-678. DOI: 10.1007/s10592-010-0173-x

[25] El-Shatnawi MKJ, Ereifej KI. Chemical composition and livestock ingestion of Carob (*Ceratonia siliqua* L.) seeds. Journal of Range Management Archives. 2001;**54**(6):669-673. DOI: 10.2307/4003669

[26] Du Hyun K, Han SH. Direct effects on seed germination of 17 tree species under elevated temperature and CO_2 conditions. Open Life Sciences. 2018;**13**(1):137-148. DOI: 10.1515/biol-2018-0019

[27] Trindade LA, de Jesus da CPH, Dantas BF, Vinicius MM. Does discontinuous hydration of Senna

spectabilis (DC.) HS Irwin & Barneby var. excelsa (Schrad.) HS Irwin & Barneby (Fabaceae) seeds confer tolerance to water stress during seed germination? Journal of Seed Science. 2018;**40**:36-43. DOI: 10.1590/2317-1545v40n1182838

[28] Funes G, Venier P. Dormancy and germination in three *Acacia* (Fabaceae) species from Central Argentina. Seed Science Research. 2006;**16**:77-82. DOI: 10.1079/SSR2005229

[29] Auld T. Dormancy and viability in *Acacia suaveolens (Sm.) Willd*. Australian Journal of Botany. 1986;**34**(4):463-472. DOI: 10.1071/BT9860463

[30] Malavasi U, Malavasi M. *Dormancy breaking and germination of Enterolobium contortisiliquum (Vell.) Morong seed*. Brazilian Archives of Biology and Technology. 2004;**47**:851-854. DOI: 10.1590/S1516-89132004000600003

[31] Pound L, Ainsley P, Facelli J. Dormancy-breaking and germination requirements for seeds of *Acacia papyrocarpa*, *Acacia oswaldii* and *Senna artemisioides* ssp. x coriacea, three Australian arid-zone *Fabaceae* species. Australian Journal of Botany. 2015;**62**:546-557. DOI: 10.1071/BT14077

[32] Conafor. *Ebenopsis ebano (Berl.) Britton et Rose*. Available from: http://www.conafor.gob.mx:8080/documentos/docs/13/919Ebenopsis%20ebano%20.pdf

[33] Adji BI, Akaffou SD, Sabatier S-A. Ecological environment effects on germination and seedling morphology in Parkia biglobosa in nursery (Côte d'Ivoire) and greenhouse (France). International Journal of Horticulture, Agriculture and Food science. 2021;**5**(5):01-13. Available from: https://hal.inrae.fr/hal-03423951/file/1IJHAF-SEP20214-Ecological.pdf

[34] Medina L, Eloy S, Rivero G, Efraín A. Efecto del acondicionamiento osmótico en la germinación de semillas de Caesalpinia spinosa (Feuillée ex Molina) Kuntze (Fabaceae) 'taya'. Arnaldoa. 2017;**24**(1):333-342. DOI: 10.22497/arnaldoa.241.24115

[35] Rodrigues A, Osuna J, Queiroz SR, Rios A. Efeito do substrato e luminosidade na germinação de Anadenanthera colubrina (Fabaceae, Mimosoideae). Revista Arvore. 2007;**31**:187-193. DOI: 10.1590/S0100-67622007000200001

[36] Tadros MJ, Samarah NH, Alqudah AM. Effect of different pre-sowing seed treatments on the germination of *Leucaena leucocephala (Lam.)* and *Acacia farnesiana (L.)*. New Forests. 2011;**42**:397. DOI: 10.1007/s11056-011-9260-1

[37] Merou T, Takos I, Konstantinidou E, Galatsidas S, Varsamis G. Effect of different pretreatment methods on germination of *Albizia julibrissin* seeds. Seed Science and Technology. 2011;**39**(1):248-252. DOI: 10.15258/sst.2011.39.1.26

[38] Rawat M, Panwar A. Effect of different seed treatments on seed germination of *Cassia fistula*. Balawala, Dehradun, Uttarakhand: Sardar Bhagwan Singh University; 2022. DOI: 10.13140/RG.2.2.15233.06240

[39] Idu M, Omonhinmin AC. Effect of oven-heat and boiling on the germination and seedling development of *Dichrostachys cinerea* (L) Wight and Arn (Fabaceae). Agronomie. 1999;**19**(8):671-676. Available from: https://hal.science/hal-00885960/document

[40] Park EH, Cho MS. Effect of pretreatments on seed germination of *Acacia* species. Flower Research Journal. 2020;**28**(3):147-155. Available from: https://db.koreascholar.com/article/Detail/400616

[41] Dhanai CS, Uniyal AK. Effect of provenance, temperature and pretreatment on germination of *Albizia chinensis*. Seed Technology. 2009;**31**(2):164-176

[42] Sikuku PA, Musyimi DM, Amusolo M. Effect of seed depth on germination, growth and chlorophyll contents of *Senna spectabilis*. Discovery Science. 2018;**14**:84-92. Available from: http://edocs.maseno.ac.ke/handle/123456789/1963

[43] Missanjo E, Maya C, Kapira D, Banda H, Kamanga-Thole G. Effect of seed size and pretreatment methods on germination of Albizia lebbeck. ISRN Botany. 2013;**2013**(1):1-4. DOI: 10.1155/2013/969026

[44] Samreen S, Shaukat S. Effect of seed size on germination, emergence, growth and seedling survival of Senna occidentalis Link Pakistan Journal of Biological Sciences. 2000;**3**(2):292-295. DOI: 10.3923/pjbs.2000.292.295

[45] Reichman S, Bellairs S, Mulligan D. Effect of seed treatment on the emergence of *Cassia brewsteri* and *Lysiphyllum carronii* seeds stored in soil. The Rangeland Journal. 2007;**29**:133-137. DOI: 10.1071/RJ07004

[46] Omari MA. Effect of temperature and seed treatment on germination of five *Acacia species*. Dirasat (Jordan). 1992;**19**(1):297-315. Available from: https://agris.fao.org/agris-search/search.do?recordID=JO9300031

[47] Mojeremane W, Mathowa T, Teketay D. Effects of different pre-sowing treatments on germination of *Peltophorum africanum Sond*. Seeds from two provenances in Botswana. Journal of Biodiversity and Environmental Sciences. 2018;**12**(2):230-236

[48] Diallo I, Danthu P, Sambou B, Dione D, Goudiaby A, Poulsen K. Effects of different pretreatments on the germination of *Faidherbia albida (Del.) A. Chev*. seeds. International Tree Crops Journal. 1996;**9**(1):31-36. DOI: 10.1080/01435698.1996.9752957

[49] Jeffery DJ, Holmes PM, Rebelo AG. Effects of dry heat on seed germination in selected indigenous and alien legume species in South Africa. South African Journal of Botany. 1988;**54**(1):28-34. DOI: 10.1016/S0254-6299(16)31358-8

[50] Stuart-Hill GC. Effects of elephants and goats on the Kaffrarian succulent thicket of the Eastern Cape, South Africa. Journal of Applied Ecology. 1992;**29**(3):699-710

[51] Walters M, Midgley JJ, Somers MJ. Effects of fire and fire intensity on the germination and establishment of *Acacia karroo*, *Acacia nilotica*, *Acacia luederitzii* and *Dichrostachys cinerea* in the field. BMC Ecology. 2004;**4**(3):1-13. DOI: 10.1186/1472-6785-4-3

[52] Salazar A, Ramirez, Claudia. Effects of mechanical and acid scarification on germination performance of *Schizolobium parahyba* (*Fabaceae - Caesalpinioideae*) seeds. Journal of Tropical Biology & Conservation (JTBC). 2019;**16**:209-223. DOI: 10.51200/jtbc.v16i0.2040

[53] Gilani M, Ahmad I, Farooq T, Wu P, Yousaf M, Tarin M, et al. Effects of pre-sowing treatments on seed germination and morphological growth of *Acacia nilotica* and *Faidherbia albida*. Scientia Forestalis/Forest Sciences.

2019;**47**(122):1-9. DOI: 10.18671/scifor.v47n122.20

[54] Oke DO. Effects of propagation method, planting medium and nursery technique on early growth of *Albizia adianthifolia* on a humid tropical alfisol. International Journal of Applied Agricultural Research. 2009;**4**(2):155-163. Available from: http://www.ripublication.com/IJAER/ijaarv4n2_8.pdf

[55] Al Hassan M, Fuertes MM, Sánchez FJR, Vicente O, Boscaiu M. Effects of salt and water stress on plant growth and on accumulation of osmolytes and antioxidant compounds in cherry tomato. Notulae Botanicae Horti Agrobotanici Cluj-Napoca. 2015;**43**(1):1-11. DOI: 10.15835/nbha4319793

[56] Luera P, Wahl-Villarreal K, Christoffersen BO, Treviño A, Soti P, Gabler CA. Effects of scarification, phytohormones, soil type, and warming on the germination and/or seedling performance of three tamaulipan thornscrub forest species. Plants. 2021;**10**(8):1489. Available from: https://www.mdpi.com/2223-7747/10/8/1489/pdf?version=1626840052

[57] Marunda CT et al. Effects of seed pretreatments on the development of *Acacia auriculiformis* and *A. Holosericea* seedlings. In: Tropical Tree, Seed Research. Canberra, Australia: ACIAR; 1990. pp. 33-36. Available from: https://www.iufro.org/download/file/3506/4376/20903-gympie1989-1.pdf#page=35

[58] Soliman A, Abbas M. Effects of sulfuric acid and hot water pre-treatments on seed germination and seedlings growth of *Cassia fistula L*. American-Eurasian Journal of Agricultural & Environmental Sciences. 2013;**13**:7-15

[59] Aliero BL. Effects of sulphuric acid, mechanical scarification and wet heat treatments on germination of seeds of African locust bean tree, *Parkia biglobosa*. African Journal of Biotechnology. 2004;**3**(3):179-181. Available from: https://www.ajol.info/index.php/ajb/article/view/14937/58847

[60] Bell DT, Rokich DP, McChesney CJ, Plummer JA. Effects of temperature, light and gibberellic acid on the germination of seeds of 43 species native to Western Australia. Journal of Vegetation Sience. 1995;**6**(6):797-806. DOI: 10.2307/3236393

[61] Jeller H, Perez SC, Gualtierres J, de Andrade. Effects of water and salt stress and the gibberellin action in *Senna spectabilis* seeds. Ciência Florestal. 2001;**11**:93-104. DOI: 10.5902/19805098498

[62] Rampart M, Teketay D, Kamogelo M, Mojeremane W, Mmolotsi R, Makoi T, et al. Enhancing germination of seeds of *Cassia abbreviata* and *Senegalia nigrescens* using pre-sowing seed treatments in Botswana. Current Agriculture Research Journal. 2021;**9**:179-187. DOI: 10.12944/CARJ.9.3.05

[63] Puri S, Kumar A. Establishment of *Prosopis cineraria (L.)* Druce in the hot deserts of India. New Forests. 1995;**9**:21-33. DOI: 10.1007/BF00028923

[64] Pieterse PJ, Cairns ALP. Factors affecting the reproductive success of *Acacia longifolia (Andr.) Willd*. in the Banhoek Valley, South-Western Cape, Republic of South Africa. South African Journal of Botany. 1988;**54**(5):461-464. DOI: 10.1016/S0254-6299(16)31278-9

[65] Factsheets. Greening Australia. Available from: https://www.greeningaustralia.org.au/publications/

[66] Riveiro SF, Cruz Ó, Casal M, Reyes O. Fire and seed maturity drive the viability, dormancy and germination of two invasive species: *Acacia longifolia (Andrews) Willd* and *Acacia mearnsii* De Wild. Annals of Forest Science. 2020;**77**:60. DOI: 10.1007/s13595-020-00965-x

[67] Emery SM, Uwimbabazi J, Flory SL. Fire intensity effects on seed germination of native and invasive Eastern deciduous forest understory plants. Forest Ecology and Management. 2011;**261**(8):1401-1408. DOI: 10.1016/j.foreco.2011.01.024

[68] Palmer HD, Denham AJ, Ooi MKJ. Fire severity drives variation in post-fire recruitment and residual seed bank of *Acacia* species. Plant Ecology. 2018;**219**:527-537. DOI: 10.1007/s11258-018-0815-5

[69] León M, Ibacache E, Sandoval A, Pañitrur C, Espejo M, Mundaca G, et al. Germinación, estadios de plántulas y manejo del crecimiento aéreo temprano en el arbusto endémico Erythrostemon angulatus (Hook. & Arn.) E. Gagnon & G. P. Lewis. Chloris Chilensis. 2021;**24**(2):48-66

[70] Bhatt A, Caró MM, Verheyen K, Elsarrag E, Alhorr Y. Germination and seedling performance of five native legumes of the Arabian Desert. Flora-Morphology, Distribution, Functional Ecology of Plants. 2016;**220**:125-133. DOI: 10.1016/j.flora.2016.03.002

[71] Ferro J, Rocha D, Luciano J, Junior A, Araujo Neto J, Silva J, et al. Germination and viability of seeds of *Caesalpinia pulcherrima* newly harvested and stored. Journal of Agricultural Science. 2019;**11**:73-85. DOI: 10.5539/jas.v11n7p73

[72] Bargali K, Singh SP. Germination behaviour of some leguminous and actinorhizal plants of Himalaya: Effect of temperature and medium. Tropical Ecology. 2007;**48**(1):99. ResearchGate: 238087379

[73] Wright BR, Clarke PJ. Germination biologies and seedbank dynamics of *Acacia* shrubs in the Western Desert: Implications for fire season impacts on recruitment. Australian Journal of Botany. 2018;**66**:278-285. DOI: 10.1071/BT18003

[74] Parolin P, Ferreira L, Junk W. Germination characteristics and establishment of trees from central Amazonian flood plains. Tropical Ecology. 2003;**44**(2):155-168

[75] Pèrez-Garcìa F. Germination characteristics and intrapopulation variation in Carob (*Ceratonia siliqua L.*) seeds. Spanish Journal of Agricultural Research. 2009;**7**(2):398-406

[76] Ghassali F, Salkini AK, Petersen SL, Niane AA, Louhaichi M. Germination dynamics of *Acacia* species under different seed treatments. Range Management & Agroforestry. 2012;**33**(1):37-42

[77] Souza E, Pacheco M, Matos V, Luiz R. Germination of *Adenanthera pavonina L.* seeds as a function of different temperatures and substrates. Revista Árvore. 2006;**31**(3):437-443

[78] Mitchell E. Germination of seeds of plants native to Dutchess County, New York. Botanical Gazette. 1926;**81**(1):108-112

[79] Delachiave MEA, de Pinho SZ. Germination of Senna occidentalis link: Seed at different osmotic potential levels. Brazilian Archives of Biology and Technology. 2003;**46**:163-166. DOI: 10.1590/S1516-89132003000200004

[80] Gunn BV. Germination pretreatments for selected *Acacia* species from the Pilbara region of Western Australia. Tropical Tree Seed Research. 1989;**28**:46-50. Available from: https://www.iufro.org/download/file/3506/4376/20903-gympie1989-1_pdf/

[81] Penman TD, Binns DL, Allen RM, Shiels RJ, Plummer SH. Germination responses of a dry sclerophyll forest soil-stored seedbank to fire related cues. Cunninghamia. 2008;**10**(4):547-555. Available from: https://ro.uow.edu.au/cgi/viewcontent.cgi?referer=&httpsredir=1&article=1760&context=scipapers

[82] Shekafandeh A, Shibani B. Germination studies on *Pistacia terebinthus L*. Iran Agricultural Research. 1996;**5**(1):13-20. Available from: https://iar.shirazu.ac.ir/article_4516_9f9b6fffd6a1b912748ab66cc193d18d.pdf

[83] Witness M, Kamogelo M, Demel T, Melusi R, Ronnie M, Ismael K, et al. *Germination studies on seeds of Burkea africana and Erythrophleum africanum from Kazuma Forest Reserve, Northern Botswana*. African Journal of Biotechnology. 2020;**19**(9):675-683. DOI: 10.5897/AJB2020.17213

[84] Hauser TP. Germination, predation and dispersal of *Acacia albida* seeds. Oikos. 1994;**71**:421-426

[85] Carvajal D, Loayza A, López R, Toro P, Squeo F. Growth and early seedling survival of four *Atacama Desert* shrub species under experimental light and water availability regimes. Revista Chilena de Historia Natural. 2014;**87**:1-9. DOI: 10.1186/s40693-014-0028-9

[86] Kube PD. Growth rates, establishment techniques, and propagation of some central Australian *Acacias*. Australian Acacias in Developing Countries. 1987;**16**:77-80. ISBN: 9780949511263

[87] Hall S, Newton R, Holmes P, Gaertner M, Esler K. Heat and smoke pre-treatment of seeds to improve restoration of an endangered Mediterranean climate vegetation type. Austral Ecology. 2017;**42**:354-366. DOI: 10.1111/aec.12449

[88] Hanley ME, Lamont BB. Heat pre-treatment and the germination of soil- and canopy-stored seeds of south-western Australian species. Acta Oecologica. 2000;**21**(6):315-321. DOI: 10.1016/S1146-609X(00)01087-0

[89] Danthu P, Ndongo M, Diaou M, Thiam O, Sarr A, Dedhiou B, et al. Impact of bush fire on germination of some West African acacias. Forest Ecology and Management. 2003;**173**(1-3):1-10. DOI: 10.1016/S0378-1127(01)00822-2

[90] Bush JK, Van Auken OW. Importance of time of germination and soil depth on growth of *Prosopis glandulosa* (Leguminosae) seedlings in the presence of a C4 grass. American Journal of Botany. 1991;**78**(12):1732-1739

[91] Humtsoe C, Khare N, Rout S, debbarma, Ronald. Influence of different pre-treatments methods on seed germination and seedling growth performance of golden shower tree (*Cassia fistula L.*). Journal of Plant Development Sciences. 2018;**10**(11):617-621

[92] Roeder M, Yang W, Tomlinson K. Influence of smoke, heat and fire on germination of woody species occurring in the dry valleys of Southwest China. Journal of Plant Ecology. 2019;**12**(6):931-940. DOI: 10.1093/jpe/rtz030

[93] Villacis J, Armas C, Hang S, Casanoves F. Initial selection of forest

species for the rehabilitation of soils disturbed by petroleum extractions in the Ecuadorian Amazon. 2016. pp. 1-2

[94] Leal C, Torres S, de Freitas RM, Nogueira N, Farias R. Light intensity and type of container on producing *Cassia grandis L. f.* seedlings. Revista Brasileira de Engenharia Agrícola e Ambiental. 2015;**19**:939-945. DOI: 10.1590/1807-1929/agriambi.v19n10p939-945

[95] Bacchetta G, Ballesteros D, Baaziz K, Bou Dagher Kharrat M, El Hamrouni K. et al. Manual for the propagation of selected mediterranean native plant species. 2015. Available from: http://www.enpicbcmed.eu/sites/default/files/english_version_0.pdf

[96] Burrows G, Alden R, Robinson W. Markedly different patterns of imbibition in seeds of 48 *Acacia* species. Seed Science Research. 2019;**29**(4):270-282. DOI: 10.1017/S0960258519000242

[97] Ferraz D, Ramalho WP, Vale LS, Rodrigues. Methods for overcoming dormancy and seed storage of Enterolobium contortisiliquum. Acta Scientiarum. Agronomy. 2019;**41**:e42602. DOI: 10.4025/actasciagron.v41i1.42602

[98] Okunlola AI, Adebayo RA, Orimogunje AD, et al. Methods of braking seed dormancy on germination and early seedling growth of African locust bean (*Parkia biglobosa*) (*Jacq.*) Benth. Journal of Horticulture and Forestry. 2011;**3**(1):1-6

[99] L'Huillier L, Jaffré T, Wulff A, Lebrun M, Maggia L, et al. Mines et environnement en Nouvelle-Calédonie : Les Milieux Sur Substrats Ultramafiques et Leur Restauration. 2010. Available from: https://agritrop.cirad.fr/561101/

[100] Varela O, Albornoz P. Morpho-anatomy, imbibition, viability and germination of the seed of *Anadenanthera colubrina var. cebil (Fabaceae)*. Revista de Biología Tropical. 2013;**61**:1109-1118

[101] Nery F, Nery M, Prudente D, Alvarenga A, Paiva R. Morphological and physiological germination aspects of *Anadenanthera colubrina (Vell.) Brenan*. Notulae Botanicae Horti Agrobotanici Cluj-Napoca. 2018;**46**:570. DOI: 10.15835/nbha46211094

[102] McDonald MW, Harwood CE, Whitfeld SJ. Morphological variation in *Acacia tumida* and implications for its utilisation. Recent Developments in Acacia Planting. 1998;**82**:341-346. Available from: https://www.aciar.gov.au/sites/default/files/legacy/node/13704/pr82_pdf_41862.pdf

[103] Boulet F. Mycorrhizal symbiosis as a strategy for survival in ultramafic soils. 2003. Available from: https://research-repository.uwa.edu.au/en/publications/mycorrhizal-symbiosis-as-a-strategy-for-survival-in-ultramafic-so

[104] Bowers JE. Natural conditions for seedling emergence of three woody species in the northern Sonoran desert. Madrono. 1994;**41**(2):73-84

[105] Emery N, Lacey E. Optimizing the application of smoke water to maximize germination of the flannel flower, Actinotus helianthi. Seed Science and Technology. 2010;**38**(3):797-801. DOI: 10.15258/sst.2010.38.3.29

[106] Erickson TE, Merritt DJ, Turner SR. Overcoming physical seed dormancy in priority native species for use in arid-zone restoration programs. Australian Journal of Botany. 2016;**64**(5):401-416. DOI: 10.1071/BT16059

[107] Khan D, Sahito ZA, Javed Zaki M. Parental investment of biomass in pod,

seed and seed packaging in a tree of wiry wattle (*Acacia coriacea subsp*. Pendens) growing in Karachi, Pakistan. International Journal of Biology and Biotechnology. 2013;**10**(4):515-536

[108] Morrison D, Auld T, Rish S, Porter C, Mcclay K. Patterns of testa-imposed seed dormancy in native Australian legumes. Annals of Botany. 1992;**70**:157-163

[109] Gaol ML, Fox JED. Pengaruh variasi ukuran biji terhadap perkecambahan *Acacia fauntleroyi*. Journal of Biological Research. 2009;**14**(2):153-160. DOI: 10.23869/bphjbr.14.2.20095

[110] Harwood CE, Kha LD, Dien PQ, Van Thang L. Performance of Australian dry-zone *Acacia* species on white sandy soil in dry, Southeastern Vietnam. Recent Developments in Acacia Planting. 1998;**82**:29-35. Available from: https://www.aciar.gov.au/sites/default/files/legacy/node/13704/pr82_pdf_41862.pdf

[111] Thinh RR, Kha LD, Searle SD, Tung RV. Performance of Australian temperate *Acacias* on subtropical highlands of Vietnam. Recent Developments in Acacia Planting. 1998;**82**:51-59. Available from: https://www.aciar.gov.au/sites/default/files/legacy/node/13704/pr82_pdf_41862.pdf

[112] Bragante R, Hell A, Silva J, Centeno D, Figueiredo-Ribeiro R, Barbedo C. Physiological and metabolic responses of immature and mature seeds of *Libidibia ferrea ((Mart. Ex Tul.) L.P. Queiroz)* under contrasting storage temperatures. Brazilian Journal of Botany. 2018;**41**:43-55. DOI: 10.1007/s40415-018-0442-3

[113] Erickson TE, Barrett RL, Merritt DJ, Dixon KW. Pilbara Seed Atlas and Field Guide. Clayton, Australia: Csiro Publishing; 2016. ISBN: 9781486305520

[114] Alicia Krige. Albizia adianthifolia var. adianthifolia. In: 2017 National Herbarium. Pretoria, PlantZafrica; 2017

[115] Griesser SS, Prakash S, Griesser HJ. Plasma discharge treatment for improved germination of seeds and killing of fungal spores on seed coats. Final Project Report to the Australian Flora Foundation. 2011;**5095**:1-9. Available from: http://aff.org.au/results/grant-summaries/aff-griesser-plasma/

[116] Auld T. Population dynamics of the shrub *Acacia suaveolens* (Sm.) Willd.: Fire and the transition to seedlings. Australian Journal of Ecology. 1986;**11**:235-254. DOI: 10.1111/j.1442-9993.1986.tb01395.x

[117] Lavallee K, Soti P, Rodrigo H, Kariyat R, Racelis A. Pre-sowing treatments improve germinability of South Texas native plant seeds. Plants. 2021;**10**(11):25-45. DOI: 10. 10.3390/plants10112545

[118] Fordham AJ. Propagation of *Albizia julibrissin*. Arnoldia. 1968;**28**:4-5. Available from: http://arnoldia.arboretum.harvard.edu/pdf/articles/1968-28--propagation-of-albizia-julibrissin.pdf

[119] Pound L, Obst C, How T, Bickerton D. Recovery plan for *Acacia pinguifolia* (fat-leaved wattle). 2011. Available from: https://www.agriculture.gov.au/sites/default/files/documents/acacia-pinguifolia-recovery-plan.pdf

[120] Bremner M, Goeth A. Recovery plan for *Acacia* terminalis Subsp. terminalis (sunshine wattle). 2010. Available from: https://www.awe.gov.au/sites/default/files/documents/acacia-terminalis-terminalis.pdf

[121] Berry E, Wheeler J, Darnowski C. Reproductive biology of northern wild senna (*Senna hebecarpa*): Flowering phenology, fruit set, and a test for inbreeding depression. Journal of the Torrey Botanical Society. 2013;**140**:9-19. DOI: 10.3159/TORREY-D-12-00061.1

[122] Gaol ML, Fox JED. Reproductive potential of *Acacia* species in the central wheatbelt: Variation between years. Conservation Science Western Australia. 2002;**4**(3):147-157. Available from: https://library.dbca.wa.gov.au/static/Journals/080559/080559-04.021.pdf

[123] Sabina PD, Dorin C, et al. Research regarding the influence of the preparing methods on seed germination on *Gleditsia triacanthos L.* Romanian. Biotechnological Letters. 2015;**20**(6):11035. Available from: https://rombio.unibuc.ro/wp-content/uploads/2022/05/20-6-16.pdf

[124] Spooner PG. Response of *Acacia* species to disturbance by roadworks in roadside environments in southern New South Wales, Australia. Biological Conservation. 2005;**122**(2):231-242. DOI: 10.1016/j.biocon.2004.07.012

[125] Cazebonne C, Vega A, Varela DA, Cardemil LA. Salinity effects on germination and growth of *Prosopis chilensis*. Revista Chilena de Historia Natural. 1999;**72**:83-91. Available from: http://rchn.biologiachile.cl/pdfs/1999/1/Cazebonne_et_al_1999.pdf

[126] Commander L, Merritt D, Rokich D, Dixon K. Seed biology of Australian arid zone species: Germination of 18 species used for rehabilitation. Journal of Arid Environments. 2009;**73**:617-625. DOI: 10.1016/j.jaridenv.2009.01.007

[127] Sajeevukumar B, Sudhakara K, Ashokan PK, Gopikumar K. Seed dormancy and germination in *Albizia falcataria* and *Albizia procera*. Journal of Tropical Forest Science. 1995;**7**(3):371-382

[128] Kannan CS, Sudhakara K, Augustine A, Ashokan PK. Seed dormancy and pre-treatments to enhance germination in selected *Albizia* species. Journal of Tropical Forest Science. 1996;**8**(3):369-380

[129] Weatherall-Thomas CR. Seed dynamics and seedling survival in mainland thicket of the Eastern Cape. 2009. Available from: https://core.ac.uk/download/pdf/145051121.pdf

[130] Smith NJC, Zahid DM, Ashwath N, Midmore DJ. Seed ecology and successional status of 27 tropical rainforest cabinet timber species from Queensland. Forest Ecology and Management. 2008;**256**:1031-1038. DOI: 10.1016/j.foreco.2008.06.005

[131] Nogueira FCB, Pinheiro CL, Sebastiã MF. Seed germination and seedling development of *Anadenanthera colubrina* in response to weight and temperature conditions. Journal of Plant Sciences. 2014;**2**(1):37-42. Available from: https://article.sciencepg.com/pdf/10.11648.j.jps.20140201.17.pdf

[132] Ranal MA, Ferreira WR. Seed germination and seedling emergence measurements of Brazilian native species. Pesquisa Florestal Brasileira. 2023;**43**:1-39. DOI: 10.4336/2023.pfb.43e202002166

[133] Bell DT, Plummer JA, Taylor SK. Seed germination ecology in southwestern Western Australia. The Botanical Review. 1993;**59**:24-73

[134] Sorensen B, Jusaitis M. Seed germination of endangered south Australian plants. Proceedings of the International Plant Propagators' Society. 1995;**45**:86-92

[135] Zuloaga-Aguilar S, Briones O, Orozco-Segovia A. Seed germination of montane forest species in response to ash, smoke and heat shock in Mexico. Acta Oecologica. 2011;**37**(3):256-262. DOI: 10.1016/j.actao.2011.02.009

[136] Sy A, Grouzis M, Danthu P. Seed germination of seven *Sahelian legume species*. Journal of Arid Environments. 2001;**49**:875-882. DOI: 10.1006/jare.2001.0818

[137] Higgisson W, Reynolds B, Cross Y, Dyer F. Seed germination requirements of an Australian semi-arid floodplain *Acacia species, Acacia stenophylla*. Marine and Freshwater Research. 2022;**73**(5):615-623. DOI: 10.1071/MF21226

[138] Ngulube MR. Seed germination, seedling growth and biomass production of eight Central-American multipurpose trees under nursery conditions in Zomba, Malawi. Forest Ecology and Management. 1989;**27**(1):21-27. DOI: 10.1016/0378-1127(89)90079-0

[139] Nrownn J, Enrightb J, Miller P. Seed production and germination in two rare and three common co-occurring *Acacia species* from south-East Australia. Austral Ecology. 2003;**28**(3):271-280. DOI: 10.1046/j.1442-9993.2003.t01-4-01287.x

[140] Khan D, Zaki M. Seedling characteristics of *Cassia fistula L. (Caesalpiniaceae)*. International Journal of Biology and Biotechnology. 2019;**16**:231-253

[141] Baskin C, Baskin JM. Seeds: Ecology, Biogeography, and, Evolution of Dormancy and Germination. Elsevier; 2014. pp. 1-1586. DOI: 10.1016/C2013-0-00597-X

[142] Clark FH. Senna hebecarpa (Fern.) Irwin & Barneby Northern Wild Senna. New England Plant Conservation Program Conservation and Research Plan. 2000. pp. 1-21. Available from: http://www.nativeplanttrust.org/documents/120/Sennahebecarpa.pdf

[143] Stone L, Fryirs K, Leishman M. Simulating the effect of environmental flow duration on seedling emergence from riparian seed banks of the Upper Hunter River, New South Wales. Environmental Engineering Ecology. 2020;**36**:607-619. DOI: 10.1002/rra.3376

[144] Chou Y-F, Cox RD, Wester DB. Smoke water and heat shock influence germination of shortgrass prairie species. Rangeland Ecology & Management. 2012;**65**(3):260-267. Available from: https://journals.uair.arizona.edu/index.php/jrm/article/download/22771/21563

[145] Penman T, Towerton A. Soil temperatures during autumn prescribed burning: Implications for the germination of fire responsive species? International Journal of Wildland Fire. 2008;**17**(5):572-578. DOI: 10.1071/WF07092

[146] Broadhurst LM, Young AG, Thrall PH, et al. Sourcing seed for *Acacia acinacea*, a key revegetation species in south eastern Australia. Conservation Genetics. 2006;**7**:49-63. DOI: 10.1007/s10592-005-7855-9

[147] Portlock CC, Shea SR, Majer JD, Bell DT. Stimulation of germination of *Acacia pulchella*: Laboratory basis for forest management options. Journal of Applied Ecology. 1990;**27**(1):319-324. DOI: 10.2307/2403588

[148] MacPhee E. Studies of the germination of seven Australian alpine and subalpine shrub species. 1998. Available from: http://hdl.handle.net/11343/39397

[149] Morgan A, Hamdoun A, Bashir N. Studies on seed germination and seedling emergence of Mesquite, Prosopis juliflora (Swartz) DC. in Sudan Gezira. Universal Journal of Agricultural Research. 2017;**5**:159-163. DOI: 10.13189/ujar.2017.050211

[150] Oduor N, Ngugi W, Gathui T. Sustainable tree management for charcoal production *Acacia species* in Kenya. Acacia Pocketbook. 2012. pp. 1-24

[151] Fangqiu Z, Chen Zuxu SD, Searle LX, Junj Z, Qiang L. Temperate Australian *Acacia species* elimination trials in Southern China. Recent Developments in Acacia Planting. 1998;**82**:36-44. Available from: https://www.aciar.gov.au/sites/default/files/legacy/node/13704/pr82_pdf_41862.pdf

[152] Melo L, Joao L, Magalhães I, Medeiros A, Sebastião MJ, Jairo J, et al. Temperature and substrate effects on the germination of *Caesalpinia ferrea Mart. Ex Tul*. African Journal of Agricultural Research. 2017;**12**:3348-3354. DOI: 10.5897/AJAR2017.12748

[153] Sean B, Bell DT. Temperature effects on the seed germination of ten Kwongan species from Eneabba, Western Australia. Australian Journal of Botany. 1990;**38**(5):451-458. DOI: 10.1071/BT9900451

[154] Zoghi Z, Azadfar D, Kooch Y. The effect of different treatments on seeds dormancy breaking and germination of Caspian Locust (*Gleditschia caspica*) tree. Annals of Biological Research. 2011;**2**(5):400-406. Available from: https://www.scholarsresearchlibrary.com/articles/the-effect-of-different-treatments-on-seeds-dormancy-breaking-and-germination-of-caspian-locust-gleditschia-caspica-tree.pdf

[155] Vilela AE, Ravetta DA. The effect of seed scarification and soil-media on germination, growth, storage, and survival of seedlings of five species of Prosopis L.(Mimosaceae). Journal of Arid Environments. 2001;**48**(2):171-184

[156] Al-Menaie H, Al-Ragam O, Al-Shatti A, Mathew M, Suresh N. The effects of different treatments on seed germination of the *Cassia fistula L.* and Cassia nodosa Buch.-Ham. Ex Roxb. in Kuwait. African Journal of Agricultural Research. 2010;**5**(3):230-325

[157] Aref I. The effects of light intensity on seed germination and seedling growth of Cassia fistula (Linn.), Enterolobium saman (Jacq.) Prain ex King. and Delonix regia (Boj) Raf. Riyadh, Saudi Arabia: Plant Production Department, College of Agriculture, King Saud University; 2002. pag. 1-12

[158] Brenan JPM. The genus Adenopodia (Leguminosae). Kew Bulletin. 1986;**41**(1):73-90

[159] Lynch AJJ, Gilfedder L, Kirkpatrick JB. The Tasmanian endemic shrub *Acacia axillaris:* Conservation ecology applied to the question of rarity or vulnerability. Australian Journal of Botany. 1999;**47**(1):97-109. DOI: 10.1071/BT96137

[160] Oliveira L, Bruno R, Jr A, Gonçalves E. Tratamentos pré-germinativos em sementes de Caesalpinia pulcherrima (L.) Sw. - Leguminosae. Revista Caatinga. 2010;**23**:71-76

[161] Paula D'Agostini Contreiras Rodrigues A, Oliveira A, Laura V, Yamamoto C, Chermouth K, Freitas M. Treatments for *Adenanthera pavonina L.* seed dormancy overcoming. Revista Árvore. 2009;**33**(4):617-623. DOI: 10.1590/S0100-67622009000400004

[162] Johnson LAS. Treatments to promote seed germination in Australian Acacias. Australian Acacias in Developing Countries. 1989;**16**:57-63. ISBN: 9780949511263

[163] Clarke B, McLeod I, Vercoe T. Trees for farm forestry: 22 promising species. Land & Water Australia/FWPRDC Joint Venture Agroforestry Program. 2009;**9**(15):1-20. ISBN: 1440-6845

[164] Marcar N, Crawford D, Leppert P, Jovanovic T, Floyd R, Farrow R. Trees for Saltland, a Guide to Selecting Native Species for Australia. Csiro Publishing; 1995. ISBN: 9780643101425

[165] Burns RM, Honkala BH. USDA Forest Service's Silvics of North America: Conifers. Vol. 1. US Department of Agriculture; 1990. Available from: https://www.srs.fs.usda.gov/pubs/misc/ag_654_vol1.pdf

[166] Burns RM, Honkala BH. USDA Forest Service's Silvics of North America: Hardwoods. Vol. 2. US Department of Agriculture; 1990. Available from: https://www.srs.fs.usda.gov/pubs/misc/ag_654_vol2.pdf

[167] Francis JK. USDA forest service's tropical tree seed manual. US Department of Agriculture; 2002. Available from: https://rngr.net/publications/ttsm

[168] Francis JK, Whitesell CD. USDA forest service's woody plant seed manual. US Department of Agriculture; 2008. Available from: https://www.fs.usda.gov/nsl/nsl_wpsm.html

[169] Houck MJ, Row JM. Row. Plant Guide for Partridge Pea (Chamaecrista fasciculata). USDA-Natural Resources Conservation Service, James E. "Bud" Smith Plant Materials Center, Knox City, Texas, 79529, and Manhattan Plant Materials Center. Manhattan, Kansas, 66502. Chamaecrista fasciculata

[170] Francis JK. Wildland Shrubs of the United States and Its Territories: Thamnic descriptions. USDA Forest Service, International Institute of Tropical Forestry, San Juan PR; Rocky Mountain Research Station, Fort Collins CO, General Technical Report. 2004;**1**:529-531. DOI: 10.2737/IITF-GTR-26

[171] Temperate Plants Database, Ken Fern. temperate.theferns. Chamaecrista fasciculata

[172] de Souza TV, Voltolini CH, Santos M, Paulilo S, Terezinha M. Water absorption and dormancy-breaking requirements of physically dormant seeds of *Schizolobium parahyba* (*Fabaceae - Caesalpinioideae*). Seed Science Research. 2012;**22**(3):169-176. DOI: 10.1017/S0960258512000013

[173] Native Plants of North America. Wildflower.org (University of Texas). Available from: https://www.wildflower.org/

Chapter 3

Trends in Germination Technology of Edible Seeds and Applications for Functional Food

Junzhou Ding, Hanieh Sadeghi, Hao Feng and Boris Nemzer

Abstract

This chapter summarizes recent progress in the application of novel germination technologies and their impact on seed germination and the improvement of the nutritional quality of germinated seeds and sprouts. Specifically, the effects of oxygen regulation, power ultrasound, pulsed electric fields, magnetic fields, high-pressure processing, and non-thermal plasma treatment on germination performance and nutritional properties are highlighted. Additionally, efforts are made to introduce the industrialization of sprouted grains, microgreens, and vegetable sprouts. The chapter also discusses the use of traditional methods, such as advanced rotating drums, fermenting tanks, and automatic germinating tray systems in vertical farming practices. Finally, current market trends of functional foods derived from germinated seeds are discussed.

Keywords: germination, edible seeds, abiotic stress, power ultrasound, plant growth regulators, functional food

1. Introduction

1.1 Edible seeds and germinating process

Until now, there has been no formal definition of "edible seeds." Generally, edible seeds include grains (rice, wheat, barley, oat, millet, etc.), legumes (mung beans, soybeans, peas, snow peas, kidney beans, chickpeas, lentils, etc.), vegetable seeds (broccoli, alfalfa, mustard greens, etc.), and nuts (sunflower seeds, pumpkin seeds, peanuts, almonds, etc.).

The germination and sprouting processes have been widely used in the food industry, such as malting barley for beer production, germinating brown rice, and sprouting beans in Asian countries. Researchers worldwide have reported the benefits of the germination process on the physicochemical properties and sensory acceptance of grain-based foods (bread, cereal, pasta, tortillas, bread crisps, etc.), including increased levels of oligosaccharides, amino acids, vitamins, and antioxidants, as well as improved digestibility and absorptivity [1, 2]. In recent years, germination-based products have emerged as a new addition to the food industry.

1.2 Controlled germination and environmental stresses

Since the germination process activates dormant enzymes in seeds, it reactivates seed metabolism and leads to the degradation of anti-nutrient and macro-nutrient compounds. The degree of germination affects the quality of the end-products. In addition to the enhanced nutritional value of germinated grain flour and sprouts, controlled germination is needed as a low-carbon green technique for soilless agriculture. The germination of seeds is triggered by external environmental factors, including temperature, moisture, oxygen, and lighting [2]. Controlled germination involves managing these environmental factors and the germination time. Furthermore, controlled germination also refers to the application of physical energy forms to stimulate germinating seeds and enhance the accumulation of health-promoting compounds [3, 4].

Plants have natural physiological and chemical responses to environmental stresses. These stresses trigger the elevated production of metabolites that can protect plant cells against damage. These responses provide practical pathways to enhance the health-promoting compounds in plants. For example, gamma-amobutyric acid (GABA) is a compound with known health benefits that function in plant stress resistance [5]; thus, the enhancement of GABA in germinating seeds has been widely reported. Some flavonoids are also found in grains and are proven to play a role in plant defense [6]. Accumulation of phenolic compounds is another typical response to environmental stress [7]. Therefore, the application of environmental stress treatment provides an option for improving functional biochemical compositions (GABA, etc.) and antioxidant activities during sprouting.

In recent years, various environmental stresses (also known as abiotic stresses or elicitors) have been applied and evaluated for their effects on the concentrations of nutritional compounds in sprouting seeds. While researchers have summarized the enhancement of bioactive components in sprouted grains using novel techniques such as ultrasound, cold plasma, microwave, and pulsed electric fields, further research is still needed to understand how these treatments enhance the germination process [8].

1.3 Technical application of physical field treatments on germination

In recent years, the application of various forms of physical energy, such as microwave radiation, ultrasound, high-pressure processing, ultraviolet, electric field energy, magnetic fields, and plasma, has emerged as an innovative approach to improve seed germination and the nutritional value [9, 10]. These non-chemical, physical treatments also offer a sustainable alternative with potential benefits in health-promoting metabolites [11]. The novel technologies discussed in the following sections have proven effective as efficient techniques in breaking seed dormancy and improving germination characteristics. Hence, the application of these novel technologies including power ultrasound, non-thermal plasma, pulsed electric fields, and magnetic fields is expected to significantly enhance the germination rate (**Table 1**).

1.3.1 Power ultrasound

Ultrasound, a high-efficiency, non-toxic, and environmentally friendly method of physical stimulation, involves mechanical sound waves produced by the oscillation of molecules within a medium. These waves propagate at frequencies exceeding the upper limit of human hearing (>20 kHz). Typically, ultrasound applications are classified based on their frequency, with distinctions like power ultrasound and

Treatment	Grain	Positive effect	References
Power Ultrasound	Seeds of watermelon, melon, leek, pepper, carrot, tomato and aubergine	Increased germination; increased the seedling quality of vegetable seeds	[12]
	Cucumber, rice, barley, pepper and watermelon seeds, Tartary buckwheat seeds	Increased germination;	[13–15]
Non-thermal Plasma	Barely	Increased germination percentage (93.3%)	[16]
	Wheat	Enhanced the germination rate; positive effects on seedling growth	[17]
		Modify flour functionality; increase the flour hydration properties and depolymerization of starch	[18, 19]
Pulsed Electric Fields (PEF)	Wheat	Increased water uptake, germination of seeds, and growth parameters of seedlings; A significant increase in total phenolic contents, chlorophylls, carotenoids, soluble proteins, and amino acids	[20]
Magnetic Fields	Wheat (10 mT for 10 min/15 min and 15 mT for 15 min)	Increased germination rate, the growth of the wheat plant and improved their fruit yield and yield parameters	[21]
	Corn	Increased germination; fresh mass, dry biomass, and contents of chlorophyll a, chlorophyll b, and total chlorophyll	[22]

Table 1.
Application of physical field energy stimulation for enhancing the germination of grains and health-promoting components.

diagnostic ultrasound [23]. The power ultrasound used in food treatment typically involves sound waves with frequencies ranging from 20 to 100 kHz and sound intensities of 10–1000 W/cm^2 [24].

Ultrasound's application spans a wide range of processes, depending on its intensity. It is widely used as a supplementary extraction and drying method [10] and is also employed for enzyme activation or deactivation, mixing, homogenization, emulsification, dispersion, preservation, stabilization, dissolution, and crystallization [25]. Additionally, it exhibits the capacity to enhance germination rates, induce specific physiological and biochemical alterations, and reduce germination time [26]. This is because the energy generated by ultrasound boosts the vitality of plant cells by enhancing the activity of enzymes like amylase and peroxidase, which stimulate cell division and accelerate growth [27]. They also showed that the ultrasonic treatment effectively reduces the germination period by 30–45% [28]. Ultrasound, as an emerging acoustic processing technology, has also been explored for its potential to enhance various physiological and biochemical changes in cereal seeds, thereby increasing their digestibility and bioavailability while enhancing the bioactive compound content and nutritional value, flavor, and quality of cereal sprouts [29].

Wang et al. used ultrasound to enhance the metabolites and polyphenol compounds including flavonoids, isoflavones, phenols, and coumarins in mung bean [27]. In other research studies, ultrasound treatment of seeds has been shown to promote germination rates, enhance seedling growth, and increase the production of beneficial active compounds [10]. Consequently, this technique has garnered escalating interest among researchers.

1.3.2 Non-thermal plasma

Non-thermal plasma, also known as cold plasma, is a partially ionized gas containing ions, electrons, and neutral particles. It can be produced under atmospheric and low-pressure conditions using radio frequency or microwave sources [30, 31]. Recent studies highlight that cold plasma technology is a potent and beneficial advancement for the food industry. This technology provides numerous advantageous including effective microbial decontamination of food products, processing of packaging materials, modification of food component functionalities, altering the hydrophilic/hydrophobic properties, and degradation of agrochemical residues [32].

In recent years, it is considered a rapid, cost-effective, and environmentally friendly method for enhancing seed performance, plant growth, and overall plant production. This technique significantly impacts various aspects of plant development and physiology, including the stimulation of seed germination and seedling growth, activation of photosynthesis, and regulation of carbon and nitrogen metabolism [31]. Treatment of wheat seeds with 80 W cold plasma significantly improved seed germination potential (6.0%) and germination rate (6.7%) compared to the control group. Also, the yield of treated wheat was 5.89% more than that of the control [33]. Argon plasma treatment has been linked to the enhancement of soybean seed germination and sprout growth through the upregulation of ATP demethylation levels. Similarly, cold plasma treatment of germinating mung beans has been shown to elevate the activity of hydrolytic enzymes such as amylase, protease, and phytase, thereby enhancing the germination process [30].

Guragain et al. [34] indicated that the final germination percentage, germination index, germination value, coefficient of velocity of germination, vigor index, and chlorophyll content were all improved in the case of cold plasma-treated radish (*Raphanus sativus*) and carrot (*Daucus carota sativus L.*) seeds.

1.3.3 Pulsed electric fields

Pulsed Electric Fields (PEF) is a non-thermal processing technology that involves the application of short, high-voltage pulses to food or biological materials placed between two electrodes. These pulses create an electric field within the material, causing permeabilization or structural changes in cell membranes. In a PEF system, there is a high-voltage source, capacitor bank, switch, and treatment chamber. When the sample is treated, it experiences a force per unit charge that can break down microorganism cell membranes for good. Therefore, PEF has been used for the inactivation of microorganisms in liquid foods by exposing them to high voltage short pulses. The dose is adjusted by electric field intensity, amount of the pulse, and treatment time.

While many past studies have explored the use of PEF to deactivate microorganisms in liquids and enhance plant cell permeability for compound extraction, recent research has begun investigating its application at lower energy levels for non-inactivation purposes, including seed treatment [35]. It is shown that treatment of soybean seeds at 0.1, 1.0, 10.0, and 100.0 Hz for 5 h per day for 20 days significantly increased the rate of seed germination, while 10 and 100 Hz PMFs showed the most effective response. In addition, activities of alpha-amylase, acid phosphatase, alkaline phosphatase, nitrate reductase, peroxidase, and polyphenol oxidase were increased in PMF (10 Hz)-exposed soybean plants [36]. Rezaei-Zarchi et al. [37] reported that exposure of alfalfa seed to PEF (at four voltages V1 = 1, V2 = 3, V3 = 5 and V4 = 7 as a major factor in both h1 = 2.5 and h2 = 5 control sample) significantly increased germination and seedling rates, especially at the treatment of V4h1 and V3h2. In the PEF-treated oat seed (99 monopolar, rectangular pulses lasting 10 μs each, with a frequency of 13 Hz and a nominal electric field strength of 2250 V/cm), Al-Khafaji et al. reported an increase in the growth of the root and shoot of oat seedlings [38].

1.3.4 Magnetic fields

Magnetic fields are regions around a magnetic material or a moving electric charge where the force of magnetism acts. They are characterized by the direction and strength of the force they exert. Magnetic fields play crucial roles in various natural phenomena. To conduct static magnetic field treatment, an electromagnet is utilized [20]. This electromagnet setup involves two sets of cylindrical coils made of enameled copper wire, each wound around an iron bar. These bars are then positioned vertically, one over the other, and held in place at their ends by metallic supports.

The application of magnetic field treatment has proven effective in promoting the growth and productivity of a wide range of crops, spanning barley, oat, rice, wheat, and maize, along with other significant crops like chickpea, sunflower, and tomato [39]. Exposure of bean or wheat seeds to static magnetic fields (4 or 7 mT) promoted the germination ratios, especially at 7 mT MF [40]. Static magnetic field strengths (from 0 to 300 mT in steps of 50 mT for 30, 60, and 90 min) on the seeds of soybean) was reported to have a wide range of positive physiological effects on germination-related parameters like water uptake, speed of germination, seedling length, fresh weight, dry weight, and vigor indices [41]. Magnetic field (from 0 to 250 mT in steps of 50 mT for 1-4 h) application significantly enhanced chickpea germination performance [9].

2. Industrialization of germination process

2.1 Soaking and washing system of seeds

Soaking seeds is a necessary step to enhance germination rates and speed up the seedling establishment, which can contribute to softening seed coats, facilitating water hydration, and providing operational possibility to create favorable conditions and moderate stress for controlled germination.

Based on the industrial sprouting experience, the key point of safe sprout production is "Cleaning Seeds". One of important considerations is that some damaged seeds are prone to fungal infections during soaking. Therefore, actions are needed to ensure proper hygiene and food safety during the soaking process, such as cleaning containers, sterilizing tools, and adding food-grade hydrogen peroxide solution to the soaking water.

Additionally, power ultrasound-assisted soaking and washing system can be an eco-friendly green technology since the ultrasound treatment can enhance the removal of microorganisms and dirt on rough surface of seeds. In the other hand, ultrasound-assisted soaking may create modification on enzyme and substrate structure thus facilitation the enzymatic hydrolysis process [42]. Ultrasound treatment during soaking not only significantly enhanced rice sprouting speed and starch hydrolysis, but also reduced energy use in germination period [43].

Numerous studies have proven that high-intensity ultrasound has shown effectiveness in the inactivation of pathogenic microorganisms in various food systems. Feng's lab developed a pilot-scale continuous-flow washing system with ultrasonic capability for sanitation [44]. Ultrasound transducer boxes with three frequencies (25, 40, and 75 kHz) were used in the unit, so that this ultrasound-assisted washing system provides spatial uniform ultrasound treatment. Nowadays, many manufacturers released new design of ultrasonic soaking and washing system, mostly for cleaning of hard objects, as shown in **Figure 1**.

2.2 Tanks for steeping and malting

In the technological history of malting processing, there are four major types: malting on floor, malting in a germination box/tank, utilizing separate germination drum and kilning vessels, and a combined germination and kilning system. **Figure 2a** shows grains in a large open germinating box. **Figure 2b** shows a typical steeping tank and drum-style system.

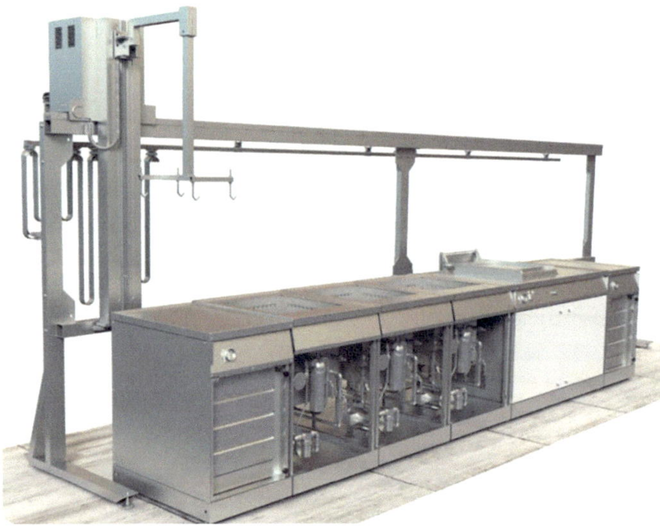

Figure 1.
Ultrasonic cleaning automatic system (emerson.com).

Figure 2.
(a) Malting in a large open germinating box (ukmalt.com), (b) steeping tank and drum-style system (maltersadvantage.com), and (c) grains in steeping tank (https://canadamalting.com/process/).

As disclosure information in malting and beer brewing industry, the malting process starts by moving grains into steeping tanks and absorbing water (**Figure 2c**). Then germination takes place in controlled condition, including temperature, humidity, and carbon dioxide.

2.3 Advanced rotary drums for sprouting

The Sentrex Sprout Equipment with cylindrical hydroponic growing System is a rotating system (**Figure 3**) and becomes one of the manufacturers for commercial sprout growing. Clean seeds are placed inside rotary drums, which incorporate cleaning, sanitation, and lighting systems. After rinsing, soaking, and germinating, sprouts grow to maturity, and then the wash system carries sprouts through a stainless steel tank, where they are cleaned, de-hulled, and de-rooted.

2.4 Vertical farming systems

As we know, developing sustainable technology to grow fresh produce under a controlled environment is an important strategy for practicing urban agriculture. Sprouts and microgreens are ideal fresh produce for indoor farming due to their rich nutrients and short growth cycle [45]. Vertical farming is an efficient way to provide a controlled environment by stacking plant beds vertically on shelves in modern greenhouses, which increases times of the yield on a given land area, reduces electricity costs with LED lights, and reduces water consumption with water recycling [46]. Therefore, vertical farming is also a sustainable way to grow ready-to-eat and pesticide-free sprouts since the indoor space is free of pests and dirt. On the other hand, vertical farming allowed the simultaneous investigation of multiple environmental stresses that affect sprout growth [47].

Various cultivation methods have been applied to grow sprouts and microgreens, including the traditional soil-based media and hydroponics. Furthermore, Du et al. shared their perspectives on developing biopolymer-based hydrogels as novel media (**Figure 4**) [45].

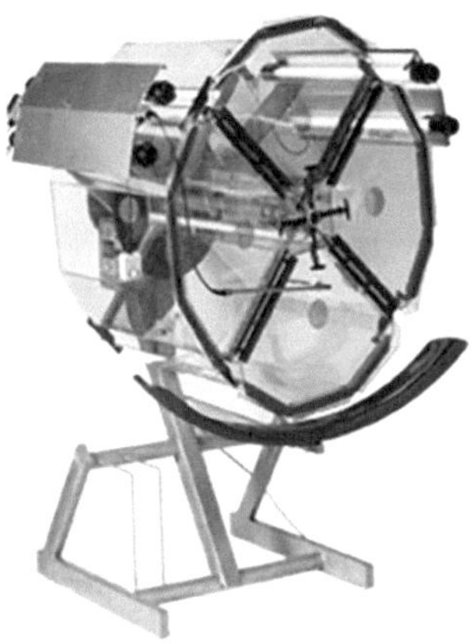

Figure 3.
Rota-tech rotary drum. ISS Rota-Tech Rotary Drum - International Specialty Supply (sproutnet.com).

Figure 4.
Vertical farming microgreens (modern farmer).

3. Functional foods trends

3.1 Germinated seeds products

Germination, known as sprouting, is versatile, suitable for implementation in home, traditional, and industrial environments [48]. The process of germination has gained widespread use in the manufacturing of diverse food products [8]. Beyond malting, a specific form of germination used in alcoholic beverage creation, edible seedlings can be consumed as ready-to-eat sprouts or subjected to further processing such as drying or roasting [49]. In recent years, numerous germinated grain-based products have been developed and companies globally are increasingly manufacturing sprouted grain products, deepening our understanding of germination and production techniques. **Figure 5** shows some packaged germinated/sprouted products available in the global market.

An alternative approach to enhance the nutritional content of bread involves supplementing flours with sprout wheat powder, especially sprouts derived from *Triticum* species, which have been identified as rich sources of bioactive compounds, including phytochemicals, phospholipids, reducing glycosides, and low molecular weight peptides [50]. Studies have shown improvements in the bread-making properties of sprouted wheat flour, the physicochemical traits and consumer acceptance of bread produced from germinated brown rice flour [1, 51], and the dough characteristics and bread-baking performance of wheat flour mixed with germinated quinoa flour [52]. Based on a Google search, the United States, Canada, and the United Kingdom are the most significant consumers of sprouted grain bread. In recent years, sprouted grain products have become popular in the food industry for their improved nutritional value and enhanced nutrient absorption [2]. They were found to possess high concentrations of bioactive compounds (polyphenols), GABA, oligosaccharides, amino acids, and antioxidants [53, 54]. This approach also tackles malnutrition by enriching foods with protein, calories, vitamins, and minerals [48]. Furthermore, it triggers the synthesis of secondary metabolites like vitamin C and polyphenols, which are valuable in nutraceuticals and functional foods. These changes enhance both the health and nutritional content of food products [55]. Sprouting effectively addresses grain processing challenges such as dehulling, lengthy cooking times, and undesirable taste profiles [56].

Figure 5.
Figure 5. Packaged germinated sprouted seed products available in the global market (Image source: cdnilbedhp.nitrocdn.com; foodsalive.com; essentialeating.com; bio-kinetics.com; goraw.com; silverhillsbakery.ca; mannaorganicbakery.com; purelivingorganic.com; m.cjcatalog.co.kr;).

Brand	Commercial sprouted grains products	Country
Oat House	Sprouted oatmeal	China
CJ Foods	Cooked budding sprouted brown rice	Korea
Essential Eating	Sprouted wheat pasta; sprouted whole grain flours	The United States
Manna Organics	Sprouted grain bread/cereal/gluten-free pasta	The United States
Silver Hills Bakery	Sprouted grain bread	The United Kingdom
Angelic Bakehouse	Sprouted grain bread; sprouted seven-grain bread crisps	Canada
One Degree Organics	Sprouted grain cereals, bread, and granolas	The United States
Bio-Kinetics	Sprouted grain bread, crackers, and cereal bars	Canada
Power Super Foods	Sprouted chia seeds, flaxseeds, and quinoa, which can be used in smoothies, baking, or sprinkled	South Africa
Gopal's Healthfoods	Sprouted grain and seed snacks including sprouted flaxseed crackers and sprouted mung bean snacks	Australia
Food for Life	Sprouted grain bread; sprouted grain end-products (bread, English muffins, tortillas, pasta, cereals, waffles)	India
Shiloh Farms	Sprouted grains breads; sprouted grain flours; sprouted nuts	The United States
Pastabilities	Sprouted grain pasta	The United States
Go Raw	Sprouted grain snacks including sprouted flaxseed crackers; Sprouted pumpkin/watermelon seeds	The United States
Unique Snacks	Sprouted grain snacks	The United States

Table 2.
Sprouted grains products retail brands.

A list of some food products formulated with different germinated seeds is given in **Table 2**, including sprouted wheat/barley bread, sprouted wheat pasta, sprouted grain flours (wheat, brown rice, sorghum, millet), and processed grain-based products (snacks, cereals, oatmeal, muffins, tortillas, etc.).

3.2 Germination-based marketing insights

As a potential source of numerous functional ingredients, germinated edible foods have attracted more and more attention for its physiological and sensory benefits. Therefore, germinated grain and legume seeds are expected to be functional dietary supplements with a variety of health benefits and medicinal values. Appropriate germination treatments can enrich bioactive components, which is conducive to the efficient production of bioactive functional food.

3.2.1 Germinated grains

Germinated grains-based foods have attracted much attention because of their unique physiological effects. Germinated whole grains come from wheat, barley, rice, or other cereal grains that are allowed to germinate slightly and grow a small sprout (**Figure 6**). The germinated grains can be processed into three forms such as dry and cooked rice or wheat berries. Dry and ground into flour as food ingredients, wet and blended into puree for future processing, then become a variety of breads, pastas, noodles, granolas, cookies, crackers, and more delicious grain-based food.

3.2.2 Malting grains for beer-producing

China has overtaken the United States to become the world's largest beer producer and has become the #1 consumption market. In 2023, China generated more than 125 billion U.S. dollars in revenues from its beer industry, while U.S. beer market generated $117–$119 billion in 2023 (Data from Statista & Brewers Association). Malt is referred to as the soul of beer, which provides enzymes, carbohydrates, and sugars necessary for fermentation. Two-row barley and other malting grains (wheat, rice, etc.) are selected by brewers and farmers to ensure high-quality ingredients for craft beer. The first step is soaking (steeping) to awaken the dormant grains. The second is germinating and sprouting. The third step is heating (kilning) the malting grains to create the final color (ranging from golden to red to black) and flavor (malty, sweet caramel, nutty, woody, burnt, etc.). Nowadays, the malting system consists of an external conical steeping vessel and germination-kilning combo drum with a connected aeration system to improve the beer production process more efficiently. The germination-kilning drum looks like a rotating cylinder unit (**Figure 7**).

3.2.3 Fresh sprouts of edible seeds

As shown in **Figure 8**, general edible sprouts can be counted into four categories: grain sprouts (wheatgrass, buckwheat and quinoa sprouts, etc.), legume sprouts (mung bean, soybean, pea, kidney sprouts, etc.), vegetable seeds sprouts (similar with microgreens, including broccoli, alfalfa, mustard greens, red clover sprouts, etc.), and nut sprouts (pumpkin seed, sunflower seed, peanuts sprouts, etc.).

Figure 6.
Germinated whole-grain rice and wheat berries. (A) www.homestatepng.com, (B) depositphotos.com, (C) foodtolive.com, and (D) sprouting.com.

Figure 7.
The SCHULZ malting system's germination-kilning combo drum Kaspar Schulz | Malting in detail (kaspar-schulz.de).

Figure 8.
Fresh sprouts of edible seeds. (A) Alfalfa (myemeraldhealth.com), (B) wheatgrass (thrivemarket.com), (C) broccoli (www.spsidahoinc.com), and (D) mung-bean (www.premierseedsdirect.com).

3.3 Current trends and future perspectives

Germination-based foods are becoming increasingly popular worldwide due to growing concerns about health. Firstly, developing germinated products with local grains and edible seeds can attract more consumers by catering to diverse tastes. Secondly, the germination process is now fully controlled using advanced technologies to enhance food safety and enrich nutrient content, resulting in high-quality end products. Thirdly, modern biotechnology offers great potential for the development of fermentation and extraction technologies, creating novel ingredients and improving the nutritional function of germinated seed-based foods. The nutritional and health benefits of germination-based foods are well-known among food scientists, dietitians, chefs, food manufacturers, and others. Increasing evidence from human studies is proving their positive effects on human health.

4. Concluding remarks

The germination process significantly enhances the nutritional profile of sprouts through enzymatic activation, which hydrolyzes complex macromolecules into simpler, more digestible forms. This reduction in anti-nutritional factors such as phytic acid enhances mineral bioavailability and increases concentrations of essential vitamins and minerals, including vitamins C and E, B vitamins, magnesium, zinc, and iron. Germination also induces the biosynthesis of bioactive compounds like antioxidants, flavonoids, and phenolic acids, which contribute to health benefits. Enhanced dietary fiber content and improved digestibility further elevate the nutritional value, while beneficial amino acids like GABA are synthesized. These improvements make germinated seeds highly nutritious and health-promoting. Germination-based foods are gaining global traction due to rising health consciousness. Developing products with local grains and seeds caters to diverse palates, while advanced technologies ensure precise control over germination, boosting food safety and nutrient density. Modern biotechnological applications facilitate the development of fermentation and extraction processes, creating novel bioactive ingredients and enhancing the nutritional functions of germinated foods. The benefits of germination-based foods are well-recognized by food technologists, nutritionists, culinary professionals, and food manufacturers, with growing evidence supporting their positive health impacts. Physical and chemical treatments further optimize germination efficiency and nutrient profiles. Technologies like oxygen regulation, electrolyzed oxidizing water, ultrasonication, pulsed electric fields, magnetic fields, high-pressure processing, and non-thermal plasma play crucial roles in enhancing germination. The commercialization of sprouted grains, microgreens, and vegetable sprouts is driven by advanced systems like rotating drum sprouts, fermenters, and automated germination trays in vertical farming setups. These technological advancements foster the development of innovative, nutrient-dense food products, addressing the increasing consumer demand for health-enhancing diets. It is foreseeable that in the near future, germination-based food products may become a common feature in health-conscious diets worldwide.

Conflict of interest

The authors declare no conflict of interest.

Author details

Junzhou Ding[1], Hanieh Sadeghi[1], Hao Feng[1] and Boris Nemzer[2,3]*

1 North Carolina A&T State University, Greensboro, NC, USA

2 VDF FutureCeuticals, Momence, IL, USA

3 University of Illinois at Urbana-Champaigne, Urbana, IL, USA

*Address all correspondence to: bnemzer@futureceuticals.com

References

[1] Charoenthaikij P, Jangchud K, Jangchud A, Prinyawiwatkul W, No HK. Composite wheat-germinated brown rice flours: Selected physicochemical properties and bread application. International Journal of Food Science and Technology. 2012;**47**(1):75-82

[2] Nelson K, Stojanovska L, Vasiljevic T, Mathai M. Germinated grains: A superior whole grain functional food? Canadian Journal of Physiology and Pharmacology. 2013;**91**(6):429-441

[3] Hübner F, Arendt EK. Germination of cereal grains as a way to improve the nutritional value: A review. Critical Reviews in Food Science and Nutrition. 2013;**53**(8):853-861

[4] Singh A, Sharma S. Bioactive components and functional properties of biologically activated cereal grains: A bibliographic review. Critical Reviews in Food Science and Nutrition. 2017;**14**(57):3051-3071

[5] Fait A, Yellin A, Fromm H. GABA shunt deficiencies and accumulation of reactive oxygen intermediates: Insight from Arabidopsis mutants. FEBS Letters. 2005;**579**(2):415-420

[6] Moheb A, Agharbaoui Z, Kanapathy F, Ibrahim RK, Roy R, Sarhan F. Tricin biosynthesis during growth of wheat under different abiotic stresses. Plant Science. 2013;**201**(1):115-120

[7] Cheynier V, Comte G, Davies KM, Lattanzio V, Martens S. Plant phenolics: Recent advances on their biosynthesis, genetics, andecophysiology. Plant Physiology and Biochemistry. 2013;**72**:1-20

[8] Deepika K, Hamid H, Chavan P, Jaiswal AK, Thakur A, Dhiman AK. A comprehensive review on sprouted seeds bioactives, the impact of novel processing techniques and health benefits. Food Reviews International. 2024;**40**:370-398

[9] Vashisth A, Nagarajan S. Exposure of seeds to static magnetic field enhances germination and early growth characteristics in chickpea (*Cicer arietinum* L.). Bioelectromagnetics. 2008;**29**:571-578

[10] Wang J, Ma H, Wang S. Application of ultrasound, microwaves, and magnetic fields techniques in the germination of cereals. Food Science and Technology Research. 2019;**25**(4):489-497

[11] Rifna EJ, Ratish Ramanan K, Mahendran R. Emerging technology applications for improving seed germination. Trends in Food Science and Technology. 2019;**86**:95-108

[12] Memis N, Njie ES, Muhie S, Sahin G, Demir I. Anadolu Tarim Bilimleri Dergisi 2022. Effect of ultrasonic treatment on seed germination and seedling emergence in seven vegetable species. Anadolu Journal of Agricultural Sciences. 2022;7(1):57-66

[13] Gossous SJ, Samarah NS, Alquadah AM, Othman MO. Enhancing seed germination of four crop species using an ultrasonic technique. Experimental Agriculture. 2010;**46**(2):231-242

[14] Miano AC, Forti VA, Abud HF, Gomes-Junior FG, Cicero SM, Augusto PED. Effect of ultrasound technology on barley seed germination and vigour. Seed Science and Technology. 2015;**43**(2):297-302

[15] Wang J, Bian Z, Wang S, Zhang L. Effects of ultrasonic waves, microwaves, and thermal stress treatment on the germination of Tartary buckwheat seeds. Journal of Food Process Engineering. 2022;**43**:e13494

[16] Feizollahi E, Iqdiam B, Tasanthan A, Thilakarathna MS, Roopesh MS. Effects of atmospheric-pressure cold plasma treatment on deoxynivalenol degradation, quality parameters, and germination of barley grains. Applied Sciences. 2020;**10**:3530

[17] Los A, Ziuzina D, Boehm D, Patrick J, Bourke P. Investigation of mechanisms involved in germination enhancement of wheat (*Triticum aestivum*) by cold plasma: Effects on seed surface chemistry and characteristics. Plasma Processes and Polymers. 2019;**16**:e1800148

[18] Bahrami N, Bayliss D, Chope G, Penson S, Perehinec T, Fisk D. Cold plasma: A new technology to modify wheat flour functionality. Food Chemistry. 2016;**202**:247-253

[19] Chaple S, Sarangapani C, Jones J, Carey E, Causeret L, Genson A, et al. Effect of atmospheric cold plasma on the functional properties of whole wheat (*Triticum aestivum L.)* grain and wheat flour. Innovative Food Science and Emerging Technologies. 2020;**66**:102529

[20] Ahmed Z, Manzoor MF, Ahmad N, Zeng X, Din ZU, Roobab U, et al. Impact of pulsed electric field treatments on the growth parameters of wheat seeds and nutritional properties of their wheat plantlets juice. Food Science and Nutrition. 2020;**8**:2490-2500

[21] Sabir Hussain M, Dastgeer G, Afzal AM, Hussain S, Kanwar RR. Eco-friendly magnetic field treatment to enhance wheat yield and seed germination growth. Environmental Nanotechnology, Monitoring and Management. 2020;**14**:100299

[22] Ortiz J, Suarez D, Puentes A, Velasquez P, Santis Navarro A. Comparison of the effects in the germination and growth of corn seeds (*Zea Mays L.*) by exposure to magnetic, electrical and electromagnetic fields. Chemical Engineering Transactions. 2015;**43**:169-174

[23] Feng H, Yang W, Hielscher T. Power ultrasound. Food Science and Technology International. 2008;**14**(5):433-436

[24] Kentish S, Feng H. Applications of power ultrasound in food processing. Annual Review of Food Science and Technology. 2014;**5**:14.1-14.22

[25] Gallo M, Ferrara L, Naviglio D. Application of ultrasound in food science. Food. 2018, 2018;7(10):164

[26] Naumenko N, Potoroko I, Kalinina I. Stimulation of antioxidant activity and γ-aminobutyric acid synthesis in germinated wheat grain *Triticum aestivum L.* by ultrasound: Increasing the nutritional value of the product. Ultrasonics Sonochemistry. 2022;**86**:106000

[27] Wang LD, Li XQ, Gao F, Liu Y, Lang SJ, Wang CY, et al. Effect of ultrasound combined with exogenous GABA treatment on polyphenolic metabolites and antioxidant activity of mung bean during germination. Ultrason Sonochem. 2023;**94**:106311

[28] Yaldagard M, Mortazavi SA, Tabatabaie F. Application of ultrasonic waves as a priming technique for accelerating and enhancing the germination of barley seed: Optimization of method by the Taguchi approach.

Journal of the Institute of Brewing. 2008;**114**(1):14-21

[29] Jaime A, Teixeira da Silva JA, Dobra'nszki J. Sonication and ultrasound: Impact on plant growth and development. Plant Cell, Tissue and Organ Culture. 2014;**117**:131-143

[30] Zhang JJ, Jo JO, Huynh DL, Mongre RK, Ghosh M, Singh AK, et al. Growth-inducing effects of argon plasma on soybean sprouts via the regulation of demethylation levels of energy metabolism-related genes. Scientific Reports. 2017;7:41917

[31] Ling L, Jiangang L, Minchong S, Chunlei Z, Yuanhua D. Cold plasma treatment enhances oilseed rape seed germination under drought stress. Scientific Reports. 2015;**5**:13033

[32] Ekezie FGC, Sun DW, Cheng JH. A review on recent advances in cold plasma technology for the food industry: Current applications and future trends. Trends in Food Science and Technology. 2017;**69**:46e58

[33] Jiang J, Xin HX, Ling LI, Jiangang LI, Hanliang S, Qilai X. Effect of cold plasma treatment on seed germination and growth of wheat. Plasma Science and Technology. 2014;**16**(1):54-58

[34] Guragain RP, Baniya HB, Shrestha B, Guragain DP, Subedi DP. Non-thermal plasma: A promising technology for the germination enhancement of radish (*Raphanus sativus*) and carrot (*Daucus carota sativus* L.). Journal of Food Quality. 2023;**4131657**:15

[35] Sitzmann W, Vorobiev E, Lebovka N. Applications of electricity and specifically pulsed electric fields in food processing: Historical backgrounds. Innovative Food Science and Emerging Technologies. 2016;**37**:302-311

[36] Radhakrishnan R, Kumari BD. Influence of pulsed magnetic field on soybean (*Glycine max* L.) seed germination, seedling growth and soil microbial population. Indian Journal of Biochemistry and Biophysics. 2013;**50**(4):312-317

[37] Rezaei-Zarchi S, Imani S, Alikhani Mehrjerdi H, Mohebbifar MR. The effect of electric field on the germination and growth of medicago sativa planet, as a native Iranian alfalfa seed. Acta Agriculturae Serbica. 2012;**34**:105-115

[38] Al-Khafaji AH, Kwao S, Galindo FG, Sajeevan RS. Germination and stress tolerance of oats treated with pulsed electric field at different phases of seedling growth. Bioelectrochemistry. 2024;**158**:108692

[39] Kang T, You Y, Jun S. Supercooling preservation technology in food and biological samples: A review focused on electric and magnetic field applications. Food Science and Biotechnology. 2020;**29**(3):303-321

[40] Cakmak T, Dumlupinar R, Erdal S. Acceleration of germination and early growth of wheat and bean seedlings grown under various magnetic field and osmotic conditions. Bioelectromagnetics. 2010;**31**:120-129

[41] Shine MB, Guruprasad KN, Anand A. Enhancement of germination, growth, and photosynthesis in soybean by pre-treatment of seeds with magnetic field. Bioelectromagnetics. 2011;**32**:474-484

[42] Xu B, Azam SMR, Feng M, Wu B, Yan W, Zhou C, et al. Application of multi-frequency power ultrasound in selected food processing using large-scale reactors: A review. Ultrasonics Sonochemistry. 2021;**81**:105855

[43] Ding J, Hou GG, Dong M, Xiong S, Zhao S, Feng H. Physicochemical properties of germinated dehulled rice flour and energy requirement in germination as affected by ultrasound treatment. Ultrasonics Sonochemistry. 2018;**41**:484-491

[44] Zhou B, Feng H, Pearlstein AJ. Continuous-flow ultrasonic washing system for fresh produce surface decontamination. Innovative Food Science and Emerging Technologies. 2012;**16**:427-435

[45] Du M, Xiao Z, Luo Y. Advances and emerging trends in cultivation substrates for growing sprouts and microgreens toward safe and sustainable agriculture. Current Opinion in Food Science. 2022;**46**:100863

[46] Oh S, Lu C. Vertical farming—Smart urban agriculture for enhancing resilience and sustainability in food security. The Journal of Horticultural Science and Biotechnology. 2023;**98**(2):133-140

[47] Farhangi H, Mozafari V, Roosta HR, Shirani H, Farhangi M. Optimizing growth conditions in vertical farming: Enhancing lettuce and basil cultivation through the application of the Taguchi method. Scientific Reports. 2023;**13**:6717

[48] Mahsa M, Ziyu W, Shahla T, Nelum P, Charles SB, Asgar F. Unlocking the potential of sprouted cereals, pseudocereals, and pulses in combating malnutrition. Food. 2023;**12**:390

[49] Paolo B, Beatrice F, Stanley L, Fabio S, Angelica G. Sprouted grains: A comprehensive review. Nutrients. 2019;**11**:421

[50] Beatrice F, Isabella C, Lilia G, Renzo T, Livia P, Vincenzo V, et al. Phenolic content and antioxidant activity of wholegrain breads from modern and old wheat (*Triticum aestivum* L.) cultivars and ancestors enriched with wheat sprout powder. Italian Journal of Agronomy. 2018;**13**:1220

[51] Charoenthaikij P, Jangchud K, Jangchud A, Prinyawiwatkul W, Tungtrakul P. Germination conditions affect selected quality of composite wheat-germinated brown rice flour and bread formulations. Journal of Food Science. 2010;**75**(6):312-319

[52] Park SH, Morita N. Dough and breadmaking properties of wheat flour substituted by 10% with germinated quinoa flour. Food Science and Technology International. 2005;**11**(6):471-476

[53] Singh AK, Rehal J, Kaur A, Jyot G. Enhancement of attributes of cereals by germination and fermentation: A review. Critical Reviews in Food Science and Nutrition. 2015;**55**(1):1575-1589

[54] Patricio JC, Elena P, Cristina MV, Patricia GM, Juana F. Development of a multifunctional yogurt-like product from germinated brown rice. LWT. 2019;**99**:306-312. DOI: 10.1016/j.lwt.2018.10.008. Available from: https://www.sciencedirect.com/science/article/pii/S0023643818308260

[55] Ali I, Farhan S, Muhammad A, Ali I, Bushra N, Tabussam T, et al. Nutritional and end-use perspectives of sprouted grains: A comprehensive review. Food Science and Nutrition. 2021;**9**:4617-4628

[56] Erba D, Angelino D, Marti A, Manini F, Faoro F, Morreale F, et al. Effect of sprouting on nutritional quality of pulses. International Journal of Food Sciences and Nutrition. 2019;**70**:30-40

Chapter 4

Utilizing a Thermogradient Device to Establish the Germination Vigor Response of Crop Seeds

Timothy L. Grey, Juliana de Souza Rodrigues and Samantha J. Bowen

Abstract

The initiation of seed germination is a critical first step in establishing many plant species associated with agronomic crop production. Crop species are uniformly tested for standard germination and sometimes vigor prior to market release but, these assessments are predominantly implanted for commercial production of registered and certified seed. Seed germination, vigor, and initial seedling growth can be evaluated using a thermogradient device that holds constant temperatures across a wide range for evaluation. Models can then be used to establish germination, vigor, and seedling growth indices related to time, temperature, and response to predict genomic stability, and phenotypic responses due to variations in climate during seed production. This will be reviewed in this chapter.

Keywords: germination, nonlinear regression, phenological development, thermal time, seedling growth

1. Introduction

Grain and legume crops form foundational cornerstones for agronomic production worldwide. In 2022, corn (*Zea mays* L.) and wheat (*Triticum aestivum* L.) were the most widely cultivated grain crops in temperate regions globally, covering over 200,720,000 and 219,580,000 hectares, respectively [1]. Corn and wheat are produced in all regions of the world with the cultivation of these crops occurring somewhere on the planet practically every month of the year. The legume dicot crops soybean [*Glycine max* (L) Merr.] and peanut (*Arachis hypogaea* L.) were grown on 136,900,000 and 28,940,000 hectares in 2022, respectively [1]. Soybean is the most widely grown oilseed crop providing two commonly used dietary products for human and animal production, oil and protein. These soybean components are obtained after chemical extraction and heating which is then followed by various processing methods (pressing, grinding) [2]. In contrast, peanut is consumed by cooking with other foods, roasted and then processed directly into peanut butter (or paste), or the oil extracted for dietary use as a food. Peanuts are also a major component used in many confectionary products, either processed or as a whole nut, such as in candy bars. Another dicot crop, cotton (*Gossypium* spp.), has been dually adapted over time, most widely grown for fiber around the world, but is also an oilseed crop [3]. In 2022, over 31,250,000 hectares of cotton were grown in tropical and subtropica-l environments [1]. Two different biotypes of cotton are commonly grown including *G. hirsutum* (L.), known as upland, and G. *barbadense* (L.) known as pima and egyptian.

One factor all these crops commonly share is the annual planting of seed for initial establishment. These species have origination from all around the world and were initially adapted from wild species. Corn (maize) has been associated with teosinte species in Mexico, domesticated around 7000 years ago [4]. Wheat has genetic associations with the *Aegilops*, *Amblyopyrum*, and *Triticum* genera, which have origins in the Mediterranean-Southwest Asia and Arctic-Temperate regions, with some of the earliest domestication occurring over 10,000 years ago [5]. Soybean is believed to have been domesticated from *Glycine soya* (Sieb. and Zucc.) from east Asia 6000–9000 years ago [6]. Peanut, associated with Arachis species from South America, has archaeological evidence suggesting domestication as far back as 8500 years ago [7]. Cotton, from the *Gossypium* species has origins in Africa, but has been domesticated worldwide over the past 5000 to 8000 years, largely due to the natural distribution of seeds by nature [8]. Domestication of these crops has removed almost all forms of seed dormancy. However, peanut still have some inherent seed dormancy traits that appear in dry seed following the after-ripening process. However, as abscisic acid-like inhibitors decrease in peanut seed over time, the dormancy variable is released which ultimately allows for germination [9–11].

1.1 Importance

Reliable seed quality is one of the most critical aspects of agronomic grain production. Crop seed germination that is consistent and vigorous is a key element for successful farming. While genotypic traits can be modified via traditional and modern breeding methods for grass crops, legumes, and lint/oil seed crops [12], genetic modification has become a major source of improving yield, altering herbicide tolerance, and developing pest resistance [13]. Field testing and evaluation of newly developing seedlings is often difficult due to soil interaction. As seeds are often buried 2 to 5 cm deep, early seedling development is difficult to measure under varying temperature conditions. To measure initial seedling development, controlled experiments are often employed to monitor growth over time.

1.2 Background information on seed production

Seed production is an international multi-billion-dollar industry. Major seed companies associated with the previously mentioned crops are often international conglomerates that utilize patents and/or via genetically modified traits to ensure continued annual grower purchases. Additionally, there are public plant breeding programs in the United States. These organizations are usually associated with peanut, soybean, and wheat cultivar development and work in conjunction with Land-Grant universities and state-affiliated seed production organizations. Standards for seed quality and purity are set by national and international organizations such as the Association of Official Seed Certifying Agencies (AOSCA) and the International Seed Testing Association (ISTA) [14, 15]. Additionally, seeds sold by private companies are tested and displayed via tag for purity and standard germination.

1.2.1 Seed production requirements

Seed vigor and crop establishment are research topics that have been investigated for multiple decades. A recent review by Finch-Savage and Bassel [16], outlined key components of crop seed production, inherited traits associated with genetic stability, and the environmental effects on seed vigor and performance. Emphasis was placed on an increase in seed vigor over time, driven by the parental plant's influence of promoting endosperm growth and development in the maturing seed. As the seed develops, the parent plant presumably dies, shifting the responsibility for maintaining vigor to the harvest and storage conditions seed are exposed too. Overall germination and vigor decline over time for crop seeds in storage.

1.2.2 Seed storage and deterioration

Morton et al. [17] reviewed the factors that contribute to seed deterioration over time during storage and stated that seed moisture content, mechanical and insect damage, pathogen attack, seed maturity, relative humidity, and temperature can have negative impacts. These impacts have been quantified by previous research on peanuts [18, 19], soybean [20–23], corn [24], cotton [25, 26], and wheat [27, 28]. Vigor testing can be utilized to evaluate seeds for successful field establishment under different environmental conditions [29]. Strong primary seedling development in standard germination testing indicates strong vigor [14], but this does not always translate into adequate field performance.

1.2.3 Seed germination modeling with base temperature

Seed germination modeling is dependent on base temperatures set for crops by years of research across multiple cultivars and seasons. These requirements vary by crop with respect to models utilizing growing degree days as a measure of thermal accumulation based on maximum and minimum temperatures [30]. These models used base temperatures of 10°C for corn [31]; 10°C for soybean [32]; 15.5°C for cotton [33]; 13°C for peanut [34]; and 0°C for wheat [35, 36]. Previous seed vigor research with peanut [37–43], soybean [44], clovers (*Trifolium* species) [45], and cotton [46] using base temperatures as a basis for modeling has demonstrated utility and consistency.

1.2.4 Seedling formation

The progression of germination is strictly linked to the rate of water uptake. Initially, there is a rapid imbibition of water by the dry seed (phase I) until the seed tissues become fully hydrated. This is followed by a slower, more controlled water uptake during phase II, and in phase III, water uptake increases again, signaling the completion of germination. Phase II is marked by increased metabolic and cellular activity and is particularly critical, as it is associated with a range of cellular and biochemical processes, including DNA repair and the translation of both stored and newly synthesized mRNAs [47].

The ability of seedlings to develop properly after emergence also depends on their metabolic efficiency in mobilizing nutrients from the cotyledons and transitioning to photosynthesis. A rapid transition from reliance on seed reserves to photosynthesis is crucial for seedlings to become independent of the seed nutrient reserves and begin proper plant establishment in the soil [48]. These factors are governed by the seeds' inherent physiological and genetic potential, as well as the environmental conditions experienced during storage.

2. Research

The root growth (cm) of five crops was measured over six constant temperatures (15°C, 20°C, 25°C, 30°C, 35°C, and 40°C) for 7 days. These measures were taken every 24, 48, 72, 96, 120, 144, and 168 hours. Certified seeds of corn (DKC 70-45), cotton (DP2127B3XF), peanut (GA06G), soybean (AG58VF3), and wheat (UNK) were evaluated for seedling growth. The experiment followed a completely randomized design and was repeated six times. Eight seeds of each crop were used per dish, and the results from the root lengths (cm) were averaged, constituting one experiment. Each experiment was considered a replicate, generating six replicates per crop.

2.1 Thermogradient apparatus

Crop seeds were placed in a 100 × 25 mm sterile Petri dish (Fisher Scientific Education, Hanover Park, IL) containing germination paper (SDB 86 mm, Anchor Paper Co., St. Paul, MN) and watered with 10 ml of deionized water to initiate germination. The Petri dishes were arranged on a thermogradient table featuring columns of cells maintained at constant temperatures of 15°C, 20°C, 25°C, 30°C, 35°C, and 40°C. The thermal gradient table was constructed from a solid aluminum block measuring 2.4 m long by 0.9 m wide by 7.6 cm thick with a mass of 470 kg (**Figure 1**).

Figure 1.
Thermal gradient table used to determine optimum crop seedling growth over time (photograph by Sidney Cromer).

On each end of the aluminum blocks, a 1.0 cm hole was drilled across the side section to allow fluid to be pumped into the table. One side contains a chiller set at 12° C and on the other side, a heating unit at 32°C. Each unit pumps a mixture of ethylene glycol plus water (1:10 mixture) at a rate of 3.8 L per minutes to generate the thermal gradient. The solutions from the chiller and heating units were independent of each other and never mixed. A grid pattern consisting of 10 cm by 10 cm cells allowed for even distribution of individual Petri dishes. This resulted in 24 increments across the established temperature gradient, with nine cells at each temperature [49, 50], resulting in 216 total cells. Thermocouples made from duplex insulated PR-T-24 wire (Omega Engineering, Inc. Stamford, CT) were mounted to the underside of the table in the center of each cell. The wire was inserted vertically into a hole on the underside of the table held in place by a washer and screw. Holes measured 8 mm wide by 7 cm deep to allow the thermocouple to be within 5 mm of the upper table surface at 10 cm intervals along the length of the table. Temperature was monitored continuously for each thermocouple and recorded at 30 minutes intervals with a Graphtec data logger (Graphtec America, Inc. Irvine CA). Data indicated a continuous temperature gradient ranging from 12°C to 40°C resulting in changes of 1.67°C in each cell along the length of the table.

2.2 Image capture

Daily images of the seeds captured 24, 48, 72, 96, 120, 144, and 168 hours after the experiment started were taken using a Canon EOS 7D camera equipped with an 18–55 mm lens. Seeds were placed on a black panel to enhance root visualization in the images (**Figure 2**). Digital images were then analyzed using ImageJ software [51–53].

To ensure accurate root length measurement, the scale in ImageJ was set to 1 cm based on a ruler placed next to the seeds (**Figure 2**). The "freehand line tool" was used to trace the roots according to their architecture and length. The scale was recalibrated when switching crops, restarting the software, and adjusting zoom resolutions during

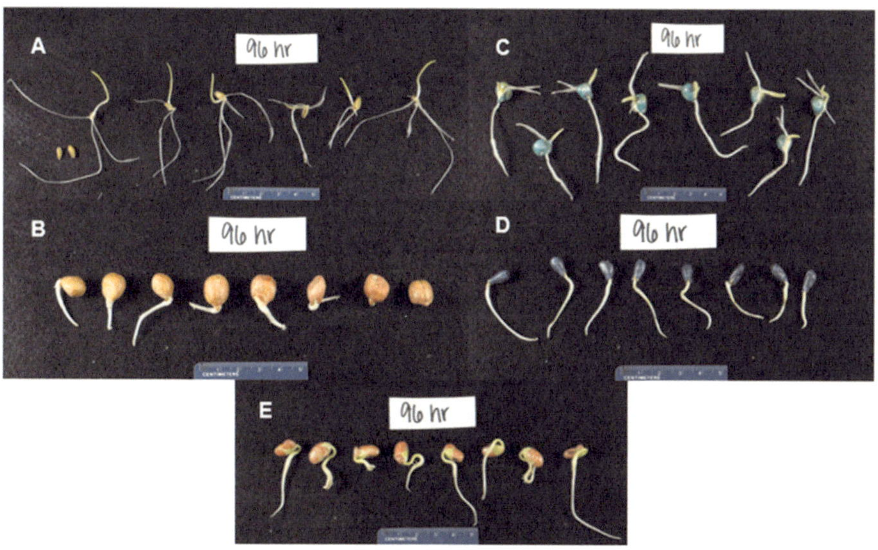

Figure 2.
A black background used to improve crop seed and root visualization for further analysis on ImageJ software. Crops included (A) wheat, (B) peanut, (C) corn, (D) cotton, (D) soybean, and (E) peanut.

image capture. Root growth rates (cm/day) were calculated by subtracting the length of roots measured at consecutive time points: 0–24, 24–48, 48–72, 72–96, 96–120, 120–144, and 144–168 hours.

2.3 Models

2.3.1 Log logistic model for plant growth temperature comparisons

Log-logistic models are widely used in plant growth modeling [54–57]. In this study, these models were applied to describe the root growth rate (cm/day) over 24, 48, 72, 96, 120, 144, and 168 hours at specific temperatures of 15°C, 20°C, 25°C, 30°C, 35°C, and 40°C. Each model was adjusted for different crops, including corn, cotton, peanut, soybean, and wheat.

For model validation, residual analysis was conducted to examine homoscedasticity (constant variance of the response variables) and to identify possible outliers. Model assumptions of normal distribution were assessed using a Q-Q plot, while homogeneous variance was evaluated with a residual versus fitted values graph [58, 59]. The models were compared using the bias-corrected Akaike information criterion (AICC) [60] and the lack-of-fit test [61]. Additionally, a transform-both-sides approach, employing the Box-Cox power family transformation [62], was used to improve the model fit [63]. After reevaluating the model fit and assumptions, the three-parameter log-logistic model was chosen.

The log-logistic model with a lower (c) limit set as 0, was defined as:

$$f(x) = 0 + \frac{d - 0}{1 + \exp(b(\log(x) - \log(e)))} \tag{1}$$

where d is the higher growth rate, c is the lower asymptote (which has a lower limit of 0 as previously described), e is X value producing a response half the maximum growth rate (between d and c), while b is the slope around the inflection point.

Model fitting and comparisons were conducted using the *drc* package in R software [64, 65]. Based on the model fit, the temperatures of 15°C, 20°C, 25°C, 30°C, 35°C, and 40°C were used to fit the curves for corn and wheat. For peanut, soybean, and cotton, data generated at 15°C did not fit this model as it resulted in nonsensical parameter estimates (data not shown). Model fit test, following Bates and Watts [61], indicated that the three-parameter log-logistic model fit the data ($p > 0.05$). The slope/temperature comparison was tested using an F-test for the extra-sum-of-squares with Bonferroni correction for multiple comparisons, utilizing the *aomisc package* [63]. A 5% error level was adopted for all statistical analyses. Data visualization was performed using R software.

2.3.2 Lorentzian modeling to predict time and temperature to maximum seedling growth

In order to establish the relationship between time, temperature, and seedling growth, a three-dimensional non-linear regression was conducted in SigmaPlot (SigmaPlot 15.0. SPSS Inc. 233 S. Wacker Dr., 11th Floor, Chicago, Illinois). The non-linear relationship between the two independent variables of time and temperature were regressed to seedling growth. A Lorentzian distribution model,

$$z = \frac{a}{\left[\left(1 + ((x - x_0)/b)^2\right) * \left(1 + ((y - y_0)/c)^2\right)\right]} \tag{2}$$

was fitted to seedling growth data. Temperature (x) and time (y) data were used to model the parameters (a, b, and c) to predict the maximum temperature (x_0) and maximum time (y_0) that produced the maximum seedling growth (z), similar to other research [55]. Data for cm total growth, and by each measure to establish the maximum, was modeled. The Lorentzian distribution belongs to the category of t-distribution with wider and more comprehensive analysis than other models when there are outlier data points [66].

3. Results and discussions

This study analyzed root growth rates of corn, wheat, peanuts, soybeans, and cotton using a three-parameter log-logistic model, supplemented by Lorentzian regression to estimate the temperature (x_0) and time (y_0) required to achieve maximum growth (**Table 1/Figure 3** and **Table 2/Figure 4**, respectively).

For corn, root growth rates increased gradually at higher temperatures (25°C, 30°C, 35°C, and 40°C), with an average of 60 hours required to reach half the maximum growth rate (*e*). However, at cooler temperatures (15°C and 20°C), the time to reach (*e*) increased significantly to approximately 375 hours and 86 hours, respectively, indicating corn's greater sensitivity to low temperatures and its preference for warmer conditions. Lorentzian regression estimated that corn achieved its maximum growth at 30.3°C, requiring approximately 124 hours, highlighting the importance of higher temperatures for optimal root development.

Wheat exhibited a similar trend but with a broader temperature tolerance. At moderate temperatures (25°C, 30°C, and 35°C), it required around 67 hours to reach (*e*). In contrast, at the extremes of 15°C and 40°C, the time to reach (*e*) increased to 124 hours and 82 hours, respectively. These findings suggest that while wheat can adapt to a wider temperature range, its growth efficiency declines at both low and high extremes. Lorentzian analysis showed that wheat reached its maximum growth at 26.1°C, requiring about 124 hours, indicating its preference for cooler conditions compared to corn.

Peanuts demonstrated a distinct growth pattern, with a statistically significant difference in slope across temperatures. At 35°C, it took 50 hours to reach (*e*), while at 20°C, only 7 hours were needed. This striking difference reflects the substantial impact of temperature on peanut growth dynamics, with lower temperatures accelerating the time to reach half the maximum growth rate. According to Lorentzian regression, peanuts achieved their maximum growth at 31.2°C, requiring around 116 hours, suggesting a balance between rapid early growth at cooler temperatures and optimal overall development near 31°C.

Soybeans, in contrast, displayed remarkable stability in growth rates across a broad temperature range. At 20°C, 25°C, 35°C, and 40°C, the time required to reach (*e*) showed no significant variation, averaging 49 hours. This consistency emphasizes the temperature tolerance of soybeans, which maintain steady growth under diverse conditions. Lorentzian analysis estimated their maximum growth at 31.3°C, also requiring about 116 hours, confirming their adaptability and resilience within this optimal temperature range.

Parameter	Temperature (°C)	Corn		Peanut		Soybean		Wheat		Cotton	
		Estimate	*P* value	Estimate	*P* value	Estimate	*P* value	Estimate	*P* value	Estimate	*P* value
Slope (b)		−6.094	<0.001***	—	—	—	—	−40.281	<0.001***	—	—
Upper (d)	15	3.689	ns	—	—	—	—	0.055	ns	—	—
e		5.927	<0.001 ***	—	—	—	—	4.823	<0.001***	—	—
Slope (b)		−10.905	<0.001***	−81.416	ns	−14.960	ns	−46.702	ns	−6.744	<0.001***
Upper (d)	20	0.503	0.016*	0.4210	<0.001***	0.744	0.002**	0.525	0.007**	1.792	0.014*
e		4.458	<0.001***	4.573	<0.001***	3.915	0.033*	3.805	<0.001***	4.812	<0.001***
Slope (b)		−11.323	<0.001***	−34.357	ns	16.918	ns	−42.015	<0.001***	−6.019	0.0017***
Upper (d)	25	1.094	0.005**	1.5788	<0.001***	2.806	<0.001***	1.323	0.004**	3.311	<0.001***
e		4.223	<0.001***	3.925	<0.001***	3.916	<0.001***	3.897	<0.001***	4.274	<0.001***
Slope (b)		−12.504	<0.001***	−36.155	<0.046*	7.596	0.0153*	−56.781	ns	−6.847	<0.001***
Upper (d)	30	0.938	0.004**	2.529	<0.001***	3.283	<0.001***	0.779	0.003**	2.778	<0.001***
e		4.094	<0.001 ***	3.902	<0.001***	4.244	<0.001***	3.701	0.033*	4.182	0.011*
Slope (b)		−13.192	<0.001***	−19.950	<0.001***	17.908	ns	−52.526	ns	−4.018	<0.001***
Upper (d)	35	0.764	0.004**	2.831	<0.001***	1.688	<0.001***	0.386	0.004**	3.298	<0.001***
e		4.030	<0.001***	3.924	<0.001***	3.850	ns	3.730	<0.001***	4.142	<0.001***
Slope (b)		−12.605	<0.001***	−31.923	ns	−14.222	ns	−30.397	<0.001***	−6.438	0.008**
Upper (d)	40	0.405	0.005**	1.042	<0.001***	0.320	0.012*	0.244	0.022*	1.001	<0.001***
e		4.021	<0.001***	3.984	<0.001***	3.9117	<0.001***	4.415	<0.001***	3.771	<0.001***

[1]*Data was box-cox transformed to improve the model fit. Please refer Eq. (1). The model lack-of-fit test [61] for the crops tested was $p > 0.05$. Signif. codes: '***' 0.001 '**' 0.01 '*' 0.05.*

Table 1.
Parameters b (slope), d (upper), and e were fitted to a three-parameter log-logistic regression model[1] for root growth rate (cm/day) under the tested temperatures. For corn and wheat, models were fitted for temperatures of 15°C, 20°C, 25°C, 30°C, 35°C, and 40°C. For peanut, soybean, and cotton, models were fitted for temperatures of 20°C, 25°C, 30°C, 35°C, and 40°C.

$$f(x) = 0 + \frac{d - 0}{1 + \exp(b(\log(x) - \log(e)))}$$

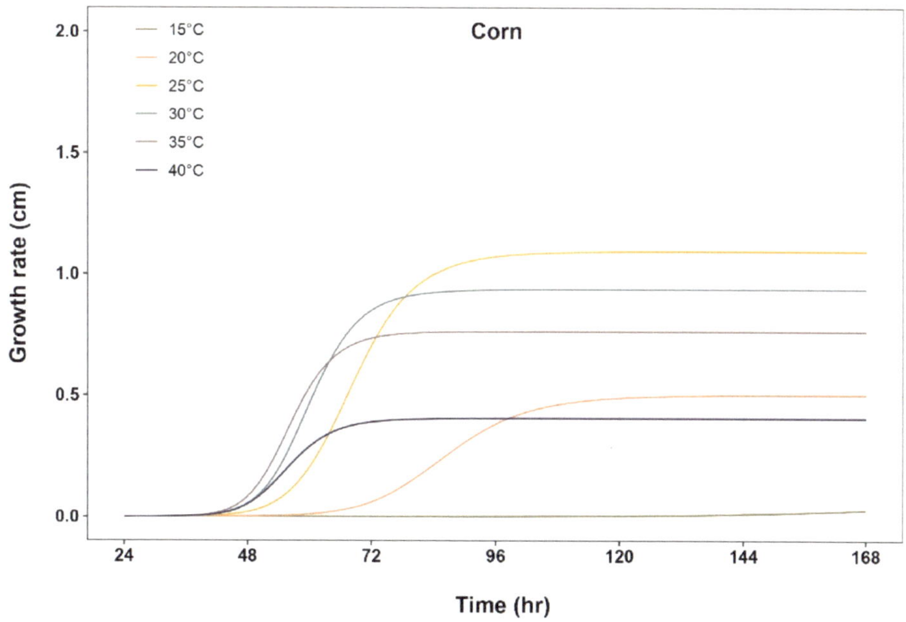

15°C

$$f(x) = 0 + \frac{3.689 - 0}{1 + \exp(-6.094(\log(x) - \log(5.927)))}$$

20°C

$$f(x) = 0 + \frac{0.503 - 0}{1 + \exp(-10.905(\log(x) - \log(4.458)))}$$

25°C

$$f(x) = 0 + \frac{1.094 - 0}{1 + \exp(-11.323(\log(x) - \log(4.223)))}$$

30°C

$$f(x) = 0 + \frac{0.938 - 0}{1 + \exp(-12.504(\log(x) - \log(4.094)))}$$

35°C

$$f(x) = 0 + \frac{0.764 - 0}{1 + \exp(-13.192(\log(x) - \log(4.030)))}$$

40°C

$$f(x) = 0 + \frac{0.405 - 0}{1 + \exp(-12.605(\log(x) - \log(4.021)))}$$

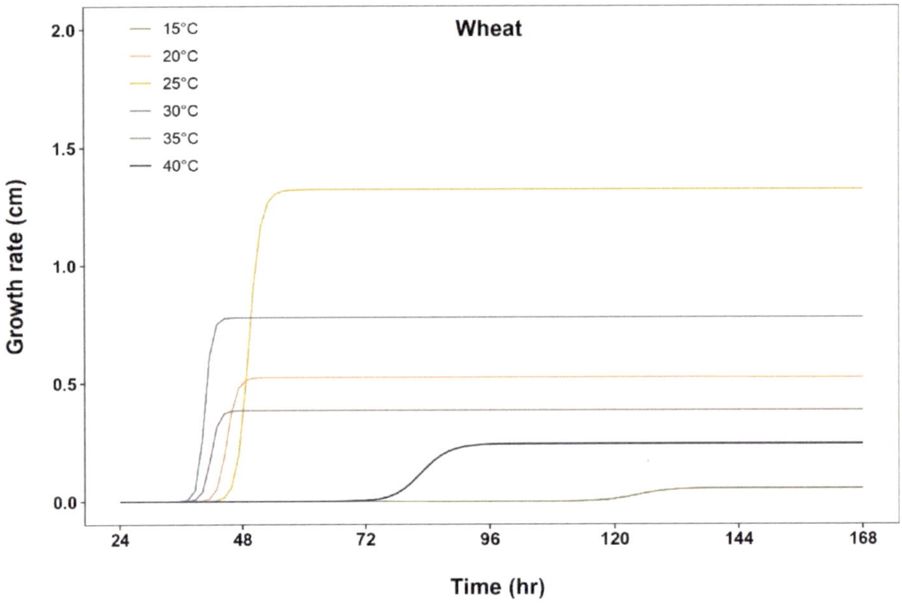

15°C

$$f(x) = 0 + \frac{0.055 - 0}{1 + \exp(-40.281(\log(x) - \log(4.823)))}$$

20°C

$$f(x) = 0 + \frac{0.525 - 0}{1 + \exp(-46.702(\log(x) - \log(3.805)))}$$

25°C

$$f(x) = 0 + \frac{1.323 - 0}{1 + \exp(-42.015(\log(x) - \log(3.897)))}$$

30°C

$$f(x) = 0 + \frac{0.779 - 0}{1 + \exp(-56.781(\log(x) - \log(3.701)))}$$

35°C

$$f(x) = 0 + \frac{0.386 - 0}{1 + \exp(-52.526(\log(x) - \log(3.730)))}$$

40°C

$$f(x) = 0 + \frac{0.244 - 0}{1 + \exp(-30.397(\log(x) - \log(4.415)))}$$

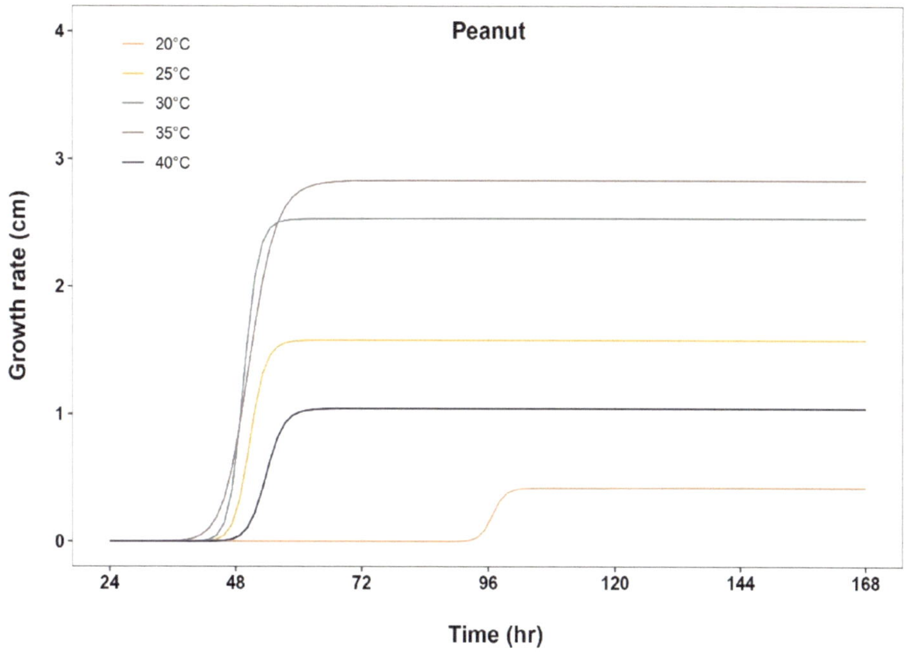

20°C

$$f(x) = 0 + \frac{0.421 - 0}{1 + \exp(-81.416(\log(x) - \log(4.573)))}$$

25°C

$$f(x) = 0 + \frac{1.578 - 0}{1 + \exp(-34.357(\log(x) - \log(3.925)))}$$

30°C

$$f(x) = 0 + \frac{2.529 - 0}{1 + \exp(-36.155(\log(x) - \log(3.902)))}$$

35°C

$$f(x) = 0 + \frac{2.831 - 0}{1 + \exp(-19.950(\log(x) - \log(3.924)))}$$

40°C

$$f(x) = 0 + \frac{1.042 - 0}{1 + \exp(-31.923(\log(x) - \log(3.984)))}$$

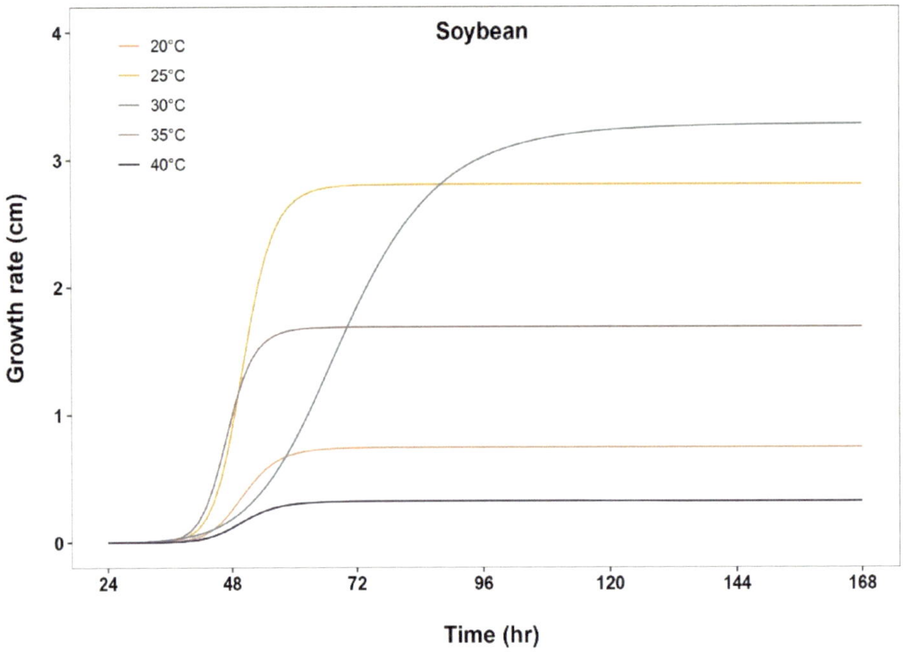

20°C

$$f(x) = 0 + \frac{0.744 - 0}{1 + \exp(-14.960(\log(x) - \log(3.915)))}$$

25°C

$$f(x) = 0 + \frac{2.806 - 0}{1 + \exp(16.918(\log(x) - \log(3.916)))}$$

30°C

$$f(x) = 0 + \frac{3.283 - 0}{1 + \exp(7.596(\log(x) - \log(4.244)))}$$

35°C

$$f(x) = 0 + \frac{1.688 - 0}{1 + \exp(17.908(\log(x) - \log(3.850)))}$$

40°C

$$f(x) = 0 + \frac{0.320 - 0}{1 + \exp(-14.222(\log(x) - \log(3.9117)))}$$

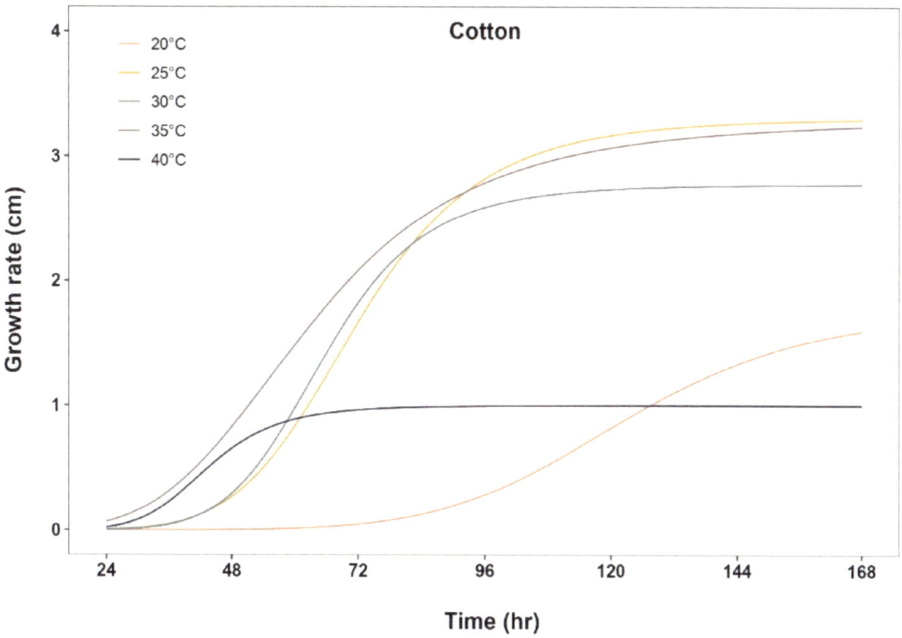

20°C

$$f(x) = 0 + \frac{1.792 - 0}{1 + \exp(-6.744(\log(x) - \log(4.812)))}$$

25°C

$$f(x) = 0 + \frac{3.311 - 0}{1 + \exp(-6.019(\log(x) - \log(4.274)))}$$

30°C

$$f(x) = 0 + \frac{2.778 - 0}{1 + \exp(-6.847(\log(x) - \log(4..182)))}$$

35°C

$$f(x) = 0 + \frac{3.298 - 0}{1 + \exp(-4.018(\log(x) - \log(4.142)))}$$

40°C

$$f(x) = 0 + \frac{1.001 - 0}{1 + \exp(-6.438(\log(x) - \log(3.771)))}$$

Figure 3.
The three-parameter log-logistic function was used to evaluate seedling growth rate for corn, wheat, peanut, soybean, and cotton. For this model, d *is the higher asymptote,* c *is the lower asymptote (which has a lower limit of 0),* e *is* X *value producing a response half-way between* d *and* c*, while* b *is the slope around the inflection point. The parameter* b *can be positive or negative and, consequently* Y *may increase or decrease as* X *increases.*

Crop	x_0‡		y_0‡		*a*		*b*		*c*	
	°C		Hours							
Corn§	30.3	(0.53)	124.2	(1.99)	3.45	(0.29)	7.3	(0.81)	26.2	(2.67)
Wheat	26.1	(0.37)	124.9	(2.18)	3.60	(0.31)	4.7	(0.50)	30.9	(3.26)
Soybean	31.3	(1.33)	114.6	(6.32)	1.43	(0.17)	12.8	(2.68)	63.2	(12.06)
Peanut	31.2	(0.71)	118.6	(4.97)	1.15	(0.11)	9.3	(1.26)	61.6	(9.31)
Cotton	32.3	(0.86)	117.4	(4.44)	1.81	(0.21)	9.1	(1.67)	49.9	(7.51)

†*Lorentizian regression equation for estimating time and temperature to maximum seedling growth after 168 hours across 15°C, 20°C, 25°C, 30°C, 35°C, and 40°C. Please refer Eq. (2). Number in () represents the ±SE.*‡x_0*, temperature to achieve seedling growth germination;* y_0*, hours to achieve maximum seedling growth.*§n = *x# seed for each crop.*

Table 2.
Temperature (x_o) *and time* (y_o) *to maximum seedling growth using nonlinear Lorentzian regression*† *for grain crops with a thermal gradient assay.*

$$z = \frac{a}{[(1+((x-x0)/b)^2) * (1+((y-y0)/c)^2)]}$$

Temperature (x) and time (y) data were used to model the parameters (a, b, and c) to predict the maximum temperature ($x0$) and maximum time ($y0$) that produced the maximum seedling growth (z). Data for root growth every 24 hours was used to establish the maximum, was modeled. The Lorentzian distribution belongs to the category of t-distribution with wider and more comprehensive analysis than other models when there are outlier data points. Legend colors, blue hue represents lower seedling growth with green to yellow to red hues indicating increased seedling growth:

Corn $$z = \frac{3.45}{[(1+((x-30.3)/7.3)^2) * (1+((y-124.2)/26.2)^2)]}$$

Wheat $$z = \frac{3.60}{[(1+((x-26.1)/4.7)^2) * (1+((y-124.9)/30.9)^2)]}$$

Soybean $$z = \frac{1.43}{[(1+((x-31.3)/12.8)^2) * (1+((y-114.6)/63.2)^2)]}$$

Peanut $$z = \frac{1.15}{[(1+((x-31.2)/9.3)^2) * (1+((y-118.6)/61.6)^2)]}$$

Cotton $$z = \frac{1.81}{[(1+((x-32.3)/9.1)^2) * (1+((y-117.4)/49.9)^2)]}$$

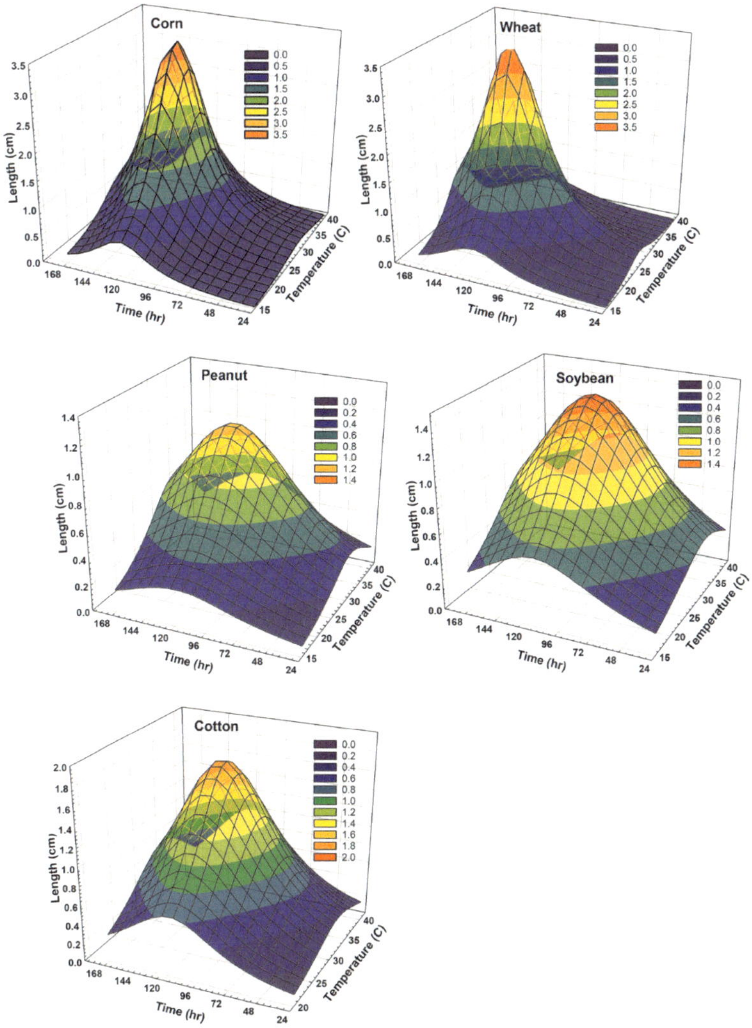

Figure 4.
Temperature and time to maximum seedling growth using Lorentzian regression for corn, wheat, soybean, peanut, and cotton with a thermal gradient assay (n = 60 per crop).

Similarly, cotton exhibited stable growth rates within the moderate temperature range of 25–35°C, requiring approximately 67 hours to reach (*e*). However, slight sensitivity to temperatures outside this range was observed. Lorentzian regression revealed that cotton reached its maximum growth at 32.3°C, requiring around 116 hours, aligning closely with the results for soybeans and peanuts and emphasizing its suitability for similar thermal environments. The results highlighted species-specific thermal preferences, with corn and wheat favoring distinct temperature ranges for maximum growth, while soybeans, peanuts, and cotton demonstrated consistent growth rates and similar optimal conditions around 31–32°C.

Studies suggest that the interaction between environmental factors (i.e., light, temperature, and water availability) and growth hormones (including abscisic acid, gibberellic acid, and ethylene) plays a crucial role in regulating dormancy and germination [67]. Temperature is one of the most influential factors affecting seed imbibition and germination, as it impacts water uptake and the reactivation of metabolic processes [68, 69]. Under optimal conditions, rapid water uptake activates respiration within hours of imbibition, supplying energy for biochemical reactions like hydrolysis and biosynthesis, which drive plant growth [70–72]. Deviations from optimal germination temperatures, whether too low or too high, can hinder metabolic activities, resulting in slower growth rates [72].

Root length growth in different crops is also affected by the structure and dynamics of cell walls. Petrova et al. [73] discussed the role of cellulose microfibril distribution in roots and how it varies between monocots and dicots, influencing their growth. In monocots like rye and maize, internal tissues primarily restrict root growth, whereas, in dicots such as soybeans, external tissues play an important role. Additionally, hormonal interactions, genetic factors, and environmental conditions are crucial in regulating seedling root growth [74, 75].

4. Conclusion

Comparing data generated from the thermal gradient using these growth curve models provided maximum germination rates with optimal temperatures for cotton, corn, soybean, wheat, and peanut (**Tables 1** and **2**). Cold germination testing can be used as a measure to stress seeds to evaluate vigor [11, 14]. The thermal gradient apparatus in this study established variation in seed vigor across a wide range of temperatures simultaneously. This method of seed evaluation provided an indication of seedling vigor across different species and temperatures and will assist scientists in the future in understanding the differences in germination and vigor with respect to the $Germ_{80}$, and the GDD needed to reach maximum germination (*a*) among these crops. Phenotypic differences will always be noted when these crops are grown around the globe, but similarity in genetics will always confer consistency. This information will help to broaden the knowledge base on how crop seed can be evaluated based on these vigor testing methods.

Author details

Timothy L. Grey*, Juliana de Souza Rodrigues and Samantha J. Bowen
Crop and Soil Sciences Department, University of Georgia, Tifton, Georgia, United States

*Address all correspondence to: tgrey@uga.edu

References

[1] Foreign Agricultural Service/USDA. 2024 World Agricultural Production. Circular Series WAP7-24. 2024. Available from: https://apps.fas.usda.gov/psdonline/circulars/production.pdf [Accessed: November 13, 2024]

[2] Preece KE, Hooshyar N, Zuidam NJ. Whole soybean protein extraction process: A review. Innovative Food Science and Emerging Technologies. 2017;**43**:163-172

[3] Riaz T, Iqbal MW, Mahmood S, Yasmin I, Leghari AA, Rehman A. Cottonseed oil: A review of extraction techniques, physicochemical, functional, and nutritional properties. Critical Reviews in Food Science and Nutrition. 2023;**63**:1219-1237

[4] Sánchez González JJ, Ruiz Corral JA, García GM, Ojeda GR, Larios LDC, Holland JB. Ecogeography of teosinte. PLoS ONE. 2018;**13**(2):e0192676. DOI: 10.1371/journal.pone.0192676 [Accessed: November 13, 2024]

[5] Feldman M, Levy AA. Origin and evolution of wheat and related triticeae species. In: Molnár-Láng M, Ceoloni C, Doležel J, editors. Alien Introgression in Wheat. Cham: Springer; 2015. DOI: 10.1007/978-3-319-23494-6_2 [Accessed: November 13, 2024]

[6] Sedivy EJ, Wu F, Hanzawa Y. Soybean domestication: The origin, genetic architecture and molecular bases. New Phytologist Foundation. 2017;**214**:539-553

[7] Hamons RO, Herman D, Stalker HT. Chapter 1—Origin and early history of the peanut. In: Stalker HT, Wilson RF, editors. Peanuts. AOCS Press; 2016. pp. 1-26. ISBN 9781630670382. DOI: 10.1016/B978-1-63067-038-2.00001-0 [Accessed: November 13, 2024]

[8] Viot CR, Wendel JF. Evolution of the cotton genus, *Gossypium*, and its domestication in the Americas. Critical Reviews in Plant Sciences. 2023;**42**:1-33. DOI: 10.1080/07352689.2022.2156061 [Accessed: November 13, 2024]

[9] Moss JP, Rao VR. The peanut—Reproductive development to plant maturity. In: Pattee HE, Stalker HT, editors. Advances in Peanut Science. Tifton, GA: American Peanut Research and Education Society, Inc; 1995. p. 31793

[10] Narasimhareddy SB, Swamy PM. Abscisic acid-like inhibitors and cytokinins during after-ripening of dormant peanut seeds (*Arachis hypogaea*). Physiologia Plantarum. 1979; **46**:191-193

[11] Bradnock W. Seed testing: The international perspective. Journal of Seed Technology. 1990;**14**:111-116

[12] Breseghello F, Coelho G, Siqueria A. Traditional and modern plant breeding with examples in rice. Journal of Agricultural and Food Chemistry. 2013; **61**:8277-8286

[13] Kniss AR. Genetically engineered herbicide-resistant crops and herbicide-resistant weed evolution in the United States. Weed Science. 2018;**66**:260-273

[14] Association of Official Seed Certifying Agencies (AOSCA). Seed Certification. 2024. Available from: https://aosca.org/programs-services/seed-certification/ [Accessed: November 13, 2024]

[15] International Seed Testing Association (ISTA). International rules

for seed testing. 2025. Available from: https://www.seedtest.org/en/publications/international-rules-seed-testing.html [Accessed: January 29, 2025]

[16] Finch-Savage WE, Bassel GW. Seed vigor and crop establishment: Extending performance beyond adaptation. Journal of Experimental Botany. 2016;**67**:567-591

[17] Morton BR, Tillman BL, Gorbet DW, Boote KJ. Impact of seed storage environment on field emergence of peanut cultivars. Peanut Science. 2008; **35**:108-155

[18] Isleib TG, Day JL, Coy AE, Beasley JP, Branch WD. Genotype-by-irrigation interaction in the Georgia peanut official variety test. Peanut Science. 2014;**41**:1-7

[19] Tai PY, Hammons RO. Genotype-environment interaction effects in peanut variety evaluation. Peanut Science. 1978;**5**:72-74

[20] Mbofung GCY, Goggi AS, Leandro LFS, Mullen RE. Effects of storage temperature and relative humidity on viability and vigor of treated soybean seeds. Crop Science. 2013;**53**(3):1086-1095. DOI: 10.2135/cropsci2012.09.0530 [Accessed: November 13, 2024]

[21] Moreano TB, Braccini AL, Scapim CA, Krzyzanowski FC, França-Neto JB, Marques OJ. Changes in the effects of weathering and mechanical damage on soybean seed during storage. Seed Science and Technology. 2011;**39**(3): 604-611. Available from: https://www.ingentaconnect.com/content/ista/sst/2011/00000039/00000003/art00007 [Accessed: November 13, 2024]

[22] Egli DB, White GM, TeKrony DM. Relationship between seed vigor and the storability of soybean seed. Journal of Seed Technology. 1978;**3**:1-11

[23] Byrd HW, Delouche JC. Deterioration of soybean seed in storage. Proceedings of the Association of Official Seed Analysts. 1971;**61**:41-57

[24] Tang S, TeKrony DM, Egli DB, Cornelius PL. Survival characteristics of corn seed during storage: II. Rate of seed deterioration. Crop Science. 1999;**39**: 1400-1406. DOI: 10.2135/cropsci1999.3951400x [Accessed: November 13, 2024]

[25] Afzal I, Kamran M, Basra SMAB, Khan SHU, Mahmood A, Farooq M, et al. Harvesting and post-harvest management approaches for preserving cottonseed quality. Industrial Crops and Products. 2020;**155**:112842

[26] Dharmalingam C, Basu RN. Control of seed deterioration in cotton. Current Science. 1978;**47**:484-487

[27] Ellis RH, Filho CP. The development of seed quality in spring and winter cultivars of barley and wheat. Seed Science Research. 1992;**2**(1):9-15. DOI: 10.1017/S0960258500001057

[28] Pomeranz Y, Zeleny L. Biochemical and functional changes in stored cereal grains. Critical Reviews in Food Technology. 1971;**2**(1):45-80. DOI: 10.1080/10408397109527116 [Accessed: November 13, 2024]

[29] Balota M, Jordan D, Langston D, Shortridge J, Taylor S. Peanut Production Guide. In: Balota M, editor. Virginia Cooperative Extension. Available from: https://sites.ext.vt.edu/newsletter-archive/peanut-production/2023.pdf [Accessed: January 29, 2025]

[30] Baskerville GL, Emin P. Rapid estimation of heat accumulation from

maximum and minimum temperatures. Ecology. 1969;**50**:514-517

[31] Noblet A, Leymarie J, Bailly C. Chilling temperature remodels phospholipidome of *Zea mays* seeds during imbibition. Scientific Reports. 2017;**7**:8886. DOI: 10.1038/s41598-017-08904-z [Accessed: November 13, 2024]

[32] Alsafri FA, Wijewardana C, Krutz LJ, Irby JT, Golden B, Reddy KR. Quantifying and validating soybean seed emergence model as a function of temperature. American Journal of Plant Sciences. 2019;**10**:111-124

[33] Raphael JPA, Gazola B, Nunes JGS, Macedo GC, Rosolem CA. Cotton germination and emergence under high diurnal temperatures. Crop Science. 2017;**57**:2761-2769

[34] Ketring DL, Wheless TG. Thermal time requirements for phonological development of peanut. Agronomy Journal. 2018;**81**:910-917

[35] Cao W, Moss DN. Temperature effect on leaf emergence and phyllochron in wheat and barley. Crop Science. 1989;**29**:1018-1021

[36] Gallagher JL. Field studies of cereal leaf growth: I. Initiation and expansion in relation to temperature and ontogeny. Journal of Experimental Botany. 1979;**30**: 625-636

[37] Grey TL, Chen CY, Nuti R, Monfort WS, Cutts G III. Characterization of genotype by planting date effects on Runner-Type peanut seed germination and vigor response to temperature. In InTech eBooks. Advances in Seed Biology. 2017. DOI: 10.5772/intechopen.70584

[38] Hurdle NL, Grey TL, Branch WD, Monfort WS. High and normal oleic runner-type peanut cultivar by year effects on seed germination and vigor response to temperature. Agronomy Journal. 2022;**114**:2364-2374. Available from: https://acsess.onlinelibrary.wiley.com/doi/epdf/10.1002/agj2.21126 [Accessed: November 13, 2024]

[39] Suassuna TMF, Pilon C, Grey TL, Culbreath AK. Physiological quality of peanut seed affected by tomato spotted wilt disease. Agronomy Journal. 2022; **114**:1-9. DOI: 10.1002/agj2.20978

[40] Chamberlin KD, Grey TL, Puppala N, Holbrook CC, Isleib TG, Dunne J, et al. Comparison of field emergence and thermal gradient table germination rates of seed from high oleic and low oleic near isogenic peanut lines. Peanut Science. 2021;**48**:131-143

[41] Weaver C, Monfort WS, Grey TL, Pilon C, Tubbs RS. Effect of post shelling storage environments on germination and vigor of peanut seed. Agronomy Journal. 2021;**113**:4116-4131. DOI: 10.1002/agj2.20795

[42] Hurdle NL, Grey TL, Pilon C, Monfort WS, Prostko EP. Peanut seed germination and radicle development response to direct exposure of flumioxazin across multiple temperatures. Peanut Science. 2020;**47**: 89-93

[43] Grey TL, Branch WD, Tubbs RS, Snider JL, Webster TM, Arnold J, et al. The impact of genotype by environment effects on runner-type peanut seed vigor response to temperature. Agronomy Journal. 2016;**108**:1424-1433

[44] Bachleda N, Grey TL, Li Z. Effects of modified fatty acid composition on yield, protein and oil contents, and seed germination in soybean. Plant Breeding. 2017;**137**:539-547. DOI: 10.1111/pbr.12497

[45] Chaudhari S, Jordan DL, Grey TL, Prostko EP, Jennings K. Weed control and peanut response to acetochlor alone and in combination with various herbicides. Peanut Science. 2018;**45**: 45-55

[46] Snider JL, Collins GD, Whitaker J, Chapman KD, Horn P, Grey TL. Seed size and oil content are key determinants of seedling vigor in *Gossypium hirsutum*. Journal of Cotton Science. 2014;**18**:1-9

[47] Weitbrecht K, Müller K, Leubner-Metzger G. First off the mark: Early seed germination. Journal of Experimental Botany. 2011;**62**(10):3289-3309. DOI: 10.1093/jxb/err030

[48] Rajjou L, Duval MM, Gallardo-Guerrero KK, Catusse JJ, Bally JJ. Seed germination and vigor. Annual Review of Plant Biology. 2012;**63**:507-533. DOI: 10.1146/annurev-arplant-042811-105550

[49] Cardina J, Hook JE. Factors influencing germination and emergence of Florida beggarweed (*Desmodium totruosum*). Weed Technology. 1989;**3**: 402-407

[50] Chatterton N, Kadish A. A temperature gradient germinator. Agronomy Journal. 1969;**61**:643-644

[51] Abramoff M, Magalhaes PJ, Ram SJ. Image processing with ImageJ. Biophotonics International. 2004;**11**:36-42

[52] Rasband WS. ImageJ. Bethesda, Maryland, USA: National Institutes of Health; 2024. Available from: https://imagej.net/ [Accessed: November 13, 2024]

[53] Schneider C, Rasband W, Eliceiri K. NIH image to ImageJ: 25 years of image analysis. Nature Methods. 2012;**9**: 671-675. DOI: 10.1038/nmeth.2089 [Accessed: November 13, 2024]

[54] Gachoki P, Muraya M, Njoroge G. Modelling plant growth based on gompertz, logistic curve, extreme gradient boosting and light gradient boosting models using high dimensional image derived maize (*Zea mays* L.) phenomic data. American Journal of Applied Mathematics and Statistics. 2022; **10**(2):52-64. DOI: 10.12691/ajams-10-2-3

[55] Jiao X, Zhang H, Zheng J, et al. Comparative analysis of nonlinear growth curve models for *Arabidopsis thaliana* rosette leaves. Acta Physiologiae Plantarum. 2018;**40**:114. DOI: 10.1007/s11738-018-2686-8 [Accessed: November 13, 2024]

[56] Paine CET, Marthews TR, Vogt DR, Purves D, Rees M, Hector A, et al. How to fit nonlinear plant growth models and calculate growth rates: An update for ecologists. Methods in Ecology and Evolution. 2011;**3**(2):245-256. DOI: 10.1111/j.2041-210X.2011.00155.x

[57] Reynolds WD, Drury CF, Phillips LA, Yang X, Agomoh IV. An adapted Weibull function for agricultural applications. Canadian Journal of Soil Science. 2021;**101**(4): 680-702. DOI: 10.1139/cjss-2021-0046

[58] Archontoulis SV, Miguez FE. Nonlinear regression models and applications in agricultural research. Agronomy Journal. 2014;**107**:786-798

[59] Ritz C, Streibig JC. Nonlinear Regression with R. New York, NY, USA: Springer; 2008. pp. 4-6

[60] Yang FQ, Feng HK, Li ZH, Gao L, Yang GJ, Dai HY. Hyperspectral estimation of leaf area index for winter wheat based on Akaike's information criterion. Transactions of the Chinese

Society of Agricultural Engineering. 2016;**3**:163-168

[61] Bates DM, Watts DG. Nonlinear Regression Analysis and Its Applications. New York, NY, USA: Wiley & Sons; 1988. pp. 103-104

[62] Box GEP, Cox DR. An analysis of transformations. Journal of the Royal Statistical Society: Series B: Methodological. 1964, 1964;**26**(2): 211-252. Available from: http://www.jstor.org/stable/2984418

[63] Onofri A. The Broken Bridge between Biologists and Statisticians: A Blog and R Package, Statforbiology, IT, 2020. Available from: https://www.statforbiology.com [Accessed: November 13, 2024]

[64] R Core Team. R: A Language and Environment for Statistical Computing. Vienna, Austria: R Foundation for Statistical Computing; 2023. Available from: https://www.R-project.org/ [Accessed: November 13, 2024]

[65] Ritz C, Baty F, Streibig JC, Gerhard D. Dose-response analysis using R. PLoS ONE. 2015;**10**(12):e0146021. DOI: 10.1371/journal.pone.0146021 [Accessed: November 13, 2024]

[66] Motulsky HJ, Christopoulos A. Fitting Models to Biological Data Using Linear and Nonlinear Regression. A Practical Guide to Curve Fitting. San Diego CA, USA. Available from: www.graphpad.com: GraphPad Software Inc.; 2003 [Accessed November 13, 2024]

[67] Nautiyal PC, Sivasubramaniam K, Dadlani M. Seed dormancy and regulation of germination. In: Dadlani M, Yadava DK, editors. Seed Science and Technology. Singapore: Springer; 2023. pp. 39-66. DOI: 10.1007/978-981-19-5888-5_3

[68] Ali AS, Elozeiri AA. Metabolic Processes during Seed Germination. In: InTech eBooks. 2017. DOI: 10.5772/intechopen.70653

[69] Ghaleb W, Ahmed LQ, Escobar-Gutiérrez AJ, Julier B. The history of domestication and selection of lucerne: A new perspective from the genetic diversity for seed germination in response to temperature and scarification. Frontiers in Plant Science. 2021;**11**:578121. DOI: 10.3389/fpls.2020.578121

[70] Xia Q, Ponnaiah M, Cueff G, Rajjou L, Prodhomme D, Gibon Y, et al. Integrating proteomics and enzymatic profiling to decipher seed metabolism affected by temperature in seed dormancy and germination. Plant Science. 2018;**269**:118-125. DOI: 10.1016/j.plantsci.2018.01.014

[71] El-Maarouf-Bouteau H. The seed and the metabolism regulation. Biology. 2022;**11**(2):168. DOI: 10.3390/biology11020168

[72] Bewley JD, Bradford KJ, Hilhorst HWM, Nonogaki H. Environmental regulation of dormancy and germination. In: Seeds. New York, NY: Springer; 2013. pp. 299-339. DOI: 10.1007/978-1-4614-4693-4_7

[73] Petrova A, Ageeva M, Kozlova L. Root growth of monocotyledons and dicotyledons is limited by different tissues. The Plant Journal. 2023;**116**: 1462-1476. DOI: 10.1111/tpj.16440

[74] Ramachandran V, Tobimatsu Y, Masaomi Y, Ryosuke S, mezawa T, Demura T, Ohtani M. Plant-specific Dof transcription factors VASCULAR-RELATED DOF1 and VASCULAR-

RELATE DOF2 regulate vascular cell differentiation and lignin biosynthesis in *Arabidopsis*. Plant Molecular Biology. 2020;**104**:263-281

[75] Zhou JM, Zhang Y. Plant immunity: Danger perception and signaling. Cell. 2020;**181**:978-989

Chapter 5

Cultivating Resilience: Exploring Biotic Stressors and Plant Immune Responses

Kingsley K. Kanu and Vincent C. Kanu

Abstract

Plants encounter various biotic stressors, including pathogens, pests, herbivores, and parasites, which significantly threaten their health and productivity. To combat these challenges, plants have evolved sophisticated immune systems that rely on pattern recognition receptors (PRRs) to detect microbe-associated molecular patterns (MAMPs) or damage-associated molecular patterns (DAMPs), initiating pattern-triggered immunity (PTI). A more robust defense mechanism, effector-triggered immunity (ETI), is activated when intracellular nucleotide-binding leucine-rich repeat receptors (NLRs) recognize specific pathogen effector proteins delivered into plant cells to suppress PTI. The plant immune system is further regulated by intricate signaling networks involving phytohormones like salicylic acid (SA), Jasmonic acid (JA), and ethylene (ET). These hormonal pathways interact and cross-talk, allowing plants to fine-tune their responses to various biotic and abiotic stresses while balancing growth and defense mechanisms. Components like the RALF1 peptide and its receptor FERONIA also play crucial roles in modulating immune responses by interacting with receptor-like kinases and inhibiting or amplifying specific defense pathways.

Keywords: biotic stressors, plant health, defense mechanism, interactions, plant immunity, environment interactions, PRRs, ETI, PAMPs

1. Introduction

Stress acts as a stimulus or influence that falls beyond the normal range of homeostatic control in plants, leading to the establishment of a new physiological state once mitigated. This often impacts the accumulation or depletion of phytochemicals by either enhancing or reducing the activities of key enzymes in metabolic pathways. Hence, to cope with stress, plants activate tolerance mechanisms at various levels, and understanding the molecular control mechanisms underlying stress tolerance often results in the development of molecular tools based on the expression of specific stress-related genes [1]. Fluctuating environmental conditions often impede plants from reaching their full genetic potential for growth and reproduction. Among these conditions, recurrent attacks by herbivores and microbial pathogens represent significant challenges to plant health and productivity. These biotic stressors pose significant challenges to plant health and productivity by directly damaging plant tissues and compromising their ability to photosynthesize, absorb nutrients, and carry out essential metabolic processes [2]. They overwhelm the plants, resulting in substantial reductions in crop yields and quality. Hence, "plants tend to strike a balance between their response and biotic stress to combat the deleterious effect on their survival" [2]. This delicate equilibrium involves the deployment of various defense mechanisms, while simultaneously allocating resources toward growth and reproduction, ensuring overall fitness and sustainability in the presence of biotic stressors.

Biotic stress, being a broadly defined term, presents numerous challenges, as the types of biotic stresses encountered by an organism are contingent upon the climate of its habitat and the species' inherent ability to withstand specific stresses [3]. Furthermore, the correlation between biotic stresses and plant yield influences economic decisions and practical advancements. Hence, the repercussions of biotic damage on crop yield extend to population dynamics, thereby influencing the coevolution between plants and stressors, and nutrient cycling within ecosystems [4]. This impact is not confined to agricultural contexts but also extends to horticultural plant health and natural habitat ecology. The consequences are profound for the host plants, often resulting in reduced yield or compromised quality of harvested produce [2]. Biotic stresses rarely occur in isolation, and their interactions often have synergistic or antagonistic effects on plant health and productivity. For instance, drought stress can predispose plants to pathogen attack, while herbivory can increase susceptibility to fungal diseases [4]. These interactions are complex and can vary depending on the specific combination of stresses, their timing, and the plant species involved.

1.1 Biotic stresses faced by plants

Plants struggle with various biotic stresses caused by different living organisms. Numerous pests, parasites, and pathogens are responsible for infecting plants and inciting biotic stress. These stressors can be necrotrophic or biotrophic (fungal parasites), leading to conditions such as vascular wilts, leaf spots, and cankers in plants [3, 5]. Nematodes feed on plant parts and primarily cause soil-borne diseases, resulting in nutrient deficiency, stunted growth, and wilting [6]. They pose more threats to food security and account for a major share of the chemicals used in agricultural control. Although nematodes can independently cause diseases in plants, they typically live and function in the soil, where they are frequently surrounded by fungi and bacteria, many of which are also plant pathogens. Associations often form between nematodes and these other pathogens, creating an etiological complex. This combination can result in a pathogenic potential that is significantly greater than the sum of the damage each pathogen could cause alone [7]. Similarly, viruses can inflict both local and systemic damage, causing chlorosis and stunting. However, mites and insects damage plants either by feeding (piercing and sucking) or by laying eggs [6, 8]. Insects like aphids and beetles can act as vectors for various plant pathogens, exacerbating the damage. Additionally, insect feeding can result in direct tissue damage, loss of plant sap, and secondary infections by pathogens entering through feeding wounds [6]. Hence, herbivory can also be considered one of the major causes of plant biotic stress. Herbivory not only causes immediate physical damage to plant tissues but also leads to significant physiological and biochemical changes within the plant [9]. The removal of leaf tissue reduces the plant's photosynthetic capacity, thereby limiting its energy production and growth potential. In addition, the stress induced by herbivory triggers a cascade of defense responses in plants [9, 10]. According to Alves-Silva and Del-Claro [10], herbivory in plants results in developmental instability, as evidenced by high levels of fluctuating asymmetry in mature leaves that developed from leaf buds damaged by thrips. Fluctuating asymmetry, which refers to random deviations from perfect symmetry in bilateral traits, is a common indicator of developmental instability and stress. The high levels of fluctuating asymmetry observed in leaves damaged by the herbivore suggest that the plants are experiencing significant developmental stress. This stress can arise from several factors, including the direct physical damage caused by herbivores, the loss of essential nutrients and photosynthetic capacity, and the activation of defense mechanisms that divert resources away from growth [9, 10].

1.2 Plant immune responses

Plant immune responses are crucial for safeguarding the health and survival of plants in their natural environments. The ability to respond to infections by microbial pathogens and pests is essential for their survival and overall fitness. Therefore, understanding the intricacies of these immune responses is vital for developing innovative strategies to enhance plant resilience against biotic stressors [11]. As sessile organisms, plants are unable to escape from microscopic pathogens or voracious herbivores [12]. Hence, to effectively counteract biotic threats, plants have evolved intricate pathogen defense systems, similar to the innate immune systems found in humans [12–14]. These defense mechanisms work alongside various cellular processes, together creating surveillance networks. Through these interactions, plants gain the ability to recognize and distinguish a wide range of biotic threats, such as pathogens, pests, and viruses. Therefore, understanding how plants perceive and respond to these challenges is crucial for developing strategies to improve their resilience and ensure their survival in changing environments [13]. Some of these plant immune systems include the pathogen recognition receptors, resistance proteins, systemic acquired resistance, and RNA silencing pathways. Understanding the intricate signaling networks, molecular mechanisms, and regulation of these plant immune components is crucial for developing strategies like genetic engineering, elicitor treatments, or biocontrol agents to improve crop resilience against evolving biotic threats exacerbated by climate change and globalization.

2. Importance of understanding plant immune responses

At the forefront of plant immunity are pattern recognition receptors (PRRs), which act as sentinel proteins capable of detecting conserved molecular patterns associated with potential threats [3]. These patterns, known as pathogen-associated molecular patterns (PAMPs) or damage-associated molecular patterns (DAMPs), serve as molecular signatures that trigger defense responses upon recognition by PRRs. Upon recognition of these molecular patterns, PRRs initiate a cascade of signaling events that culminate in the activation of defense responses. This includes the production of antimicrobial compounds, reinforcement of cell walls, and induction of defense-related genes. Additionally, PRR-mediated recognition of PAMPs or DAMPs can trigger the hypersensitive response (HR), a form of programmed cell death that limits the spread of pathogens [3]. Hence, it could be said that the specificity and sensitivity of PRRs enable plants to distinguish between beneficial microbes and potential threats, allowing for a tailored response to different stimuli.

Unraveling the mechanisms of plant immunity is of paramount importance for addressing several critical challenges facing modern agriculture and ensuring global food security. Plant diseases caused by pathogenic microbes, pests, and herbivores pose a significant threat, leading to substantial crop losses of up to 40% annually [11]. These biotic stresses, heightened by climate change and globalization, have far-reaching consequences for crop productivity, food availability, and economic stability. Hence, expanding our understanding of plant immune responses holds the key to developing innovative strategies that can bolster crop resilience and disease resistance. By understanding the molecular pathways, receptors, and signaling cascades involved in pathogen recognition and defense activation, crops with improved immune capabilities can be engineered [11, 15, 16]. This not only increases yields and productivity but also reduces the reliance on excessive pesticide application, minimizing the environmental impact and ecological disruption caused by agrochemicals. Moreover, insights into plant immunity mechanisms enable the development of sustainable and environmentally friendly approaches to disease management. These include the use of elicitors or plant activators that stimulate the plant's own defense responses, aligning with the principles of sustainable agriculture practices [15]. Such strategies have the potential to reduce the ecological footprint of crop production while maintaining high yields. As climate change and globalization facilitate the spread of plant pathogens and pests into new regions, the urgency to combat these emerging threats becomes increasingly apparent. Thereby necessitating a comprehensive understanding of plant immune pathways for developing effective solutions to protect crops against these evolving biotic challenges, ensuring food security for a growing global population.

2.1 Plant immune response in agriculture

Plants have evolved sophisticated immune systems to defend against various biotic stresses such as pathogens (viruses, bacteria, fungi) and pests (insects, nematodes). This immune response is crucial for plant survival and agricultural productivity. They rely on two main branches of immunity, which are the cell surface receptors that recognize conserved pathogen/pest-associated molecular patterns (PAMPs/HAMPs), triggering PAMP-triggered immunity (PTI), and intracellular immune receptors called nucleotide-binding leucine-rich repeat (NLR) proteins that detect specific pathogen effectors, leading to effector-triggered immunity (ETI) [11, 17]. PTI acts as the first line of defense, while ETI provides a more robust and prolonged immune response, often associated with localized cell death at the infection site [11]. Both PTI and ETI involve complex signaling pathways, transcriptional reprogramming, and the production of antimicrobial compounds [11, 17]. The plant immune system can also establish systemic acquired resistance (SAR) in distal tissues after local infection, priming the plant for heightened defense against subsequent attacks [17]. SAR involves signaling molecules like salicylic acid and can be transgenerational, conferring resistance to progeny.

In agriculture, understanding and manipulating the plant immune system is crucial for developing disease-resistant crops and reducing yield losses. Strategies include breeding for resistance genes (R genes encoding NLRs), using elicitors/inducers to activate immune responses, and engineering plants with modified immune receptors or signaling components. Additionally, sustainable agricultural practices like crop rotation, intercropping, and the use of biological control agents can help reduce biotic stress and promote plant health by modulating the immune system.

2.2 Plant immune responses: PRRs and ETI

Plant immunity depends on innate immune receptors present in each cell, which detect invasion signals to initiate pattern-triggered immunity (PTI) or effector-triggered immunity (ETI) [18]. With evolved sophisticated mechanisms to detect and respond to various biotic threats. Plant pattern recognition receptors (PRRs) located on the plant cell surface remain the core of this defense system. These PRRs perceive a diverse array of molecular signatures, known as elicitors, derived from pathogens or released by the plant itself during a pathogen attack [19]. Upon recognition of pathogen-associated molecular patterns (PAMPs) by plant pattern recognition receptors (PRRs), a series of signaling events is initiated to activate the plant's immune responses. However, this initial signal needs to be amplified and precisely regulated to mount defenses appropriate for the specific threat encountered [13]. Plants achieve this through intricate phosphorylation-dependent signaling cascades. Two well-studied pathways involve mitogen-activated protein kinases (MAPKs) and calcium-dependent protein kinases (CDPKs). These kinase cascades act as molecular relays, sequentially activating downstream components through phosphorylation events [13]. One extensively characterized example is the immune response triggered by the bacterial PAMP flagellin (flg22). In this pathway, the PRR FLS2 initiates a MAPK cascade involving MAPKKK3/5, which phosphorylates and activates MAPK3/4/6. This kinase relay amplifies and transmits the initial signal, ultimately leading to the activation of specific defense mechanisms tailored to counter bacterial pathogens [20]. These signaling cascades serve as molecular switches, integrating the initial PAMP recognition event with the appropriate downstream responses. By precisely regulating the specificity and amplitude of the immune signal, plants can mount effective and targeted defenses against diverse biotic threats while minimizing the metabolic costs associated with activating unnecessary defense mechanisms.

Effector-triggered immunity (ETI) represents a more specialized and robust defense mechanism. It is initiated when intracellular nucleotide-binding leucine-rich repeat receptors (NLRs) detect specific pathogen effector proteins that are delivered into the plant cell by the pathogen to suppress PTI and facilitate infection. These effectors are recognized either directly by NLRs or indirectly through modifications they cause to host proteins. The detection of these effectors by NLRs typically results in a stronger and faster immune response, often associated with localized cell death known as the hypersensitive response (HR), which restricts pathogen growth and spread [3, 18]. PTI is initiated by cell surface-localized pattern recognition receptors (PRRs), which identify molecular structures characteristic of microbes and endogenous damage, known as microbe-associated molecular patterns (MAMPs) and danger-associated molecular patterns (DAMPs), as seen in **Figure 1** [18]. When receptors are bound to ligands, they create complexes with coreceptor or adapter kinases from the same extracellular domain class. This action activates protein phosphorylation cascades through receptor-like kinases (RLKs) and receptor-like cytoplasmic kinases. Subsequently, PRR signaling engages cytosolic calcium ions (Ca^{2+}) and bursts of reactive oxygen species (ROS) in the apoplast, calcium-dependent protein kinases (CDPKs), mitogen-activated protein kinase (MAPK) cascades, defense hormone networks, and extensive reprogramming of transcriptional, translational, and metabolic processes [18].

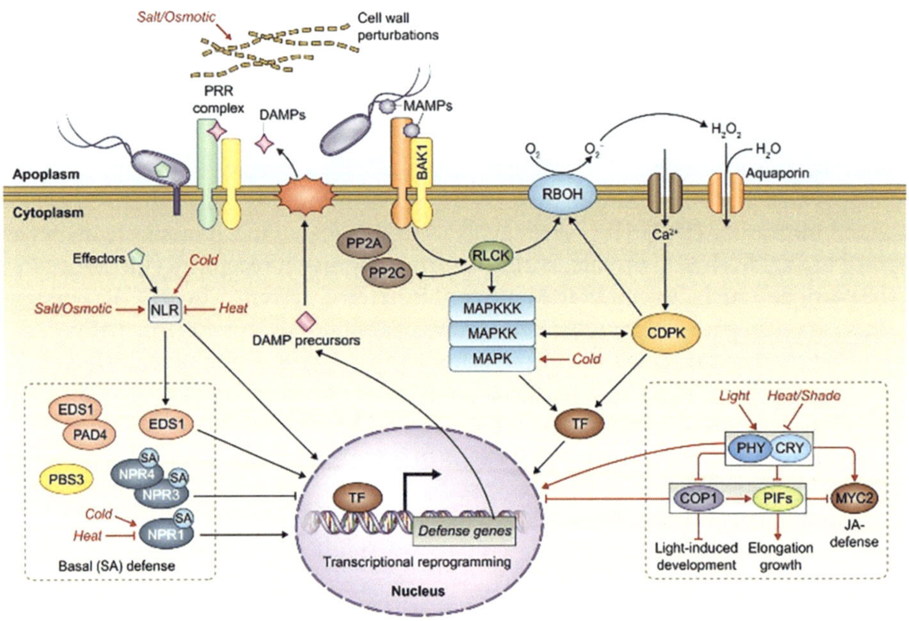

Figure 1.
[18]: A fundamental outline for plant immune signaling and its modulation by the environment involves the perception of microbe- or damage-associated molecular patterns (MAMPs/DAMPs), which initiates the formation of receptor complexes. Activation of pattern recognition receptors (PRRs) triggers intracellular signaling pathways, including calcium (Ca^{2+}) signaling, activation of Ca^{2+}-dependent protein kinases (CDPKs), mitogen-activated protein kinase (MAPK) cascades, and engagement of defense-related transcription factors (TFs), resulting in extensive transcriptional reprogramming.

2.3 RALF1 (rapid alkalinization factor 1) peptide

Another component of plant immunity is the RALF1 (Rapid alkalinization factor 1) peptide. RALF1 (Rapid Alkalinization Factor 1) is a key component in plant immunity that acts through the FERONIA (FER) receptor kinase in Arabidopsis [11]. FER associates with the malectin-like extracellular domain LLG1 to perceive and bind RALF1, a cysteine-rich peptide. Upon RALF1 recognition, FER recruits and interacts with the receptor-like cytoplasmic kinase RIPK to inhibit immune responses like PAMP-triggered ROS production. RALF peptides are widely conserved small signaling molecules in plants [11]. They play crucial roles in regulating diverse aspects of plant growth, development, reproduction, responses to environmental cues, and immunity. Interestingly, functional RALF-like peptides have also been identified in fungal pathogens like Fusarium and Verticillium, though their precise role in pathogenesis is still unclear. It is hypothesized that these pathogen-derived RALF mimics may interfere with the plant's RALF signaling pathways to suppress immunity. In addition to microbe-associated molecular pattern (MAMP) ligands, some leucine-rich repeat receptor-like kinases (LRR-RLKs) can perceive damage-associated molecular patterns (DAMPs) like Atpeps and PIPs, which are proteinaceous elicitors induced during pathogen infection or wounding [11]. For example, the LRR-RLKs PEPR1/2 recognize Atpeps (plant elicitor peptides), while the RLKs RLK7 and PCRK1 bind PIPs (PAMP-induced secreted peptides). Perception of these DAMPs activates downstream signaling cascades that amplify the plant's immune responses, such as oxidative burst, calcium influx, and defense gene expression. This highlights the complex interplay between different signaling components, including peptide hormones, receptor kinases, and downstream effectors, in regulating the multilayered plant immune system against diverse biotic threats.

3. Signaling pathway

Cell signaling pathways are mechanisms through which cells communicate to perform their functions. While these pathways may differ, they generally serve a common purpose. As sessile organisms, plants respond to various signals by altering their morphology. This process involves a complex network of interactions to initiate biochemical and physiochemical responses. Plants encounter a diverse array of microorganisms, especially at the root-soil interface level. They have the ability to detect microbial molecules, which can result in either mutualistic interactions or immune responses. Immune signaling effectively limits the invasion of various organisms, including pathogens like viruses, insects, nematodes, and even parasitic plants. Despite the apparent complexity of this process, it is coordinated by complexes of cell surface receptor kinases [21]. Among various signaling interactions, the following remains the most common in plant protection.

3.1 Salicylic acid (SA) pathway

Salicylic acid (SA) is a crucial plant hormone involved in mediating immunity, growth, and development [22]. Environmental changes directly influence various aspects of SA biosynthesis, signaling, metabolism, and transport, with these influences being positive, neutral, or negative depending on the specific environmental factor. These molecular and biochemical trends can either be conserved or vary across different species. Key components of the SA pathway are affected by abiotic factors such as temperature and water availability, as well as biotic factors like interactions with commensal and beneficial microbes [22]. Temperature regulation has revealed major thermosensitive nodes at different levels of gene and protein regulation within the SA pathway. For example, changes in temperature can impact the expression levels of genes encoding enzymes involved in SA biosynthesis or alter the stability and activity of SA signaling components. Additionally, studies are unraveling how the microbiome modulates the SA pathway at both single species and community levels, highlighting mechanisms dependent on canonical plant hormone cross-talk. Beneficial microbes can enhance SA-mediated immunity by priming plants for faster and stronger defense responses, while commensal microbes may influence SA metabolism and signaling pathways through direct interactions or by modulating the plant's hormonal balance [22].

3.2 Jasmonic acid (JA) pathway

Plant growth and development are significantly regulated by Jasmonic acid (JA) and its precursors and derivatives. They are known to play pivotal roles in regulating numerous physiological processes, particularly in mediating plant responses to both biotic and abiotic stresses [23]. Jasmonic acid is synthesized from linolenic acid *via* a series of enzymatic reactions, primarily in response to environmental cues such as herbivore attack, pathogen infection, wounding, and abiotic stressors like drought and salinity. Once synthesized, JA acts as a signaling molecule that triggers a cascade of downstream responses, coordinating various defense mechanisms and adaptive strategies to cope with stress [23]. One of the key functions of JA is its involvement in plant defense against herbivores and pathogens. Upon perception of herbivore-induced damage or pathogen invasion, plants rapidly accumulate JA, which activates defense-related genes and induces the production of secondary metabolites such as volatile organic compounds (VOCs), proteinase inhibitors, and defensive proteins [24]. These compounds can deter herbivores, inhibit pathogen growth, and attract natural enemies of herbivores, enhancing plant resistance to pests and diseases [23, 24]. Furthermore, JA interacts with other hormonal signaling pathways, such as those mediated by salicylic acid (SA), ethylene (ET), and abscisic acid (ABA), forming complex regulatory networks that integrate multiple stress responses and developmental cues. Cross-talk between JA and other hormones allows plants to fine-tune their responses to changing environmental conditions and prioritize resource allocation toward growth, reproduction, or defense as needed [9, 10, 23, 24].

3.3 Ethylene (ET) pathway

As a gas, ethylene diffuses from its production sites to where it can be perceived without being modified or metabolized [25]. Therefore, precise regulation of ethylene biosynthesis is essential. This regulation is managed through transcriptional and post-translational regulation of ethylene biosynthesis enzymes, the adjacent Yang cycle, and through the transport and conjugation of the ethylene precursor 1-aminocyclopropane-1-carboxylic acid (ACC) [25]. The biosynthesis of ethylene begins with the conversion of S-adenosyl-L-methionine (SAM) to ACC by the enzyme ACC synthase (ACS). The activity of ACS is tightly regulated at the transcriptional level, with various environmental and developmental cues influencing ACS gene expression [26]. For instance, factors such as wounding, pathogen attack, and fruit ripening can upregulate ACS genes, leading to increased ethylene production. ACS enzymes are also subject to post-translational modifications that affect their stability and activity [26–28]. Phosphorylation, for example, can either stabilize or destabilize ACS proteins, thereby modulating ethylene biosynthesis in response to specific signals. Additionally, the enzyme ACC oxidase (ACO), which converts ACC to ethylene, is regulated post-translationally to ensure that ethylene production is finely tuned [26, 27]. The Yang cycle is crucial for maintaining the levels of SAM, which is the precursor for ethylene biosynthesis. In this cycle, SAM is regenerated from methylthioadenosine (MTA), a byproduct of ethylene production. Proper functioning of the Yang cycle ensures a consistent supply of SAM for ethylene biosynthesis, highlighting its importance in the regulation of ethylene production. ACC can be transported from its site of synthesis to other parts of the plant where it can be converted to ethylene [28]. This transport is crucial for coordinating ethylene responses throughout the plant. ACC transporters facilitate the movement of ACC across cell membranes, enabling its distribution to different tissues. ACC can be conjugated to form ACC conjugates, which are inactive forms of the precursor. This process serves as a mechanism to regulate the availability of ACC for ethylene production. Conjugation acts as a buffer, preventing excessive ethylene synthesis under nonstress conditions [27, 28]. When required, ACC conjugates can be hydrolyzed back to ACC, making it available for conversion to ethylene. Once ethylene is produced, it diffuses through the plant to reach its site of action. Ethylene is perceived by a family of ethylene receptors located on the endoplasmic reticulum membrane. Upon binding to ethylene, these receptors initiate a signaling cascade that leads to various ethylene-responsive gene expressions and physiological responses as well as stress responses.

4. Conclusion

Plants encounter various biotic stressors, including pathogens, pests, herbivores, and parasites, which pose significant threats to their health and productivity. To combat these challenges, plants have evolved sophisticated immune systems that rely on pattern recognition receptors (PRRs) and effector-triggered immunity (ETI) to detect and respond to invading pathogens or herbivore-induced damage. The recognition of microbe-associated molecular patterns (MAMPs) or damage-associated molecular patterns (DAMPs) by PRRs initiates a cascade of signaling events, including calcium influx, reactive oxygen species bursts, and the activation of mitogen-activated protein kinase (MAPK) cascades. This leads to the induction of defense responses, such as the production of antimicrobial compounds, cell wall reinforcement, and the expression of defense-related genes. Additionally, components like the RALF1 peptide and its receptor FERONIA play crucial roles in modulating immune responses by interacting with receptor-like kinases and inhibiting or amplifying specific defense pathways. The plant immune system is further regulated by intricate signaling networks involving phytohormones like salicylic acid (SA), Jasmonic acid (JA), and ethylene (ET). These hormonal pathways interact and cross talk with each other, allowing plants to fine-tune their responses to various biotic and abiotic stresses while balancing growth and defense mechanisms. Understanding the molecular mechanisms underlying plant immunity, including the receptors, signaling cascades, and regulatory networks, is crucial for developing strategies to enhance crop resilience, reduce yield losses, and ensure food security in the face of evolving biotic threats exacerbated by climate change and globalization.

Acknowledgements

We would like to express our deepest appreciation to our study committee, who took it upon themselves to provide access to all relevant materials. And to miss Milena Spang for her unwavering support and encouragement. The authors acknowledge the use of QuillBot paraphrasing tool and Grammarly writing tool for language polishing and rephrasing of the major and important writeups without having to lose the points from those writeups.

Author contributions

All tasks leading to this outcome have been equally shared and handled by the authors.

Conflict of interest

The authors declare no conflict of interest.

Thanks

We would like to say a big thank you to the chief editor for giving us this opportunity.

Author details

Kingsley K. Kanu[1*] and Vincent C. Kanu[2]

1 Technical University of Munich, Freising, Germany

2 Abia State University, Uturu, Nigeria

*Address all correspondence to: kanukingsley72@gmail.com

References

[1] Peck S, Mittler R. Plant signaling in biotic and abiotic stress. Journal of Experimental Botany. 2020;**71**:1649-1651

[2] Madani B, Mirshekari A, Imahori Y. Physiological responses to stress. In: Postharvest Physiology and Biochemistry of Fruits and Vegetables. Elsevier; 2019

[3] Iqbal Z, Iqbal MS, Hashem A, Abd Allah EF, Ansari MI. Plant defense responses to biotic stress and its interplay with fluctuating dark/light conditions. Frontiers in Plant Science. 2021;**12**:631810

[4] Zhang H, Zhao Y, Zhu J-K. Thriving under stress: How plants balance growth and the stress response. Developmental Cell. 2020;**55**:529-543

[5] Sobiczewski P, Iakimova E, Mikiciński A, Węgrzynowicz-Lesiak E, Dyki B. Necrotrophic behaviour of *Erwinia amylovora* in apple and tobacco leaf tissue. Plant Pathology. 2017;**66**:842-855

[6] Osman HA, Ameen HH, Mohamed M, Elkelany US. Efficacy of integrated microorganisms in controlling root-knot nematode *Meloidogyne javanica* infecting peanut plants under field conditions. Bulletin of the National Research Centre. 2020;**44**:1-10

[7] Agrios GN. Plant diseases caused by nematodes. In: Plant Pathology. Burlington: Elsevier Academic Press; 2005. pp. 825-874

[8] Pallas V, Garcia JA. How do plant viruses induce disease? Interactions and interference with host components. Journal of General Virology. 2011;**92**: 2691-2705

[9] Mostafa S, Wang Y, Zeng W, Jin B. Plant responses to herbivory, wounding, and infection. International Journal of Molecular Sciences. 2022;**23**:7031

[10] Alves-Silva E, Del-Claro K. Herbivory-induced stress: Leaf developmental instability is caused by herbivore damage in early stages of leaf development. Ecological Indicators. 2016;**61**:359-365

[11] Bentham AR, De la Concepcion JC, Mukhi N, Zdrzałek R, Draeger M, Gorenkin D, et al. A molecular roadmap to the plant immune system. Journal of Biological Chemistry. 2020;**295**:14916-14935

[12] Bhar A, Chakraborty A, Roy A. Plant responses to biotic stress: Old memories matter. Plants. 2021;**11**:84

[13] Li P, Lu Y-J, Chen H, Day B. The lifecycle of the plant immune system. Critical Reviews in Plant Sciences. 2020;**39**:72-100

[14] Divekar PA, Narayana S, Divekar BA, Kumar R, Gadratagi BG, Ray A, et al. Plant secondary metabolites as defense tools against herbivores for sustainable crop protection. International Journal of Molecular Sciences. 2022;**23**:2690

[15] Yang B, Yang S, Zheng W, Wang Y. Plant immunity inducers: From discovery to agricultural application. Stress Biology. 2022;**2**:5

[16] Jones JD, Staskawicz BJ, Dangl JL. The plant immune system: From discovery to deployment. Cell. 2024;**187**:2095-2116

[17] Lolle S, Stevens D, Coaker G. Plant NLR-triggered immunity: From receptor

activation to downstream signaling. Current Opinion in Immunology. 2020;**62**:99-105

[18] Saijo Y, Loo EP-I. Plant immunity in signal integration between biotic and abiotic stress responses. New Phytologist. 2020;**225**:87-104

[19] Ziv C, Zhao Z, Gao YG, Xia Y. Multifunctional roles of plant cuticle during plant-pathogen interactions. Frontiers in Plant Science. 2018;**9**:1088

[20] Saijo Y, Loo EP-I, Yasuda S. Pattern recognition receptors and signaling in plant-microbe interactions. The Plant Journal. 2018;**93**:592-613

[21] Zipfel C, Oldroyd GE. Plant signalling in symbiosis and immunity. Nature. 2017;**543**:328-336

[22] Rossi CA, Marchetta EJ, Kim JH, Castroverde CDM. Molecular regulation of the salicylic acid hormone pathway in plants under changing environmental conditions. Trends in Biochemical Sciences. 2023;**48**(8):699-712

[23] Ruan J, Zhou Y, Zhou M, Yan J, Khurshid M, Weng W, et al. Jasmonic acid signaling pathway in plants. International Journal of Molecular Sciences. 2019;**20**:2479

[24] Hewedy OA, Elsheery NI, Karkour AM, Elhamouly N, Arafa RA, Mahmoud GA-E, et al. Jasmonic acid regulates plant development and orchestrates stress response during tough times. Environmental and Experimental Botany. 2023;**208**:105260

[25] Pattyn J, Vaughan-Hirsch J, Van de Poel B. The regulation of ethylene biosynthesis: A complex multilevel control circuitry. New Phytologist. 2021;**229**:770-782

[26] Yin J, Zhang X, Zhang G, Wen Y, Liang G, Chen X. Aminocyclopropane-1-carboxylic acid is a key regulator of guard mother cell terminal division in *Arabidopsis thaliana*. Journal of Experimental Botany. 2019;**70**:897-908

[27] Vanderstraeten L, Depaepe T, Bertrand S, Van Der Straeten D. The ethylene precursor ACC affects early vegetative development independently of ethylene signaling. Frontiers in Plant Science. 2019;**10**:485515

[28] Mou W, Kao Y-T, Michard E, Simon AA, Li D, Wudick MM, et al. Ethylene-independent signaling by the ethylene precursor ACC in Arabidopsis ovular pollen tube attraction. Nature Communications. 2020;**11**:4082

Chapter 6

Seed Dormancy Challenges in the Production of Medicinal and Underutilized Leafy Vegetables

Thobile Perseverance Mkhwanazi, Gugu Cynthia Maseko and Zakheleni Palane Dube

Abstract

Seed dormancy has played a significant role in the adaptation and evolution of seed plants, by ensuring germination under favorable conditions, avoiding extreme weather periods, and other unfavorable conditions. While its biological significance is clear, dormancy acts as a delaying mechanism, making it difficult to simultaneously plant and properly maintain the population of the most important indigenous high-quality plants, consequently inhibiting mass cultivation and adoption. Several genetic and environmental factors influence dormancy, and different crops and or crop varieties including those of medicinal and indigenous vegetables exhibit varying degrees of dormancy. Breaking of dormancy will make a significant contribution towards ensuring consistent germination and cultivation of these crops. It is also important to observe and understand the types of dormancy exhibited by these as this can provide a guide for effective methods of breaking it. This book chapter will comprehensively discuss the types and challenges of seed dormancy associated with wild medicinal plants and indigenous vegetables, with special mention of cancer bush and jute mallow, as well as some pre-sowing treatments that can be used to break their dormancy. It further examines the potential of technological advances such as gene editing, genome engineering, and epigenesis regulation in addressing these challenges and improving cultivation.

Keywords: dormancy mechanisms, dormancy management, seed dormancy, seed germination, seed priming, underutilized crops

1. Introduction

Medicinal plants and indigenous vegetables have long been used by traditional cultures as alternative approaches to healthcare and nutritional needs [1]. About 75–80% of the world's population, mostly from developing countries, are reported to be dependent on these plants [2]. It is estimated that there are between 10,000 and 53,000 species of medicinal and underutilized crops worldwide, with the majority

IntechOpen

still being harvested from the wild [3]. In recent years, as the prices of medicine and food have increased, the use of both medicinal plants and edible herbs as vegetables has become a topic of global importance [4]. To achieve effective and sustainable food production, there is a need for research and cultivation of all underutilized indigenous crops [5, 6].

The higher demand for medicinal plant-based products has resulted in increased overharvesting of plants from wild populations [1]. Together with overexploitation, habitat destruction, climate change, and illegal trading of wild populations over the years have threatened medicinal plants with extinction [7]. According to Julsing et al. [8], only 10% of medicinal species are cultivated, and harvesting these wild species in high volumes without implementing mitigation measures increasingly pressures wild populations, threatening biodiversity. In South Africa, over 700 wild species are actively traded, mostly illegally [9]. The cultivation of most medicinal plants has been proven challenging due to low germination rates that require a specific ecological requirement [10]. It is noted that seed dormancy contributes significantly to these low germination rates [11, 12].

Seed dormancy acts as a delaying mechanism that inhibits germination, making it difficult to plant and properly maintain crop populations in the field simultaneously [13]. It also inhibits mass cultivation, growth, yield, and adoption as viable seeds can remain dormant in the soil for extended periods [14]. Germination-promoting stimuli such as scarification have been found to assist seeds in breaking their dormancy [15]. Previous studies on medicinal and indigenous crops have focused on various agronomical tactics to improve cultivation, including pre-sowing treatments (i.e., seed scarification) to break seed dormancy and stimulate germination [11, 16, 17]. Shaik et al. [18] explored biotechnological tactics such as micropropagation of cancer bush from vegetative plant parts to reduce wild harvesting while improving ex-situ cultivation and resources for acclimatized plants. Investigation on the effect of cultural practices (pruning and fertilizer application) on the growth, biological activities, and chemical properties of cancer bush, finding significant improvements in plant growth [1]. Masenya et al. [19] studied the effect of rhizobia inoculation (both native and commercial strains) on the growth and chemical composition of cancer bush.

Currently, there is limited knowledge concerning the specific requirements for pollination, seed germination, and growth of medicinal plants and underutilized vegetable crops, as most remain in their wild stages with dormant seeds. Farahani et al. [20] reported that dormancy is the main problem preventing the sustainable use of medicinal plants that can germinate in their native arid lands but fail to germinate well under laboratory conditions or when cultivated in the field. However, research suggested that if information on cultivation and economics were available and medicinal plants were properly phased into cultivation, the economic rewards for small-scale farmers could exceed those from maize or sugarcane [2]. The lack of knowledge on the cultivation and economics of medicinal plants is considered a major limiting factor in commercializing traditional medicinal plants [21]. To effectively domesticate and cultivate any plant species, having comprehensive information on seed germination and overcoming dormancy is imperative.

This book chapter explores the challenges of dormancy associated with medicinal and underutilized crop plants, identifies reported mechanisms, and proposes potential actions to overcome them. Finally, it examines the prospects of biotechnology in improving the cultivation of medicinal and underutilized vegetables.

2. Drives to commercialize underutilized medicinal and leafy vegetables

Neglected and underutilized crops as shown in **Tables 1–3** represent an important component of agro-biodiversity with potential to contribute to climate change adaptation, food security, and human health [46]. Over the years, much attention has been paid to exotic crops due to their known inherent agronomic, ecological, economic, and nutritional value. Given the known value of exotic crops, a majority of plant breeders, researchers, and policymakers have constantly ignored the development potential of underutilized crops, which led to their poor value chain. In South Africa, there is a wide range of underutilized crops that are historically popular and used by rural communities such as cancer bush (*Sutherlandia frutescens*), jute mallow (*Corchorus olitorius* L.), *Amaranthus* spp., *Chenopodium album*, and many others as listed in **Tables 1–3** [35]. Some of these crops have been incorporated into human diets since ancient times, especially in sub-Saharan Africa and many Asian countries where they greatly contribute to food and nutrition security and medicinal needs [40]. Their well-documented nutritional quality and climate adaptability compared to the exotic plants have led them being considered as one way to curb the "hidden hunger" that is most prevalent in developing countries, and as a result contribute to the achievement of some of the UN's Sustainable Development Goals SDG-1 (no poverty), SDG-2 (zero hunger), and SDG-3 (good health and wellbeing) [40]. However, since the beginning of the Green Revolution, many of these local, traditional, and underutilized crops have been replaced by high-yielding staple crops or cultivars developed through modern breeding programs [47]. Typically, underutilized crops do not meet modern standards for uniformity and other characteristics as they have been neglected by

African leafy vegetable	Harvested from wild	Cultivated	Growth season
Abelmoschus esculentus Moench		X	Summer
Amaranthus spp.	X		Summer
Bidens spinosa L.	X		Summer
Brassica rapa L. subsp. *chinensis*		X	Winter
Chenopodium album L.	X		Summer
Citrullus lanatus		X	Summer
Cleome gynandra L.	X		Summer
Corchorus olitorius L.	X		Summer
Cucurbita spp.		X	Summer
Vigna unguiculata (L.) Walp		X	Summer
Solanum retroflexum Dun		X	Winter
Portulaca oleracea L.	X		Summer
Momordica balsamina L.	X		Summer
Galinsoga parviflora Cav	X		Summer

Source: Ref. [22].

Table 1.
African leafy vegetable commonly harvested from the wild or obtained through cultivation in South Africa.

Common name	Scientific name	Crop type[a]	Plant part used[b]	Human nutrition	Pharmaceutical and nutraceutical properties	Reference
Cancer bush	*Sutherlandia frutescens* (L.) or *Lessertia frutescens* (L.) Goldblatt and J.C. Manning	Legume, herb, shrub	St, Lv, F, Pd, R	N/A	Used to treat chickenpox, diabetes, cancer, menopausal symptoms, influenza, rheumatoid arthritis, peptic ulcers, anxiety, clinical depression, HIV infection, and external wounds. Caner bush is amino acids, proline, and alanine.	[19, 23]
Wild ginger	*Siphonochilus aethiopicus* (Schweinf.) B.L.Burtt	Root and tuber	F	Contains fat, sodium, carbohydrates, sugars, protein, and calories.	Used to treat intestinal ailments, relieve stomach aches and cramps, and reduce stress, pain, and anxiety. The rhizomes and roots are chewed fresh to treat asthma, hysteria, colds, coughs and flu, malaria, vaginal thrush, and headache, and it is chewed by women during menstruation.	[24]
Carob	*Ceratonia siliqua* L.	Tree	Lv, Pd,	Pods are a rich source of carbohydrates (providing animals with a readily available source of energy to fuel their daily activities and metabolic processes) and protein (contributes to muscle development, tissue repair, and overall animal health) and are high in fiber (which plays an important role in maintaining digestive health).	Carob possesses various pharmacological activities, such as antioxidative, anti-diarrhea, antibacterial, anti-ulcer, anti-inflammatory, and anti-diabetic effects.	[25, 26]

Common name	Scientific name	Crop type[a]	Plant part used[b]	Human nutrition	Pharmaceutical and nutraceutical properties	Reference
Drumstick	*Moringa oleifera*	Herb	Lv, F, F, B, R	Almost all tree parts are eaten or used as ingredients in traditional herbal medicines. This especially applies to the leaves and pods, commonly eaten in parts of India and Africa. Moringa leaves are excellent source of calcium, potassium, iron, magnesium, phosphorus, zinc, vitamin A, B1 (thiamine), B2, (riboflavin), B3 (niacin), B-6, folate, and ascorbic acid (vitamin C), oils, fatty acids, micro-macro mineral elements, and various phenolics.	Provide treatments for inflammation, paralysis and hypertension, rheumatism, arthritis, diabetes and high blood pressure, relieve menstrual pain, stomach pain, heals burned skin and wounds.	[27–29]
Bush tea	*Athrixia phylicoides* DC.	Herb	St, Lv, F, R	N/A	Used to clean or purify the blood, treating boils, headaches, infected wounds, and cuts, and the solution may also be used as a foam bath. Treatment of various ailments such as boils, acne, colds, loss of voice, and throat infection as a gargle. Significantly high polyphenols, tannins, antioxidants, quercetin, flavonoids, alkaloids, polysaccharides, amino acids, lipids, vitamins, and inorganic elements.	[30, 31]
Bitter gourd	*Momordica charantia* (L.)	Herb	Fr	N/A	Provides proteins, potassium, iron, and fiber. It is used in the treatment of cancer and as an aphrodisiac.	[32, 33]

Common name	Scientific name	Crop type[a]	Plant part used[b]	Human nutrition	Pharmaceutical and nutraceutical properties	Reference
Hyacinthus	Hyacinthaceae	Lv	Fr, R, Sh	Contains sodium, potassium, carbohydrates, protein, and vitamin C. Also, Hyacinthus is a good source of crude lipids, ash, fiber, proteins and minerals, potassium, and sodium	Used to treat rheumatism, cardiac, urinary infection, dermatological problems, stomach, hemorrhoid, and prostate disease.	[34]
Honeybush	*Cyclopia* (Vent.) spp.	Herb, legume	Lv	N/A	Helps in reducing digestive problems, arthritis and to treat diabetes, stress relief and relaxation remedy, treat hypertension and hypotension, chest ailments, diarrhea, immune-boosting, blood circulation and blood cleanser, kidney ailments, diabetes, eczema (internally), stomach ailments, constipation, and appetite stimulant.	[35]

[a]*Crop type -Legume (L), Herb (H), Cereal (C), Shrub (S), Leafy vegetable (LV), Tree (T).*
[b]*Plant part—Root (R), Shoot (Sh), Seed (Sd), Stem (St), Flower (Fr), Pod (Pd), Leaves (Lv), Bark (B).*

Table 2.
Potential medicinal benefits of a few selected underutilized medicinal plants of South Africa.

Common name	Scientific name	Crop type[a]	Plant part used[b]	Human nutrition	Pharmaceutical and nutraceutical properties	Reference
Jute mallow	*Corchorus carinata C. olitorius* L.	LV	Lv, Sd	High in beta-carotene, folate, calcium, protein, B vitamins, iron, vitamins C and E, lipids, carbohydrate, and dietary fiber.	The leaves may have antibacterial, anticancer, and anti-inflammatory properties which may prevent the common cold, asthma, acne, arthritis, cure gonorrhea, pain, fever, and tumors. The Lycopene on the leaves has an antioxidant that protects cells from oxidative damage, which elevates disease risk.	[35–39]
Leafy Amaranthus	*Amaranthus* spp.	LV	LV	The leafy vegetable is rich in fiber, protein, calories, protein, carbohydrates, fat, manganese, magnesium, phosphorus, iron, selenium, and copper.	Treat diarrhea, ulcers, swollen mouth, and throat.	[22, 32, 40]
Bambara groundnut	*Vigna subterranea* (L.) Verdc.	Legume	Sd	Source of moisture, protein, carbohydrate, energy, crude fiber, calcium, potassium, magnesium, sodium, phosphate, iron, zinc, copper, ascorbic acid, β-carotene, lysine, methionine, and thiamine.	N/A	[41–43]
Purslane	*Portulaca oleracea*	LV	Lv	It is rich in β-carotene, folic acid, vitamin C, and essential fatty acids.	Heal headache, stomach ache, painful urination, enteritis, mastitis, lack of milk flow in nursing mothers.	[44, 45]

[a]*Crop type - Legume (L), Herb (H), Cereal (C), Shrub (S), Leafy vegetable (LV), Tree (T).*
[b]*Plant part - Root (R), Shoot (Sh), Seed (Sd), Stem (St), Flower (Fr), Pod (Pd), Leaves (Lv), Bark (B).*

Table 3.
Potential nutritional and medicinal benefits of a few selected underutilized leafy vegetables of South Africa.

breeders from the private and public sectors [48]. This rendered them less competitive in the market as compared to the exotic or commercial cultivars.

In 2017, the African Nutrition Society reported that the food pricing system in Africa delivers food at a cost that makes nutritious food unaffordable to the majority of the population, particularly the rural communities. Consequently, a resultant high disease burden associated with child malnutrition, micronutrient deficiencies, high body mass, and dietary risk factors was later reported in a study [49]. All these challenges were associated with the fact that many African families cannot afford the expensive nutritional exotic crops and therefore rely on low-cost underutilized vegetables. As a result, the preference for normal vegetables shifted extensively away from exotics and toward the growing underutilized crop market [50].

Currently, underutilized vegetable crops are receiving a lot of attention from plant breeders, researchers, farmers, and other stakeholders including government, nutritionists, and consumers due to their recognized potential contribution toward nutritional quality and climate adaptability [40]. Apart from their commercial, medicinal, and cultural value, these crops are also considered important for sustainable food production as they reduce the impact of production systems on the environment, as many of these crops are hardy, adapted to specific marginal soil and climatic conditions, and can be grown with minimal external inputs [51, 52]. The majority of the underutilized vegetable crops are regarded as of high nutritional value in relation to global vegetables like tomato and cabbage [53]. Underutilized legume crops such as mung bean have the potential to contribute significantly as sources of essential vitamins, micronutrients, protein, and other phytonutrients toward strategies aimed at attaining nutritional security [47]. Similarly, Jew's mallow is a very nutritious vegetable that is high in beta-carotene, folate, calcium, protein, B vitamins, iron, vitamins C and E, lipids, carbohydrates, and dietary fiber in its leaves [36]. Moreover, it provides 70% and 25% of the recommended daily amount value of vitamins C and A, respectively [54]. About 87 g of cooked Jew's mallow contains about 0.021 g of tryptophan, 0.113 g of threonine, 0.152 g of isoleucine, 0.266 g of leucine, 0.151 g of lysine, 0.044 g of methionine, 94 micrograms of vitamin K, 2.73 milligrams of iron, 0.496 milligrams of vitamin B6, 225 micrograms of vitamin A, 28.7 milligrams of vitamin C, and 0.222 milligrams of copper [54]. Jew's mallow is more nutritious in contrast to cabbage and spinach [55]. According to Zeghichi et al. [56], Jew's mallow is a better source of vitamins C and E, glutathione, carotenoids, minerals, and fatty acids than most other cultivated vegetables. In addition, it can be easily integrated into children's diets as it is reputed to taste much better in many culinary dishes than spinach [56]. A larger number of the rural populations in Africa depend on underutilized leafy vegetable crops such as Jew's mallow for nutrition [57]. Most of the underutilized crops serve as an essential source of vitamins, micronutrients, and protein, thus, a valuable component to attain nutritional security. Vegetables in general are of considerable commercial value and therefore an important source of household income, particularly the small-scale farmer who rely greatly on the cultivation of underutilized crops for most of their nutritional needs.

Moreover, the increased health consciousness and dietary shifts towards healthier foods in society contribute towards the growing popularity, production, and marketing of functional food crops such as amaranth (*Amaranthus tricolor* L.), cancer bush (*Sutherlandia frutescens* L.), honeybush tea (*Cyclopia vent*), mint (*Mentha* spp), and ginger (*Siphonochilus aethiopicus* (Schweinf.) B.L. Burtt) [47]. All these crops are dual-purpose crops functioning as food and herbal medicinal crops or plant-based dietary compounds for therapeutic, nutraceutical, and pharmaceutical benefits [35]. The majority of these crop species have the potential to be distributed at a global scale but are restricted to a more local production and consumption system. Underutilized crops constitute an important part of the local diet of communities providing valuable nutritional components, which is often lacking in staple crops [47]. Hence, a wider use of neglected and undervalued crops, either intercropped with main staples in cereal-based systems or as stand-alone crops, would provide multiple options to build temporal and spatial heterogeneity into the cropping systems and, as a result, enhance the resilience of crops/ farms to biotic and abiotic stress factors and ultimately leading to a more sustainable supply of diverse and nutritious food as well as providing for medicinal needs.

3. Overview of a few selected underutilized medicinal and leafy vegetable plants of South Africa and their potential uses

Underutilized plants are found in numerous agricultural ecosystems and often survive in marginal areas or fragile environments. Underutilized plant species are essential to the livelihoods of millions of people worldwide, due to their numerous

Figure 1.
A few selected underutilized crops of Southern Africa (a) Sutherlandia frutescens*; (b)* Amaranthus *sp.; (c)* Corchorus olitorius*, and (d)* Vigna subterranea *(L.) Verdc.*

nutrition and medicinal advantages contributing significantly to poverty elimination through employment opportunities and income generation, contribute to sustainable livelihood through household food security as they add nutritional value to diets, and also are convenient food source for low-income people. Okigbo and Anyaegbu [58] emphasized that, for these plants to be called underutilized species and prioritized for selection, they must be proven able to best address challenges of food security, poverty elimination, and environmental sustainability as mentioned above. Moreover, underutilized species should be proven able to be cultivated either in the past or currently cultivated less than comparable plants. A summary of published information related to the uses of a selected underutilized medicinal and leafy vegetable plants of South Africa (see **Figure 1** and **Tables 2** and **3**) is given below.

4. Dormancy challenges associated with unlocking the potential of underutilized plants

4.1 Seed dormancy in relation to germination

Seeds play an important role in plant propagation and species maintenance within an ecosystem. The germination of seeds serves as the foundational step thereof, emphasizing the need to investigate the physiological facets of seed germination

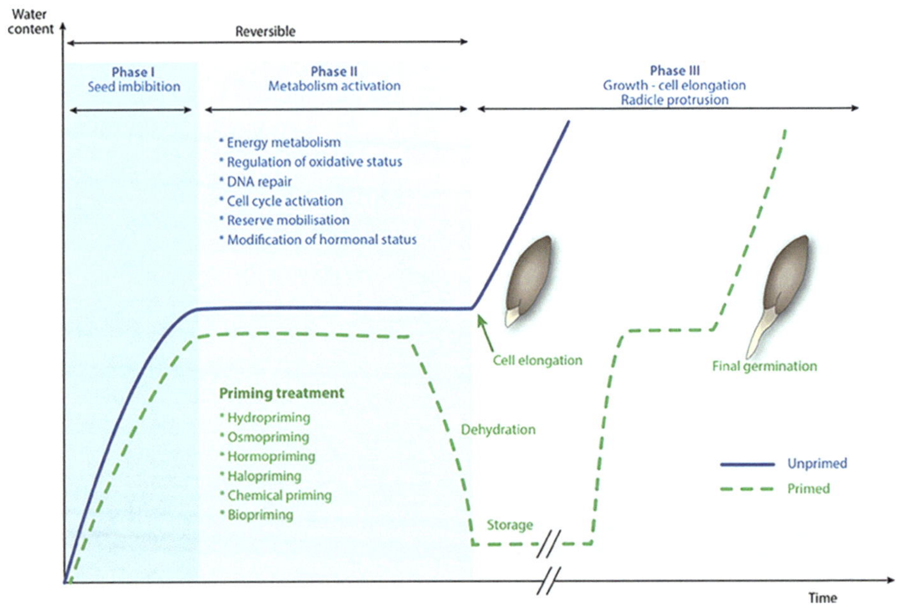

Figure 2.
Seed hydration and germination process in primed and unprimed seeds. The three distinct phases: (1) Phase I: the passive imbibition process related to rapid water uptake by the hard coat seed through hydration. Phase II is associated with an increase in embryo respiration after the establishment of metabolic activities. The increase of enzymes synthesis results in hormone release from the embryo. Lastly, Phase III demonstrates the completion of germination related to visible embryo growth processes and radicle protrusion (root and plumule) that will develop into shoot [62, 63]. Both I and III stages are associated with water uptake and increasing water content while phase II has stable water content. With primed seeds, it is common that before the end of Phase II, the seeds may dry up again and germination becomes reversible. The seed may remain alive again during storage and able to subsequently re-initiate germination under favorable conditions.

[59, 60]. Seed germination is regulated by the precise balance between endogenous (phytohormones and endosperm decay) and exogenous factors (environmental factors such as temperature and light) [61]. The transition from dormancy to germination as shown in **Figure 2** may occur as a result of the dry seed coming into contact with water and absorb water by imbibition and finally radicle emergence [62]. This encounter activates the internal metabolic process, involving the careful equilibrium phytohormones [64], in the presence of optimum light and temperature to overcome the seed dormancy [65]. Any imbalance that may occur between both factors will result in failure of seed to complete germination which is also perceived as dormancy. To understand dormancy or improve seed germination potential in plants, it requires a deeper and sequential understanding of the environmental stimuli around the seed and the interaction of intrinsic factors with extrinsic factors.

Seed dormancy is a natural process that delays seeds from germinating even when conditions are suitable for it. Several scholars define it as an innate constraint on germination under conditions that would otherwise promote germination in non-dormant seeds [66–68]. In simpler terms, dormancy must not only be associated with or defined as the absence of germination but rather the environmental conditions and characteristics of the seed that determine the conditions required for germination. When dormancy is considered in this way, any environmental cues that alter the conditions required for germination are by definition altering dormancy [69]. Other scholars define dormancy as "an innate state of arrested growth that occurs across all life forms" [69, 70]. According to Amen [71], dormancy is an internally controlled process (by enzymes, chemical inhibitors/promotors), and externally induced (by factors such as water, light, or temperature) temporal inhibition of development that is associated with reduced metabolic activity. Seed dormancy is a physiological phenomenon in wild and crop plants, more common in wild plants than crop plants [72].

During the dormancy period, seeds remain in an inactive/dormant state, often protected by mechanisms that prevent early germination. Dormancy allows seeds to germinate only when conditions become favorable. This is an adaptive feature that optimizes the distribution of seed germination over an extended period through varying degrees of dormancy [73] and bet-hedge against unpredictable variable environments (such as water content, temperature, light exposure, oxygen availability, and genetic attributes (plant hormones abscisic acid and gibberellins)) as shown in **Figure 1** [69, 74, 75]. This varied germination timing plays a pivotal role in ensuring species' survival, especially in demanding environmental circumstances [76]. By controlling the timing of germination, dormancy can strongly affect plant survival and adaptation [77]. Although, dormancy is the major determinant of species diversification by allowing colonization of new and different sites, however, this can only be possible under appropriate seasonal conditions with dormant seeds. Contrarily, non-dormant seeds that lack germination inhibitors, thus are better able to explore new and different environments because their germination is independent of specific dormancy-breaking cues that might be absent in that new environment [78]. This promotes diversification by fostering divergence and allopatric speciation [78].

Seed dormancy may be viewed as an important ecological trait ensuring survival for wild species, however, it remains an unfavorable trait in agriculture (crop species), as the main objective is to promote rapid seed germination and growth [79]. There are situations in crop production where seed dormancy can offer significant advantages, particularly during the seed development stage [79]. This advantage often benefits the production of cereal crops, which possess dormancy mechanisms that restrict germination while grains are still attached to the parent plant's ear. There it acts as a safeguarding mechanism for the plant which prevents germination, particularly during the rainfall period during harvest (also known as preharvest sprouting) to avoid agricultural losses [79]. Moreover, dormancy contributes a significant challenge for agriculture especially when it concerns issues of weed problems. Weed seeds equally maintain their inherent dormancy mechanisms as they mature and persist in the soil for many years, until the right conditions for germination arise. This may pose a threat to crop cultivation, as these seeds can rapidly multiply when favorable conditions finally occur.

4.2 Classification of dormancies

Seed dormancy classification is important in identifying the correct type of method to use to overcome any specific kind of dormancy [65]. Misidentifying or misinterpreting the dormancy of the seed may lead to failure to overcome the dormancy. Hence, the need to always direct methods of breaking dormancy towards the specific kind of dormancy. Owing to that fact, several researchers have elucidated some of the different classification systems of dormancy [80–82].

4.2.1 Classification based on barrier factors

This kind of classification categorizes dormancy in terms of barrier factors that exist within the seed. It is the earliest dormancy class discovered by Nikolaeva [82]. The barrier factors may exist inside the seed (endogenous dormancy) or outside the seed (exogenous dormancy). Exogenous dormancy is caused by conditions outside seed embryo that can prevent the seed from germination. While endogenous dormancy includes characteristics of the seed embryo that prevent the seed to germination [83]. Endogenous dormancy can be a result of an incomplete embryo development after ripening period or chemical inhibitors inside the seed. In the case of underdeveloped embryo, the seeds may require enough time to develop to their normal size [83]. The endogenous dormancy comprises the physiological, morphological, and morphophysiological characteristics of the seed embryo which may prevent germination. Within this classification, there is combinational dormancy which is activated by the combination of both exogenous and endogenous dormancy which exist in complementary fashion.

4.2.2 Classification based on time of induction

Seed dormancy is further categorized or grouped by Hilhorst [81] into primary and secondary dormancies.

4.2.2.1 Primary seed dormancy

It occurs when freshly and newly harvested seed from the mother plant is dispersed already in a dormant state [83]. Seed may be exposed to favorable conditions yet fail to complete germination, see **Figure 3** below. Primary dormancy occurs during the seed maturation phase due to the accumulation of abscisic acid (ABA). The abscisic acids prevent the loosening of the embryo cell wall which impedes water uptake and inhibits endosperm rupture instead of testa rupture. The induction of primary level of seed dormancy is regulated by several factors: genetic and non-genetic regions [83]. These factors may cause physiological variation in the seeds which manifests in different seed

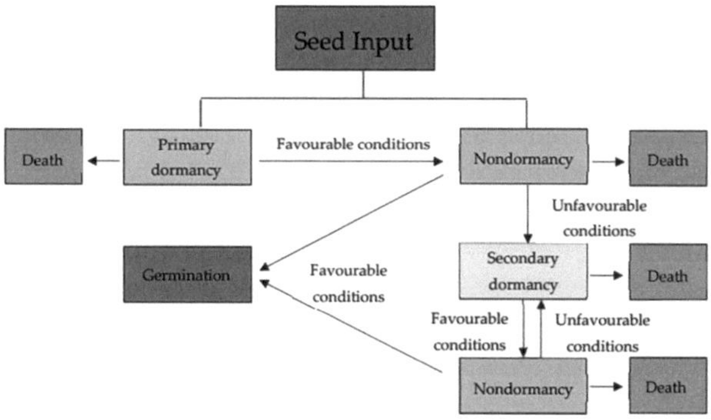

Figure 3.
A general model of the changes that happen in mature seed after being released from mother plant [84].

morphological sizes, weight, and color. Seeds with physiological variation experience the transitional period of conditional dormancy before becoming fully non-dormancy (**Figure 3**). At first, seeds gain the ability to germinate under a narrow range of environmental conditions, which increases with the loss of dormancy until they become completely non-dormant and germinate under a wide range of conditions. If non-dormant seeds are unable to germinate, because of changes in the environment, they re-enter conditional dormancy but can germinate under a full range of conditions. As time passes, the range of conditions within which germination is possible narrows to the extent that germination is not possible under any condition and the seed acquires secondary dormancy [84]. The primary dormancy is further categorized into induction seed dormancy and genetic dormancy [85].

4.2.2.2 Induction (coat-imposed) dormancy

Induction seed dormancy occurs when seeds fail to germinate due to some physical properties of the seed coats that inhibit germination [86]. Induction seed dormancy may also be what is called physical dormancy (PY) and is largely regulated by external factors (such as water, light, and temperature) regulating this kind of dormancy [87]. Hence, these three factors, water, gases, and mechanical resistance, are considered in relation to induction dormancy [88]. Physical seed dormancy is associated with some histological properties of the seed coat, such as dense epidermal palisade cells and the presence of numerous chemical compounds such as lignin, callose, lipids, phenolic deposits, cutin, wax, and suberin, in any layer of the seed coat [89]. The water-impermeable layers of palisade cells in the seed coat limit water transport, causing seed coat dormancy [90, 91].

4.2.2.3 Genetic dormancy

This is a kind of dormancy that is entirely regulated by the intrinsic factors of the seed such as embryos maturity and response to growth regulators [87]. Genetic dormancy is underpinned by three kinds of dormancies which include morphological dormancy (MD), morphophysiological dormancy (MPD), and physiological (PD).

Morphological dormant (MD) seeds have small, underdeveloped embryos that do not have a mechanism of physiological dormancy; hence, they do not require a dormancy-breaking intervention, rather sufficient time to further develop to full size to sprout [69, 80, 92].

Morphophysiological dormant seeds are also evidence of immature or underdeveloped embryos; however, they possess a component of physiological dormancy [80], as there may be hormone imbalance or the embryo's inability to push through the hard endocarp [93]. Such seeds therefore need pre-sowing treatment for dormancy breaking [80]. The time required for embryo growth or radicle protrusion to occur in seeds with morpho physiological dormancy is much longer compared to seeds with morphological dormancy [80].

Physiological dormancy (PD) is the most abundant form of dormancy found in seeds, gymnosperms, and all kinds of angiosperm clades. Due to a number of inhibitors, germination, growth-promoting enzymes and hormones can be inhibited as a result prevent complete germination of the seed. Any imbalance between inhibitors such as abscisic and promotors such as gibberellic acid may influence germination [90, 91, 94]. This suggests that if the two acids are not balanced; then, the balance or ratio needs to be tipped in the favor of those that will allow germination to proceed. Sufficient levels of abscisic inhibitors may counteract growth-promoting enzymes such as gibberellins since the two enzymes have opposite functions. These two enzymes present in seed endosperm, cotyledons, and sometimes on the outer coverings of the seed or fruit. Most of these chemicals are water soluble and can easily be leached from the seed which may help in shifting the balance towards the growth-promoting enzymes allowing germination [87]. Some of these chemicals must be degraded into other forms or reduced concentration. Inhibitors that are found in the seed embryonic axis are mostly controlled by temperature, sometimes light. Temperature may also favor the production of growth-promoting hormones and enzymes in the embryonic axis. Cool temperatures generally shift the balance of promoters and inhibitors towards promoting germination [95].

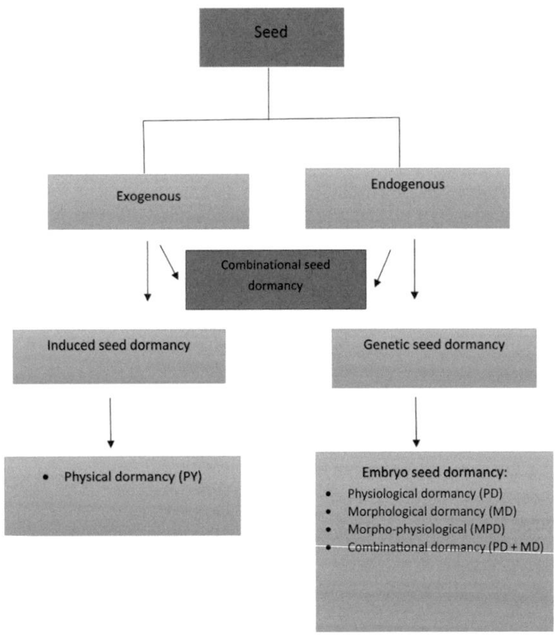

Figure 4.
An outlines the primary seed dormancy mechanisms, adopted from [83, 85].

Seeds with combinational dormancy comprised both physical dormancy and physiological dormancy [80].

4.2.2.4 Secondary dormancy

This is also called temporary or real dormancy, entirely associated with environmental conditions where seeds are planted such as light, temperature, and water; thus, when appropriate conditions are provided, the seeds will start germinating (**Figure 4**) [96].

4.3 Factors responsible for physical seed dormancy

4.3.1 Water impermeability of seed coat

The inability of seed coat to allow water penetration, which may be known as seed coat impermeability to water imbibition may be influenced by environmental and genetic factors. The interaction of environmental conditions such as climate or soil during seed development are contributing factors toward seed impermeability [87]. Seeds with impenetrable coats or hard-coated seeds may be a result of cuticle layer strong epidermal layer/cells surrounding the seed [87]. These cells render the seed impermeable to water as a result hindering germination. For instance, legumes constitute large reserves of lignin, cutin, or suberin. In some seeds, seed coat impermeability is due to large buildup of hilum, strophiole which restricts the movement of water through the seed coat.

4.3.2 Seed coat impermeability to gas exchange

Oxygen is one of the major atmospheric gases also present in soil pores spaces and is required by seeds for the metabolism process. If seeds are deprived of it either due to seeds being buried too deep into the soil or in waterlogged soils, the seed may be starved of this gas [97]. Seedlings use oxygen as the main energy source during aerobic respiration until they have grown leaves. Seed internal membrane impermeability (due to harden-seed coat) may hinder sufficient gas exchange, and such may be prevented when the seed coat is worn out enough to allow gases exchange (CO_2 and O_2) and water penetration from the environment.

4.3.3 Mechanical resistance of the seed coat

This may be associated with the development of physical limitations on the seed coat during embryo development. Although the seed may be absorbing enough water, it might not be enough to cause seed cracking and germination. Inhibitors on the seed coat may be a cause of such resistance and such is common in plantain, raspberry and cherry, and legume seeds [80, 98, 99].

4.4 Factors responsible for endogenous seed dormancy

4.4.1 Germination inhibitors

Many species' seeds fail to germinate even when the embryos are completely developed when the seed is ripe, even though the environmental conditions are excellent. In such seeds, dormancy is caused by the physiological state of the embryo [62]. The embryos of such seeds will not grow when they initially mature, even if the seed covers are removed. Abscisic acid (ABA) is one of the most commonly detected inhibitors of germination.

4.4.2 Dormant embryo

Even when the embryos are fully grown when the seed is ripe, many species' seeds fail to germinate, even when the environmental conditions are ideal. The physiological state of the embryo causes dormancy in such seeds [62]. Even if the seed coverings are removed, the embryos of such seeds will not grow when they first mature. During the period of dormancy, some physiological changes called after-ripening occur in the embryo before the seed is capable of germination.

4.5 Seed priming and breaking of dormancy

Seed priming using pre-sowing treatments is one of the approaches recommended for dormancy-breaking and enhancing germination [100]. As shown in **Table 4**, the different pre-sowing treatments use different modes of action to disrupt dormancy in plants. The majority of the pre-sowing treatments soften the water-impermeable seeds (or fruit) coat in high-order plants and enhance plant growth factors including germination rate and uniformity, as well as contributing to increased yields and plant resistance [124].

Although dormancy is an evolutionary mechanism that enables a species' long-term survival by allowing it to persist in the face of adversity [80, 91]. The ability to break dormancy is however crucial to getting seeds to germinate when necessary [91]. The purpose of breaking dormancy is to provide moisture to stimulate the seed's metabolism in repairing damages before commencing embryo development and root emergence [124].

Mechanisms of breaking dormancy are divided into seed coat treatments, also known as seed scarification techniques, and embryo treatments [87]. The embryo treatments are used to break genetic dormancy while the seed coat treatments are used to break physical dormancy which is associated with the physical properties of the seed coats.

Seed scarification techniques are used to overcome the hard, impermeable seed coat and include mechanical and chemical scarification, freeze-thaw, and cold-water soaking [87, 90, 125]. Mechanical scarification can be performed by rubbing the seed between pieces of sandpaper, abrasive, or sand, or by shaking the seed vigorously [87]. This is done to create a small hole in the seed (or fruit) forming an opening (or water gap) which will allow water to move and reach the embryo for embryo growth [126–128]. Mechanical scarifying of seeds is the most effective technique for dormancy-breaking although, it is time-consuming mostly if a large quantity of seeds requires scarification. With mechanical scarification, there are machines that allow the seeds to roll or blow seeds against abrasive surfaces (i.e. splinters or sandpaper inside containers) [128]. However, these machines do not work well for thick-coated seeds such as *Acacia* species, except on small and thin-coated seeds such as *Trifolium subterraneum* [128]. Sandpaper is used for scarification and is effective depending on the genus or species [129].

Seed priming method	Mode of action	Reference
Mechanical scarification	• Weakening/remove tissues covering the embryo surface. • Opens the micropyle.	[101, 102]
Acid scarification	• Burns/weakens the seed coat and epidermal layers.	[102, 103]
Hot treatment	• Weakening of seed coat by dissolving the lignin and pectins present in epidermal layers.	[103]
Hydropriming	• Activation of enzymes and mobilization of reserves in the aleurone layer. • Softening of hard seed coats and leaching out of chemical inhibitors (mainly ABA).	[102, 104–109]
Bio-priming	• Activation of early phases of germination. • Weakening of seed coat layers during the soaking (seed hydration) phase. • Protects seeds against the soil and seed-borne pathogens by applying antagonistic microorganism during priming.	[63, 107–110]
Gibberellic acid (GA3)	• Activates DNA in the aleurone cells. • GA counteracts the effect of ABA by promoting the embryo growth potential. • Weakening of tissues covering the embryo. • GA3 enhances cell enlargement and cell division in embryo. • Chemical activator and growth hormone. • stimulate the synthesis and production of the hydrolases enzymes resulting in the germination of seeds	[15, 102, 106–109, 111, 112]
Abscisic acid (ABA)	• Prevents loosening of the embryo cell wall which impedes water uptake. • Inhibit endosperm rupture instead of testa rupture.	[109, 113]
Indole-3-acetic acid (IAA)	• One of the prime auxins in plants, that regulate cell division, enhance photosynthetic activities, and activate the translocation of carbohydrates that enhance root initiation.	[106]
Halo-priming	• Activation of early phases of germination.	[63, 107–109]
Embryo rescue	• Break immature embryo using warm stratification. • Eliminate seed germination inhibitors.	[114]
Solid matrix priming	• The use of solid medium allows seeds to hydrate slowly and simulates natural imbibition process occurring in the soil.	[107–110, 115]
Osmo-priming (PEG, sugar, mannitol and sorbitol, NaNO3, MgCl2, NaCl, and KNO3)	• Allow pregermination metabolic activities. • Through a delayed water entry to seed reduces the ROS accumulation and thus protects the cell from oxidative injury.	[107, 109, 116]

Seed priming method	Mode of action	Reference
Priming with plant extract (such as phenolic compounds, terpenoids, flavonoids, saponins, alkaloids, and steroids)	• Saponins can enhance nutrient absorption as they are readily soluble in water. • Alkaloids, saponins, and phenolic compounds present in the leaves of various plants are involved in the production of antioxidant activities and protect the plants against pathogens.	[106, 117]
Nanoparticles (calcium-phosphate, SiO2, ZnO, and Ag)	• Allows for greater penetration of seed coat that improves nutrient and water uptake efficiency of the seed.	[109, 118, 119]
Seed priming through physical agents (magnetic field, UV radiation, gamma radiation, X-rays, and microwaves)	• Magnetic field improves germination rate, vigor, and seedling biomass as well as tolerance to various environmental stresses by means of reduction in reactive oxygen species (ROS) with increasing activities of antioxidant enzymes. • Rays interact with cellular components directly and improve the germination at lower doses. • Ultrasound priming induces mechanical pressure on seed coat that increases the seed's porosity known as acoustic cavitation and activation of enzymatic and other biological reactions due to greater water uptake in the seed.	[106, 120–123]

Table 4.
Mode of action of seed priming methods and promotion of germination.

Freeze-thaw scarification is a method of breaking the seed coat by exposing seeds to extreme cold (very low temperatures). Freeze-thaw scarification decreases hard-coated seed by creating microscopic scars on the hard seed coat and leaving the seed coat soft, which improves germination [125]. Chemical scarification involves the use of acids such as hydrochloric acid or sulfuric acid to burn the seed coating of the seed by immersing seeds in acids for a couple of minutes or seconds [130]. Hot water scarification involves the immersion of seeds in boiling water to soften their outer covering, the effectiveness of heat or hot water scarification varies with the type of device used, treatment time, and temperatures [125, 130]. However, it is necessary to identify sufficient scarification, as excessive scarification can result in the damaging of the embryo [91]. Seeds with non-deep physiological dormancy can germinate after undergoing chemical or mechanical scarification [80]. Several studies conducted on *Cassia occidentalis, C.obtusifolia, Indigofera astragalina, I. tinctoria, I. senegalensis, Tephrosia purpurea* and *Sesbania pachycarpa* [131], *Parkia biglobosa* (Jacq. Benth) [132], *Astragalus hamosus* and *Medicago orbicularis* [127], *Tylosema esculentum* (Buech) L. Schreib [133], *Senna alata* (L.) Roxb. [130], and *Senna alata* [134] have shown that when seeds were immersed in concentrated sulfuric acid, it resulted in improved final germination percentage of the dormant seed.

Hydropriming techniques may include cold water soaking and heat treatment using dry and wet heat. The dry and wet priming methods are both grouped together under heat scarification where two heating devices, which include the oven and hot water baths, are used to soften the hard seed coat [125]. The kind of heating device used, heating temperature and time, determines the effectiveness of each heat treatment. Wet heat involves the immersion of seeds that are water-impermeable in hot water so that they become permeable [126]. According to Baskin and Baskin [126], this method involves placing the seeds in cloth bag and dipped in water baths for the required period of time. The seeds are allowed to cool for a few minutes after the water bath. This method was proven effective in softening the seed coat and enhancing germination of *Sena alata* (L.) Roxb resulting in 77.7% [130]. Dry heat is the most common technique used to render seeds (of species that are water-impermeable) permeable to water by placing them in an oven at a definite temperature [130]. These techniques are effective for physical dormancy-breaking in a number of species [126].

Embryo treatments include chemicals (such as GA3 and KNO3), high temperature, and stratification treatments [125]. Stratification is the process of incubating seeds at a low temperature (also called pre-chilling) over a moist surface before transferring them to a temperature that will allow them to germinate [135]. According to Kimura and Islam [125], the force that forms scars on seed coat using this procedure is determined by the shape, size, and moisture content of the seeds, and also the duration and intensity of the treatment [125].

Bio-priming, also called the biological seed treatment, is an advanced technique of preventing stunted plant growth as a result of reduced quality of seeds physiology through the integration of biological agents (inoculation of seeds with beneficial microorganisms to protect seed) [136] or combination [137]. The treatment is applied on the seed surface, and the seed is allowed to dry. Biological seed treatment as explained in **Table 4** provides an ecological advantage to seeds by controlling several seed or soil-borne pathogens which also provide an alternative chemical treatment [136]. According to Rafi and Dawar [136], seed bio-priming enhances the initial step in the development of plants by increasing seed tolerance to different stress (seed or soil-borne pathogens), thus improving seed germination, and *Trichoderma* is the most widely used species.

4.6 Work documented on the recommended seed priming methods for breaking dormancy of some underutilized crops

Seed priming is a physiological technique that involves seed hydration and drying to enhance the metabolic process before germination in order to quicken germination, seedling growth, and crop yield under normal, as well as different biotic and abiotic stress conditions [108]. Seed priming has emerged as an effective seed treatment tool for many crops especially underutilized crops; however, treating conditions and methods of priming tend to differ with plant species (see **Table 4**) as explained below. Each of these methods can be tailored to the specific requirements of the seed species being primed as previously discussed, and different species may have different mechanisms of dormancy. Understanding the biology of the seed and dormancy mechanism, the mode of action (see **Table 4**) of the priming method is key for choosing the most effective priming method. Numerous work has been done and documented on seed priming of some underutilized medicinal and leafy vegetable plants in order to improve the final yield (see **Table 5**). Although in species such as *Athrixia phylicoides* DC. and *Siphonochilus aethiopicus* (Schweinf.) B. L. Burtt (see **Table 5**), there is no work done.

4.7 Biochemical and molecular factors regulating seed germination in plants

Dormancy release and seed germination are controlled by interconnected molecular processes regulated by different types of hormones such as abscisic acid (ABA), gibberellins (GA), ethylene, and auxin that interact with each other [94]. Furthermore, abscisic acid and gibberellins act antagonistically to each other, whereby ABA promotes induction and maintenance of dormancy during imbibition while GA promotes germination

Species name	Priming method	References
Sutherlandia frutescens L.	Scarification (mechanical and acid)	[11, 16]
Athrixia phylicoides DC.	Temperature (15, 20, and 25°C) treatments under constant light exposure	[138]
Momordica charantia (L.)	Hydropriming, halopriming (NaCl and 48 hours of ZnONP)	[106, 110, 139]
Siphonochilus aethiopicus (Schweinf.) B. L. Burtt	—	—
Cyclopia (Vent.) spp.	Scarification (H2SO4, mechanical), hot water treatment	[140]
Ceratonia siliqua L.	Scarification (mechanical and H2SO4), Hot water treatment, soaking in distilled water (24 h)	[141–143]
Portulaca oleracea	Hot water treatment, hydropriming for several hours, 100% relative humidity for several hours	[143, 144]
Amaranthus spp.	Osmo-priming, hydropriming	[145, 146]
Vigna subterranea (L.)	Osmo-priming	[147]
Moringa oleifera	Hydropriming	[148–150]
Corchorus olitorius	Hot water treatment	[14, 151]

Table 5.
List of plants and recommended priming method.

Phytohormone	Control of seed germination	Reference
Abscisic acid (ABA) and gibberellins (GAs)	• ABA and GAs balance plays a significant role in controlling seed dormancy. • ABA significantly inhibits seed germination, and a high ABA and GA ratio normally hinders germination in dormant seeds. • A decrease in ABA is mostly important for initiating germination and GA levels increased. • High levels of ABA in seeds inhibit germination by increasing the expression of dormancy-related enzymes such as NCED (9-cis-epoxycarotenoid dioxygenase) and gene DOG1 (delay of germination1) gene • And reducing growth-related GA-responsive genes such as GA3ox1 and GA3ox2	[60, 152, 153]
Ethylene	• Promotes seed germination in plants, is produced right after seed imbibition, and increases as germination continues. • Production can be increased by nitric oxide, hydrogen cyanide, low temperatures, and GA treatments, encouraging seed germination. • Arabidopsis ethylene-responsive factor ERF12 can bind to the promoter of the important dormancy gene DELAY OF GERMINATION 1 (DOG1) and recruit the transcriptional co-repressor TOPLESS (TPL), which decreases DOG1 expression and increases seed germination.	[154]

(**Table 6**) [156–158]. Ethylene is also a promoter of seed germination, having receptors such as ethylene response factor 1 (ETR1) that reduce dormancy phenotype and enhance germination [159]. Ethylene continues to be produced as germination proceeds and can be increased by nitric oxide, hydrogen cyanide, low temperatures, and GA treatments, promoting seed germination (**Table 6**) [154]. However, auxin functions negatively and positively in seed germination depending on its amount, whereby high levels of auxins promote seed dormancy by activating AB13 through auxin-responsive transcription factors (ARF) 10 and ARF16 being released [160]. Additionally, low levels result in ARF 10 and ARF16 being oppressed by AXR2/3 and failure to activate the expression of AB13 and maintain dormancy (**Table 6**) [65].

4.7.1 Phytohormone that controls dormancy and germination.

Phytohormone	Control of seed germination	Reference
Auxins	• Auxins modulate ABA positively and GA biosynthesis and signaling pathways negatively, promoting dormancy. However, auxins can also break dormancy by interacting with GA which promotes germination and is crucial for the development of the root system during germination. • While low levels of auxins result in ARF10 and ARF16 being oppressed by AXR2/3 and failure to activate the expression of AB13 and maintain dormancy. • High levels result in auxin-responsive transcription factors ARF10 and 16 being released to active AB13 transcription and maintain seed dormancy.	[65, 155]

Table 6.
Phytohormone that controls dormancy and germination.

4.7.2 Endosperm decay hinder embryo growth

Endosperm performs as a mechanical barrier to seed germination in various angiosperm clades in which a decrease in the mechanical resistance of the endosperm layer covering the radicle tip appears to be a prerequisite for radicle protrusion during seed germination [69]. Moreover, through bidirectional communication, the endosperm can affect the embryo's initial growth or even completely prevent seed germination [161], whereby endosperm weakening can be enhanced by GA hormone and inhibited ABA (**Table 7**) [164, 165].

4.7.3 Light regulation of seed germination and dormancy

Seed germination and dormancy are not only controlled by plant hormones which are internal signals but also depend on external environmental elements such as light and temperature [101]. Moreover, seeds can be categorized based on their response to light during germination, whereby the controlling effect of light on seed germination depends on the light spectrum (**Table 8**) [168]. There are two types of light-sensing systems which include blue-light (photo regulation) and red-light (phytochrome regulation) sensitive systems, in which blue light promotes dormancy [169]. In

Endosperm decay impact on embryo growth	Reference
• Endosperm break down during seed germination releasing nutrients for the growing embryo. • Slow endosperm decay can limit nutrient availability, slowing down embryo growth. • Fast endosperm decay can lead to an overabundance of nutrients, causing an imbalance that hinders embryo growth.	[162, 163]

Table 7.
Endosperm decay hinders embryo growth.

Light regulation of seed germination and dormancy	Reference
• The spectrum of light influences how seed germination is regulated. • Blue light inhibits the germination of seeds and stimulates ABA. • Red light, 600-760 nm stimulates GA biosynthesis, restricting ABA production and active seed germination. • Far-red light, 760-800 nm can inhibit seed germination.	[61, 166]
• Plants can be categorized based on their response to light during germination. • Those that require light to germinate. • Require darkness to germinate. • Plans that have a large percentage of seed neutral photoblastic.	[167]

Table 8.
Light regulation of seed germination and dormancy.

contrast, red light promotes seed germination and far-red light inhibits seed germination (**Table 8**) [166]. Furthermore, temperature elevation or lowered may impede different physiological and molecular mechanisms, hence delaying the germination of seeds causing dormancy to be disrupted while high temperatures disrupt dormancy, which causes germination (**Table 9**) [60]. Low temperatures delay germination by increasing ABA synthesis and lowering water absorption, protein breakdown, glucose metabolism, and energy production (**Table 8**) [170].

4.7.4 Temperature regulation of seed germination and dormancy

Role in regulating seed germination	Reference
• Determines germination time by influencing germination directly and regulating dormancy. • High or low temperatures can cause a delay in the germination of seeds due to the obstruction of various molecular and physiological processes. • Low temperatures delay germination by increasing ABA synthesis and decreasing water absorption, protein breakdown, carbohydrate metabolism, and energy production. • High-temperature conditions disrupt dormancy, leading to germination. • For optimal germination, most seeds require temperatures between 15 and 30°C. For instance, *Vigna subterranea* (L.) requires an optimum temperature from 30 to 35°C for germination.	[170, 171]

Table 9.
Temperature regulation of seed germination and dormancy.

5. Future prospectus in the production of underutilized medicinal and leafy vegetables

Steps towards the understanding of mechanisms of seed dormancy, identification of genes, hormones, and metabolites involved, and the role of the environment have modernized ways of addressing challenges of seed dormancy in a range of crops. Molecular biotechnology is one modern approach that has revolutionized research in biological science, resulting in the development of superior crop material [172–174].

Gene editing and genome engineering as tools have offered great advances in the study of seed dormancy [174]. In recent years, the use of CRISPR/Cas9 in gene editing has gained extensive use in plant and animal physiology [175]. Research and manipulation of seed dormancy genes such as delay of germination 1 (DOG1), seed dormancy 4 (SDR4), and mother of FT and TFL 1 (MFT) in medicinal plants and indigenous vegetables could greatly improve seed germination [176]. The applicability of the technique has been tested previously on a number of crops, including grasses, wheat, and barley [175–177]. Glison et al. [176] extensively reported on the application of genetic editing and genome engineering in the seed dormancy of grasses. The potential of marker-trait associations (MTAs) and marker-assisted selection of genes for engineering in breeding for seed dormancy in wheat was reviewed by Kulwal [177], while [175] reduced seed dormancy in rice (*Oryza sativa* L.) by knockout viviparous-1 (OsVP1) gene responsible for seed dormancy. In barley, the genetic editing of QTL for seed dormancy 1 (Qsd1) and Qsd2 genes affected seed germination of the plant [178]. Induced mutation of the Qsd1 homeoalleles was responsible for the regulation of seed germination in wheat [179].

Unlike genetic editing and genome engineering, epigenesis regulation is one modern technique exploited in the understanding and regulation of seed dormancy, that involves altering the gene expression and associated proteins without changing their sequence [180, 181]. The leading mechanisms in epigenesis include DNA methylation, modifications to chromatin, loss of imprinting, and non-coding RNA [180]. An impressive work done by Luján-Soto and Dinkova [182] in the use of epigenesis modifications in seed dormancy, highlighted the potential of the technology in detail. Work done by Sato et al. [183] on Arabidopsis demonstrated the applicability of epigenesis modification, in stimulating and inhibiting seed germination. On the other hand, [174] demonstrated that histone deacetylase HDA19 and histone methyltransferase SUVH5 both worked together in the regulation of seed dormancy in *Arabidopsis.* Recently, [184] reported also the histone modifications, acetylation, and methylation, moderating the seed dormancy of Arabidopsis *thaliana*. Secondary seed dormancy depth and germination in *Capsella bursa-pastoris* were reported to be regulated also by histone methylation [185]. Using epigenetic modification, [186] demonstrated the potential for influencing seed dormancy in *Paris polyphylla*.

Recent advances in understanding the role of environmental factors from seed development to seed dispersal have also opened a door in addressing the challenge of seed dormancy [187, 188]. Focus previously has been on the effects of temperature, light, and moisture as environmental effects, but advances in the field of metabolomics and hormonal signaling on seed dormancy and germination have recently received much attention [160, 189].

The potential of this technology in the improvement of seed dormancy and germination in medicinal and indigenous vegetables could not be overestimated. The exploitation of these modern and advanced research tools and techniques on seed dormancy regulating physiological factors could potentially be applied to the improvement and commercialization of medicinal and indigenous vegetables.

6. Conclusion

Seed dormancy is a survival trait for wild plants, but a great hindrance to the domestication of important medicinal and indigenous vegetables as it affects the crop germination rates and crop stand, resulting in reduced yield. Medicinal and indigenous vegetables have been identified as crops with potential in addressing challenges of food security and malnutrition in poor rural communities if domesticated and commercialized. This book chapter makes it clear that advances in gene editing, genome engineering, and epigenesis modification present a big opportunity in overcoming seed dormancy of these wild plants.

Acknowledgements

The authors are grateful to the University of Mpumalanga for financial support to carry out this work.

Conflict of interest

The authors declare no conflict of interest.

Acronyms and abbreviations

AUC	African Union Commission
ARF	Auxin responsive transcription factors
DOG 1	delay of germination 1
ERF	ethylene response factor 1
OsVP1	knockout viviparous 1
MTAs	Marker trait associations

Author details

Thobile Perseverance Mkhwanazi*, Gugu Cynthia Maseko and Zakheleni Palane Dube
University of Mpumalanga, Mbombela, South Africa

*Address all correspondence to: mkhwanazithobile5@gmail.com

References

[1] Raselabe MB. Effects of pruning and fertilizer on growth, phytochemistry and biological activity of *Sutherlandia frutescens* (L.) R.Br [MSc thesis]. Durban, South Africa: University of KwaZulu-Natal; 2017

[2] Tanga M, Lewu FB, Oyedeji OA, Oyedeji OO. Cultivation of medicinal plants in South Africa: A solution to quality assurance and consistent availability of medicinal plant materials for commercialization. Academia Journal of Medicinal Plants. 2018;**6**(7):168-177

[3] Yessoufou K, Daru AM, Muasya A. Phylogenetic exploration of commonly used medicinal plants in South Africa. Molecular Ecology Resources. 2015;**15**:405-413

[4] Stafford GI, Jager AK, Van Staden J. Effect of storage on the chemical composition and biological activity of several popular South African medicinal plants. Journal of Ethnopharmacology. 2005;**97**:107-115

[5] Gregory PJ, Mayes S, Hui CH, Jahanshiri E, Julkifle A. Crops for the future (CFF): An overview of research efforts in the adoption of underutilised species. Planta. 2019;**250**:979-988

[6] Knez M, Ranić M, Gurinović M. Underutilized plants increase biodiversity, improve food and nutrition security, reduce malnutrition, and enhance human health and well-being. Let's put them back on the plate. Nutrition Reviews. 2024;**82**(8):1111-1124

[7] Street RA, Stirk WA, Van Staden J. South African traditional medicinal plant trade-challenges in regulating quality, safety and efficacy. Journal of Ethnopharmacology. 2008;**119**:705-710

[8] Julsing MK, Wim JQ, Kayser O. The engineering of medicinal plants: Prospects and limitations of medicinal plant biotechnology. In: Kayser O, Quax WJ, editors. Medicinal Plant Biotechnology: From Basic Research to Industrial Applications. Weinheim: Wiley-VCH Verlag GmbH & Co. KGaA; 2007. pp. 3-8

[9] Mander M, Ntuli L, Diederichs N, Mavundla K. Economics of the Traditional Medicine Trade in South Africa [Internet] 2008. pp. 189-196. Available from: 18_SAHR_2006-2007_Section13.pdf(hst.org.za)

[10] Canter PH, Thomas H, Ernst E. Bringing medicinal plants into cultivation: Opportunities and challenges for biotechnology. Trends in Biotechnology. 2005;**23**(4):180-185

[11] Shaik S, Dewir Y, Singh N, Nicholas A. Influences of pre-sowing seed treatments on germination of the cancer bush (*Sutherlandia frutescens*), a reputed medicinal plant in arid environments. Seed Science & Technology. 2008;**36**:795-801

[12] Tsiantis M. Plant development: Multiple strategies for breaking seed dormancy. Current Biology. 2006;**16**(1):25-27

[13] Yildiz M. Seed dormancy as an obstacle in front of plant production. Trends in Horticulture. 2018;**1**(3):24-26

[14] Larnyo A, Atitsogbui PJ. Effect of temperature treatments on seed germination and seedling growth of jute mallow (*Corchorus olitorius*). International Journal of Environment, Agriculture & Biotechnology. 2020;**5**(6):1631-1640

[15] Bentsink L, Koornneef M, Hilhorst H. Seed Dormancy and Germination. In: The Arabidopsis Book [Internet]. 2008. DOI: 10.1199/tab.0119

[16] Mkhwanazi T, Ubisi R, Kgotse L, Timana M, Sebati M, Mnyambo N, et al. Seed dormancy in cancer bush (*Sutherlandia frutescens*): A major hindrance to production. Scientific Papers, Series-B, Horticulture. 2023;**67**(1):649-655

[17] Korth S. *Sutherlandia frutescens*: A Woody Perennial that Packs a Medicinal Punch [Internet]. 2021. Available from: https://hdl.handle.net/11299/225167

[18] Shaik S, Dewir YH, Singh N, Nicholas A. Micropropagation and bioreactor studies of the medicinally important plant *Lessertia* (*Sutherlandia*) *frutescens* L. South African Journal of Botany. 2010;**76**:180-186

[19] Masenya TA, Mashela PW, Pofu KM. Efficacy of rhizobia strains on growth and chemical composition of cancer bush (*Sutherlandia frutescens*). Acta Agricultural Scandinavica, Section B-Soil & Plant Science. 2022;**72**(1):358-363

[20] Farahani S, Hajibarat Z, Hajiberat Z. Effect of different treatments on breaking dormancy of *Teucrium chamaedrys* L seed. Journal of Medicinal Plants & By-products. 2014;**1**:63-67

[21] Nwafor C. Cultivation of Medicinal Plants by Smallholder Farmers in South Africa: Constraints to Commercialization [Internet]. 2020. DOI: 10.20944/preprints202012.0761.v1

[22] Maseko I, Mabhaudhi T, Tesfay S, Araya HT, Fezzehazion M, Du Plooy CP. African leafy vegetables: A review of status, production and utilization in South Africa. Sustainability. 2018;**10**:16

[23] Van Wyk BE. The potential of South African plants in the development of new medicinal products. South African Journal of Botany. 2011;**77**:812-829

[24] Xego S, Kambizi L, Nchu F. Threatened medicinal plants of South Africa: Case of the family Hyacinthaceae. African Journal of Traditional Complementary & Alternative Medicines. 2016;**13**:169-180

[25] Brassesco ME, Brandão TRS, Silva CLM, Pintado M. Carob bean (*Ceratonia siliqua* L.): A new perspective for functional food. Trends. Food Science & Technology. 2021;**114**:310-322

[26] Martins-Loucao MA, Correia PJ, Romano A. Carob: A Mediterranean resource for the future. Plants. 2024;**13**(9):1188

[27] Chuang PH, Lee CW, Chou JY, Murugan M, Shieh BJ, Chen HM. Anti-fungal activity of crude extracts and essential oil of *Moringa oleifera* Lam. Bioresource Technology. 2007;**98**:232-236

[28] Makkar HPS, Becker K. Nutritional value and antinutritional components of whole and ethanol extracted *Moringa oleifere* leaves. Animal & Feed Science Technology. 1996;**63**:211-228

[29] Seifu E, Teketay D. Introduction and expansion of *Moringa oleifera* Lam. in Botswana: Current status and potential for commercialization. South African Journal of Botany. 2020;**129**:471-479

[30] Slabbert EL, Malgas RR, Veldtman R, Addison P. Honeybush (*Cyclopia* spp.) phenology and associated arthropod diversity in the Overberg Region, South Africa. Bothalia. 2019;**49**:1-13

[31] Zonyane S, Chen L, Xu MJ, Gong ZN, Xu S, Makunga NP. Geographic-based metabolomic variation

and toxicity analysis of *Sutherlandia frutescens* L. R.Br. An emerging medicinal crop in South Africa. Industrial Crops and Products. 2019;**133**:414-423

[32] Bhattacharjee T, Sen D, Irenaeus TKS, Das SP, Chakraborty A. Diversity of underutilized vegetable crops in Tripura. Acta Horticulturae. 2019;**1241**:121-134

[33] Kutu FR, Magongwa SM. Effect of pre-treatment and preservation methods on the nutritional quality of bitter gourd leaves from two different sources. Research on Crops. 2017;**18**:675-682

[34] Masondo NA, Finnie JF, Van Staden J. Pharmacological potential and conservation prospect of the genus Eucomis (Hyacinthaceae) endemic to southern Africa. Journal of Ethnopharmacology. 2014;**151**:44-53

[35] Mudau FN, Chimonyo VGP, Modi AT, Mabhaudhi T. Neglected and underutilized crops: A systematic review of their potential as food and herbal medicinal crops in South Africa. Frontiers in Pharmacology. 2022;**12**:809-866

[36] Assi OY, Coulibaly A, Kouakou P, Konan YNG, Chatigre O, Biego HG. Vitamins contents in edible parts of some mucilaginous food plants from Côte d'Ivoire. Journal of Advances in Biology & Biotechnology. 2016;**10**(2):1-14

[37] Chan T. Jute Leaves: Nutrition, Benefits, Downsides, and Recipes [Internet]. 2021. Available from: https://www.healthline.com/nutrition/jute–leaves/

[38] Dansi A, Vodouhè R, Azokpota P, Yedomonhan H, Assogba P, Adjatin A. Diversity of the neglected and underutilized crop species of importance in Benin. The Scientific World Journal. 2012;**2012**:1-19

[39] Nyadanu D, Adu Amoah R, Kwarteng A, Akromah R, Aboagye L, Adu-Dapaah H, et al. Domestication of jute mallow (*Corchorus olitorius* L.): Ethnobotany, production constraints and phenomics of local cultivars in Ghana. Genetic Resources and Crop Evolution. 2017;**64**(6):1313-1329

[40] Imathiu S. Neglected and underutilized cultivated crops with respect to indigenous African leafy vegetables for food and nutrition security. Journal of Food Security. 2021;**9**(3):115-125

[41] Chibarabada T, Modi A, Mabhaudhi T. Expounding the value of grain legumes in the semi–And arid tropics. Sustainability. 2017;**9**:60

[42] Chibarabada TP, Modi AT, Mabhaudhi T. Nutrient content and nutritional water productivity of selected grain legumes in response to production environment. International Journal of Environmental Research & Public Health. 2017;**14**:1300

[43] Atoyebi JO, Osilesi O, Adebawo O, Abberton M. Evaluation of nutrient parameters of selected African accessions of Bambara groundnut (*Vigna Subterranea* (L.) Verdc.). Australian Jewish Fertility Network. 2017;**5**:83-89

[44] Kumar V, Uthappa AR, Srivastava M, Vijay D, Kumaranag KM, Manjunatha N, et al. Floral biology of *Grewia Flavescens* Juss. An underutilized crop. Genetics Resources and Crop Evolution. 2017;**64**:1789-1795

[45] Njume C, Goduka NI, George G. Indigenous leafy vegetables (imifino, morogo, muhuro) in South Africa: A rich and unexplored source of nutrients

and antioxidants. African Journal of Biotechnology. 2014;**13**(19):1933-1942

[46] Mabhaudhi T, Chimonyo VGP, Modi AT. Status of underutilized crops in South Africa: Opportunities for developing research capacity. Sustainability. 2017;**9**:1569

[47] Ebert AW. Potential of underutilized traditional vegetables and legume crops to contribute to food and nutritional security, income and more sustainable production systems. Sustainability. 2014;**6**:319-335

[48] Stamp P, Messmer R, Walter A. Competitive underutilized crops will depend on the state funding of breeding programmes: An opinion on the example of Europe. Plant Breeding. 2012;**131**:461-464

[49] FAO, ECA, AUC. Africa regional overview of food security and nutrition 2020: Transforming food systems for affordable healthy diets [Internet]. 2021. DOI: 10.4060/cb4831en

[50] African Nutrition Society (ANS). Traditional and indigenous foods for food systems transformation in Africa [Internet]. 2017. Available from: https://summitdialogues.org/dialogue/12901

[51] De la Peña RC, Ebert AW, Gniffke P, Hanson P, Symonds RC. Genetic adjustment to changing climates: Vegetables. In: Yadav SS, Redden RJ, Hatfield JL, Lotze-Campen H, Hall AE, editors. Crop Adaptation to Climate Change. Chichester: United Kingdom; 2011. pp. 396-410

[52] Hughes J'A, Ebert AW. Research and development of underutilized plant species; the role of vegetables in assuring food and nutritional sevurity. In: Massawe F, Mayes S, Alderson P, editors. Proceedings of the Second International Symposium on Underutilized Plant Species: Crops for the Future-beyond Food Security. Korbeek-Lo, Belgium: International Society for Horticultural Sciences (ISHS); 2018. pp. 79-91

[53] Keatinge JDH, Yang RY, Jd'A H, Easdown WJ, Holmer R. The importance of vegetables in ensuring both food and nutritional security in attainment of the millennium development goals. Food Science. 2011;**3**:491-501

[54] Bhawana D, Neetu S. Molokhia-the wealth for a better health. International Journal of Research. 2015;**2**(2):658-660

[55] Ndlovu J, Afolayan A. Nutritional analysis of the south African wild vegetable *Corchorus olitorius* L. Asian Journal of Plant Sciences. 2008;**7**(6):615-618

[56] Zeghichi S, Kallithraka S, Simopoulos A. Nutritional composition of Molokhia (*Corchorus olitorius*) and Stamnagathi (*Cichorium spinosum*). World Review of Nutrition & Dietetics. 2003;**91**:14-18

[57] Rankoana S. Indigenous plant foods of Dikgale community in South Africa. Journal of Ethnic Foods. 2021;**8**(1):1-8

[58] Okigbo RN, Anyaegbu CF. Underutilized plants of Africa. Journal of Biology & Nature. 2021;**13**(2):34-49

[59] El-Maarouf-Bouteau H. The seed and the metabolism regulation. Biology. 2022;**11**(2):168

[60] Farooq MA, Ma W, Shen S, Gu A. Underlying biochemical and molecular mechanisms for seed germination. International Journal of Molecular Science. 2022;**23**:8502

[61] Carrera-Castaño G, Calleja-Cabrera J, Pernas M, Gómez L,

Oñate-Sánchez L. An updated overview on the regulation of seed germination. Plants. 2020;**9**:703

[62] Abubakar MS, Attanda ML. Factors that Causing Seed Dormancy [Internet]. 2022. DOI: 10.5772/intechopen.107004

[63] Lutts S, Benincasa P, Wojtyla L, Szymon KS, Pace R, Lechowska K, et al. Seed priming: New comprehensive approaches for an old empirical technique. In: Araujo S, Balestrazzi A, editors. New Challenges in Seed Biology – Basic and Translational Research Driving Seed Technology [Internet]. 2016. DOI: 10.5772/64420

[64] Oracz K, Karpinski S. Phytohormones signaling pathways and ROS involvement in seed germination. Frontiers in Plant Science. 2016;**7**:864

[65] Shu K, Liu XD, Xie Q, He ZH. Two faces of one seed: Hormonal regulation of dormancy and germination. Molecular Plant. 2016;**9**:34-45

[66] Larson HJ. Breaking seed dormancy in three poaceae species. Electronic Theses and Dissertations [Internet]. 2002:3472. Available from: https://openprairie.sdstate.edu/etd/3472

[67] Simpson GM. Seed Dormancy in Grasses. Cambridge, UK: Cambridge University Press; 1990. pp. 3-59

[68] Qasema JR, Jordan A. Weed seed dormancy: The ecophysiology and survival strategies. In: Jimenez-Lopez C, editor. Seed Dormancy and Germination [Internet]. 2019. DOI: 10.5772/intechopen.88015

[69] Finch-Savage W, Leubner-Metzger G. Seed dormancy and the control of germination. New Phytologist. 2006;**171**(3):501-523

[70] Footitt S, Douterelo-Soler I, Clay H, Finch-Savage WE. Dormancy cycling in *Arabidopsis* seeds is controlled by seasonally distinct hormone-signalling pathways. Proceedings of the National Academy of Sciences, USA. 2011;**108**:20236-20241

[71] Amen R. A model of seed dormancy. The Botanical Review. 1968;**34**(1):3-6

[72] Farahani HA, Moaveni P, Maroufi K. Effect of thermopriming on germination of cowpea (*Vigna sinensis* L.). Advances in Environmental Biology. 2011;**5**(7):1668-1673

[73] Donohue KR, de Casas R, Burghardt L, Kovach K, Willis CG. Germination, post germination adaptation, and species ecological ranges. Annual Review of Ecology, Evolution, & Systematics. 2010;**41**:293-319

[74] Poisot T, Bever JD, Nemri A, Thrall PH, Hochberg ME. A conceptual framework for the evolution of ecological specialisation. Ecology Letters. 2011;**2011**(14):841-851

[75] Chandel NS, Tripathi V, Singh HB, Vaishnav A. Breaking seed dormancy for sustainable food production: Revisiting seed priming techniques and prospects. Biocatalysis & Agricultural Biotechnology. 2024;**55**:102976

[76] Bewley JD, Nonogaki H. Seed maturation and germination. In: Reference Module in Life Sciences [Internet] 2017. DOI: 10.1016/B978-0-12-809633-8.05092-5

[77] Donohue K, Dorn D, Griffith C, Kim E, Aguilera A, Polisetty C, et al. Environmental and genetic influences on the germination of *Arabidopsis thaliana* in the field. Evolution. 2005b;**59**(4):740-757

[78] Wills CG, Baskin CC, Baskin JM, Auld JR, Venable DL, Cavender-Bares J, et al. The evolution of seed dormancy: Environmental cues, evolutionary hubs, and diversification of the seed plants. New Phytologist. 2014;**203**:300-309

[79] Ermis S, Özden E, Yıldırım E. Seeds of resilience: Physiology and mechanisms of hard-seededness. In: Seed Biology - New Advances [Internet]. 2024. DOI: 10.5772/intechopen.1003847

[80] Baskin JM, Baskin CC. A classification system for seed dormancy. Seed Science Research. 2004;**14**:1-16

[81] Hilhorst HW. Standardizing seed dormancy research. Methods in Molecular Biology. 2011;**773**:43-52

[82] Nikolaeva MG. Factors controlling the seed dormancy pattern. In: Khan AA, editor. The Physiology and Biochemistry of Seed Dormancy and Germination. Netherlands: North-Holland; 1977. pp. 51-74

[83] Mbi KT, Ntsefong NG, Eugene-Lenzemo TE. Seed dormancy: Induction, maintenance and seed technology approaches to break dormancy. In: Jimenez-Lopez JC, editor. Seed Biology Updates [Internet]. 2022. DOI: 10.5772/intechopen.106153

[84] Qaderi MM. Environmental regulation of weed seed dormancy and germination. Seeds. 2023;**2**:259-277

[85] Chahtane H, Kim W, Lopez-Molina L. Primary seed dormancy: A temporally multi–layered riddle waiting to be unlocked. Journal of Experimental Botany. 2017;**68**(4):857-869

[86] Chichaghare AR, Sreejith MM, Ray RM, Abijith R, Shahina NN, Kailas BA. Seed treatment: Breaking dormancy of seeds. In: Recents Trends in Propagation of Forest and Horticultural Crops. Taran Publishers; 2021. pp. 43-47

[87] Mousavi R, Rezaei M, Mousavi A. A general overview on seed dormancy and methods of breaking it. Advances in Environmental Biology. 2011;**5**(10):3333-3337

[88] Gualbert P, Arnold H, Verkleij JAC. Endogenously induced secondary dormancy in seeds of *Striga ermonthica*. Weed Science. 2000;**48**:561-566

[89] Venier P, Funes G, Carrizo-García C. Physical dormancy and histological features of seeds of five *acacia* species (*Fabaceae*) from xerophytic forests in Central Argentina. Flora – Morphology, Distribution, Functional Ecology of Plants. 2012;**207**(1):39-45

[90] Beyaz R, Yildiz M. The use of gamma irradiation in plant mutation breeding. In: Juric S, editor. Plant Engineering. London, UK: IntechOpen; 2017. pp. 33-46

[91] Farley G, Bellairs S, Adkins S. Germination of selected Australian native grass species, with potential for minesite rehabilitation. Australian Journal of Botany. 2013;**61**(4):283-285

[92] Baskin J, Hidayati SN, Baskin CC, Walck JL, Huang Z, Chein C. Evolutionary considerations of the presence of both morphophysiological and physiological seed dormancy in the highly advanced euasterids II order dipsacales. Seed Science Research. 2006;**16**:233-242

[93] Jaganathan G. Defining correct dormancy class matters: Morphological and morphophysiological dormancy in arecaceae. Annals of Forest Science. 2020;**77**(4):1-4

[94] Kucera B, Cohn M, Leubner-Metzger G. Plant hormone interactions during seed dormancy release and germination. Seed Science Research. 2005;**15**(4):281-307

[95] Ertekin M. Effects of microorganisms, hormone treatment and stratification on seed germination of the golden rain tree (*Koelreuteria paniculata*). International Journal of Agriculture & Biology. 2011;**13**:38-42

[96] Kitis Y, Kaya D. Effects of some dormancy breaking methods on germination of jute (*Corchorus olitorius* L.). Mediterranean Agricultural Sciences. 2018;**31**(3):213-217

[97] Raven PH, Evert RF, Eichhorn SE. Biology of Plants. 7th ed. New York: W.H. Freeman & Company Publishers; 2005. pp. 504-508

[98] Sarmadnia GH. Seed Technology. Iran: University of Mashhad Publications; 1997

[99] Watkins JT, Cantliffe DJ. Mechanical resistance of the seed coat and endosperm during germination of *Capsicum annuum* at low temperature. Plant Physiology. 1983;**72**(1):146-150

[100] Dalil B. Response of medicinal plants to seed priming: A review. International Journal of Plant, Animal & Environmental Sciences. 2014;**4**(2):741-745

[101] Yang QH, Ye WH, Yin XJ. Dormancy and germination of *Areca triandra* seeds. Scientia Horticturae-Amsterdam. 2007;**113**:107-111

[102] Nautiyal PC, Sivasubramaniam K, Dadlami M. Seed dormancy and regulation of germination. In: Dadlani M, Yadava DK, editors. Seed Science and Technology. Springer Nature: Singapore; 2023. pp. 39-66

[103] Ameer JBM, Vijayakumar A, Selvaraju P. Standardization of seed dormancy breaking treatment in Senna (*Cassia auriculata*). Journal of Plant Breeding & Crop Science. 2013;**5**(11):220-223

[104] Hanson AD. The effect of imbibition drying treatment on wheat seeds. The New Phytologist. 1973;**72**:1063-1073

[105] Kumar N, Goel N. Phenolic acids: Natural versatile molecules with promising therapeutic applications. Biotechnology Reports. 2019;**24**:e00370

[106] Waqas M, Korres NE, Khan MD, Nizami AS, Ali I. In the concept and methods of seed priming. In: Hasanuzzaman M, Fotopoulos V, editors. Priming and Pretreatment of Seeds. Springer Nature: Singapore; 2019. pp. 15-22

[107] Thakur M, Tiwari S, Kataria S, Anand A. Recent advances in seed priming strategies for enhancing planting value of vegetable seeds. Scientia Horticulturae. 2022;**305**:111355

[108] Rhaman MS, Rauf F, Tania SS, Khatun M. Seed priming methods: Application in field crops and future perspectives. Asian Journal of Research in Crop Science. 2020;**5**(2):8-19

[109] Amir M, Prasad D, Khan FA, Khan A, Bela A, Belal A. Seed priming: An overview of techniques, mechanisms, and applications. Plant Science Today. 2023;**10**(10):20

[110] Adhikari B, Dhital PR, Ranabhat S, Poudel H. Effect of seed hydro-priming durations on germination and seedling growth of bitter gourd

(*Momordica charantia*). PLoS One. 2021;**16**(8):255258

[111] Ozden E, Demir I. GA3 enhanced seed germination of *Solanum torvum*. In: Works of the Faculty of Agriculture and Food Sciences. University of Sarajevo; 2016; Vol. 61(66). pp. 318-320

[112] Yücel E, Yilmaz G. Effects of different alkaline metal salts (NaCl, KNO_3), acid concentrations (H_2SO_4) and growth regulator (GA3) on the germination of *Salvia cyanescens*. Gujarat Science. 2009;**3**:123-127

[113] Schopfer P, Plachy C. Control of seed germination by abscisic acid: Effect on embryo water uptake in *Brassica napus* L. Plant Physiology. 1984;**76**:155-160

[114] Hu C, Wang P. Embryo culture: Technique and applications. In: Evans DE, Sharp WR, Ammirato PV, editors. Handbook of Plant Cell Culture. New York: Macmillan; 1986. pp. 43-96

[115] McDonald MB. Seed priming. In: Black M, Bewley JD, editors. Seed Technology and its Biological Basis. Sheffield: Sheffield Academic Press; 2000. pp. 287-325

[116] Halmer P. Methods to improve seed performance in the field. In: Handbook of Seed Physiology. 2004. pp. 125-165

[117] Satish S, Mohana DC, Ranhavendra MP, Raveesha KA. Antifungal activity of some plant extracts against important seed borne pathogens of *Aspergillus* sp. International Journal of Agricultural Technology. 2007;**3**(1):109-119

[118] Dutta P. Seed priming: New vistas and contemporary perspectives. In: Advances in Seed Priming. Singapore: Springer; 2018. pp. 3-22

[119] Ghafari H, Razmjoo J. Effect of foliar application of nano-iron oxidase, iron chelate and iron sulphate rates on yield and quality of wheat. International Journal of Agronomy & Plant Production. 2013;**4**(11):2997-3003

[120] Bilalis DJ, Katsenios N, Efthimiadou A, Karkanis A, Efthimiadis P. Investigation of pulsed electromagnetic field as a novel organic pre-sowing method on germination and initial growth stages of cotton. Electromagnetic Biology Medicine. 2012;**31**(2):143-150

[121] Bhardwaj J, Anand A, Nagarajan S. Biochemical and biophysical changes associated with magnetopriming in germinating cucumber seeds. Plant Physiology & Biochemistry. 2012;**57**:67-73

[122] Araujo SDS, Paparella S, Dondi D, Bentivoglio A, Carbonera D, Balestrazzi A. Physical methods for seed invigoration: Advantages and challenges in seed technology. Frontiers in Plant Science. 2016;**7**:646

[123] Qi W, Zhang L, Wang L, Xu H, Jin Q, Jiao Z. Pretreatment with low-dose gamma irradiation enhances tolerance to the stress of cadmium and lead in *Arabidopsis thaliana* seedlings. Ecotoxicology & Environmental Safety. 2015;**115**:243-249

[124] Waskow A, Howling A, Furno I. Mechanisms of Plasma–Seed Treatments as a Potential Seed Processing Technology [Internet]. 2021. DOI: 10.3389/fphy.2021.617345

[125] Kimura E, Islam MA. Seed scarification methods and their use in forage legumes. Research Journal of Seed Science. 2012;**5**(2):38-44

[126] Baskin CC, Baskin JM. Seeds: Ecology, Biogeography, and Evolution

of Dormancy and Germination [Internet]. 1998. Available from: https://books.google.co.za/books?hl=en&lr=&id=vXfNCgAAQBAJ&oi=fnd&pg=PP1&dq=Baskin,+C.C.+and+Baskin,+J.M.+(1998)+Seeds:+Ecology,+biogeography,+and+evolution+of+dormancy+and+germination.+San+Diego,+Academic+Press

[127] Patane C, Gresta F. Germination of *Astragalus hamosus* and *Medicago orbicularis* as affected by seed-coat dormancy breaking techniques. Journal of Arid Environments. 2006;**67**:165-173

[128] Wang Y, Hu X. Managing seed dormancy in forage legumes and grasses: An update. In: XXII International Grassland Congress Proceedings. 2013. pp. 402-404

[129] Uzun F, Aydin I. Improving germination rate of *Medicago* and *Trifolium* species. Asian Journal of Plant Science. 2004;**3**(6):714-717

[130] Arowosegbe S. Studies on methods of breaking seed dormancy and germination enhancement in *Senna alata* (L.) Roxb., a plant with great medicinal value. International Research Journal of Natural Sciences. 2016;**4**(2):31-40

[131] Sy A, Grouzis M, Danthu P. Seed germination of seven Sahelian legume species. Journal of Arid Environments. 2001;**49**:875-882

[132] Aliero BL. Effects of sulphuric acid, mechanical scarification and wet heat treatments on germination of seeds of African locust bean tree, *Parkia biglobosa*. African Journal of Biotechnology. 2003;**3**(3):179-181

[133] Travlos IS, Economou G, Karamanos AI. Germination and emergence of the hard seed coated *Tylosena esculentum* (Burch) A. Schreib in response to different pre-sowing seed treatments. Journal of Arid Environments. 2007;**68**:501-507

[134] Esan VI, Ayanbamiji TA, Abodunri AD. Physiology of breaking seed dormancy and increasing seed germination in *Senna alata* (L) Roxb seeds in Nigeria. Asian Journal of Agricultural Research. 2021;**15**:1-6

[135] Merritt DJ, Turner SR, Clarke S, Dixon KW. Seed dormancy and germination stimulation syndromes for Australian temperate species. Australian Journal of Botany. 2007;**55**(3):336-344

[136] Rafi H, Dawar S. Effect of bio-priming of leguminous and non-leguminous crop seeds in the management of root rot fungi and growth of crop plants. International Journal of Biology and Biotechnology. 2014;**11**(2-3):375-382

[137] Sahur A, Dermawan R, Sribulan H. The application of biopriming using *Trichoderma* and *Streptomyces* spp. on the germination stage of soybean. IOP Conference Series Earth and Environmental Science. 2020;**575**(1):012122

[138] Araya HT. Seed germination and vegetative propagation of bush tea (*Athrixia phylicoides*) [MSc thesis]. Gauteng, South Africa: University of Pretoria; 2005

[139] Shreevastav C, Gajurel S, Khan S, Panthi B, Dahal J. Effect of seed priming treatments on seed germination and seedling growth of bitter gourd (*Momordica charantia* L.). Asian Journal of Research in Crop Science. 2023;**8**(4):490-504

[140] Mbangcolo MM, Reinten EY, Agenbag GA. Effect of different seed

treatments on germination of *Cyclopia* spp. seeds. AgriProbe. 2022;**10**:22-24

[141] Gunes E, Gubbuk H, Ayala-Silva T, Gozlekci S, Ercisli S. Effects of various treatments on seed germination and growth of Carob (*Ceratonia siliqua* L.). Pakistan Journal of Journal. 2013;**45**(4):1173-1177

[142] Duski F. Effects of various pre-sowing treatments on *in vitro* seed germination of *Ceratonia siliqua* L. RUDN Journal of Agronomy and Animal Industries. 2022;**17**(2):158-165

[143] Akram W, Hussain A, Roomi I, Muhammad M. Effect of different priming agents on germination of purslane (*Portulaca oleracea* L.) seeds under laboratory conditions. Pakistan Journal of Botany. 2023;**55**(5):1-8

[144] Ozden E, Demir I. The effect of priming treatments on germination and seedling performance of purslane (*Portulaca oleracea*) seed lots. Bulletin of University of Agricultural Sciences and Veterinary Medicine Cluj-Napoca Horticulture. 2017;**74**(2):1843-5394

[145] Moosavi A, Afshari AT, Sharif-Zadeh F, Aynehband A. Effect of seed priming on germination characteristics, polyphenoloxidase, and peroxidase activities of four amaranth cultivars. Journal of Food, Agriculture & Environment. 2009;**7**:353-358

[146] Musa M, Lawal AA. Influence of priming concentration on the growth and yield of amaranth (*Amaranthus cruentus* L.) in Sokoto semi–arid zone of Nigeria. Journal of Plant Sciences. 2015;**3**(1):27-30

[147] Safiatou SS. Effect of different seed priming methods on germination, seedling establishment and vigour in sorghum (*Sorghum bicolor* (L.) Moench.) and Bambara groundnut (*Vigna subterrenea* (L.) Verdc.) [MSc thesis]. Ghana: Kwame Nkrumah University of Science & Technology; 2012

[148] Farahzety AM, Signgh A, Yap LV. Influence of priming treatments on seed germination and growth characteristics of *Moringa oleifera* seedlings under laboratory and greenhouse conditions. Food Research. 2023;**6**(4):71-80

[149] Nouman W, Siddiqui MT, Basra SMA, Afzal I, Rehman H. Enhancement of emergence potential and stand establishment of *Moringa oleifera* Lam. By seed priming. Turkish Journal of Agriculture & Forestry. 2012;**36**:227-235

[150] Nouman W, Basra SMA, Yasmeen A, Gull T, Bilal Hussain SB, Zubair M, et al. Seed priming improves the emergence potential, growth and antioxidant system of *Moringa oleifera* under saline conditions. Plant Growth Regulation. 2014;**73**:267-278

[151] Maina FNW, Muasya RM, Gohole LS. Dormancy breaking in *jute mallow* (*Corchorus olitorius*) seeds. Seed Technology. 2011;**33**(2):147-154

[152] Ozden E, Light ME, Demir I. Alternating temperatures increase germination and emergence in relation to endogenous hormones and enzyme activities in aubergine seeds. South African Journal of Botany. 2021;**139**:130-139

[153] Zuo Y, Xu Y. Research progress on the mechanism of GA and ABA during seed germination. Molecular Plant Breeding. 2020;**11**

[154] Zhao J, He Y, Zhang H, Wang Z. Advances in the molecular regulation of seed germination in plants. Seed Biology. 2024;**3**(1):1-13

[155] Song SQ, Liu J, Tang CF, Zhang WH, Xu HH, Zhang Q, et al. Metabolism and signaling of auxins and their roles in regulating seed dormancy and germination. Chinese Science Bulletin. 2020;**65**(34):3924-3943

[156] Bradford K, Nonogaki H. Seed Development, Dormancy and Germination [Internet]. 2008. DOI: 10.1002/9780470988848

[157] Finkelstein R, Reeves W, Ariizumi T, Steber C. Molecular aspects of seed dormancy. Annual Review of Plant Biology. 2008;**59**(1):387-415

[158] Hauvermale AL, Steber CM. GA3 signaling is essential for the embryo-to-seedling transition during arabidopsis seed germination, a ghost story. Plant Signaling & Behavior. 2020;**15**(1):1705028

[159] Sajeev N, Koornneef M, Bentsink L. A commitment for life: Decades of unraveling the molecular mechanisms behind seed dormancy and germination. The Plant Cell. 2024;**36**(5):1358-1376

[160] Matilla AJ. Seed dormancy: Molecular control of its induction and alleviation. Plants. 2020;**9**:1402

[161] Chandrasekaran U, Zhao X, Luo X, Wei S, Shu K. Endosperm weakening: The gateway to a seed's new life. Plant Physiology & Biochemistry. 2022;**178**:31-39

[162] Ingram GC. Family plot: The impact of the endosperm and other extraembryonic seed tissues on angiosperm zygotic embryogenesis. F1000Research. 2020;**9**:18

[163] Yah D, Duermeyer L, Leoveanu C, Nambara E. The functions of the endosperm during seed germination. Plant & Cell Physiology. 2014;**55**(9):1521-1533

[164] Leubner-Metzger G, Kucera B, Müller K. Emerging and established model systems for endosperm weakening. In: Seeds: Biology, Development and Ecology. Proceedings of the Eighth International Workshop on Seeds, Brisbane, Australia. Wallingford UK: CABI; 2005, 2007. pp. 195-204

[165] Sánchez-Montesino R, Bouza-Morcillo L, Marquez J, Ghita M, Duran-Nebreda S, Gómez L, et al. A regulatory module controlling GA–mediated endosperm cell expansion is critical for seed germination in arabidopsis. Molecular Plant. 2019;**12**(1):71-85

[166] Chandra RJ, Masilamani P, Suthakar B, Rajkumar P, Sivakumar SD, Manonmani V. Understanding the role of light in seed germination: A comprehensive review. Journal of Experimental Agriculture International. 2024;**46**(6):895-909

[167] Flores J, González-Salvatierra C, Jurado E. Effect of light on seed germination and seedling shape of succulent species from Mexico. Journal of Plant Ecology. 2016;**9**(2):174-179

[168] Sano N, Marion-Poll A. ABA metabolism and homeostasis in seed dormancy and germination. International Journal of Molecular Sciences. 2021;**22**(10):5069

[169] Mathews S. Phytochrome-mediated development in land plants: Red light sensing evolves to meet the challenges of changing light environments. Molecular Ecology. 2006;**15**(12):3483-3503

[170] Khan MMH, Rafii MY, Ramlee SI, Jusoh M, Al-Mamun M. Bambara groundnut (Vigna subterranea L. Verdc):

A crop for the new millennium, its genetic diversity, and improvements to mitigate future food and nutritional challenges. Sustainability. 2021;**13**(10):5530

[171] Brändel M. The role of temperature in the regulation of dormancy and germination of two related summer-annual mudflat species. Aquatic Botany. 2004;**79**(1):15-32

[172] Forti C, Gualtieri C, Pagano A, de Sousa-Araújo S, Balestrazzi A, Macovei A. Genome editing in the context of seed research: How these novel biotechnology tools can change the future face of agricultural crop development. In: Tuteja N, Tuteja R, Passricha N, Saifi SK, editors. Advancement in crop improvement techniques. Woodhead Publishing, Cambridge: United Kingdom; 2020. pp. 77-88

[173] Gao F, Ayele BT. Functional genomics of seed dormancy in wheat: Advances and prospects. Frontiers in Plant Science. 2014;**5**

[174] Kishchenko O, Zhou Y, Jatayev S, Shavrukov Y, Borisjuk N. Gene editing applications to modulate crop flowering time and seed dormancy. Biotech. 2020;**1**(4):233-245

[175] Jung YJ, Lee HJ, Bae S, Kim JH, Kim DH, Kim HK, et al. Acquisition of seed dormancy breaking in rice (*Oryza sativa* L.) via CRISPR/Cas9-targeted mutagenesis of OsVP1 gene. Plant Biotechnology Reports. 2019;**13**:511-520

[176] Glison N, Gaiero P, Monteverde E, Speranza PR. Breeding for reduced seed dormancy to domesticate new grass species. Genetics & Molecular Biology. 2024;**47**:e20230262

[177] Kulwal PL, Mir RR, Kumar S, Gupta PK. QTL analysis and molecular breeding for seed dormancy and pre-harvest sprouting tolerance in bread wheat. Journal of Plant Biology. 2010;**37**(1):1-16

[178] Hisano H, Hoffie RE, Abe F, Munemori H, Matsuura T, Endo M, et al. Regulation of germination by targeted mutagenesis of grain dormancy genes in barley. Plant Biotechnology Journal. 2022;**2010**(20):37-46

[179] Abe F, Haque E, Hisano H, Tanaka T, Kamiya Y, Mikami M, et al. Genome-edited triple-recessive mutation alters seed dormancy in wheat. Cell Reports. 2019;**28**(5):1362-1369

[180] Farsetti A, Illi B, Gaetano C. How epigenetics impacts on human diseases. European Journal of Internal Medicine. 2023;**114**:15-22

[181] Hamilton JP. Epigenetics: Principles and practice. Digestive Diseases (Basel, Switzerland). 2011;**29**(2):130-135

[182] Luján-Soto E, Dinkova TD. Time to wake up: Epigenetic and small-RNA-mediated regulation during seed germination. Plants. 2021;**10**:236

[183] Sato H, Santos-González J, Köhler C. Combinations of maternal specific repressive epigenetic marks in the endosperm control seed dormancy. eLife. 2021;**10**:e64593

[184] Amaral MN, Tognacca RS, Auge GA. Regulation of seed dormancy by histone post–translational modifications in the model plant *Arabidopsis thaliana*. Journal of Experimental Botany. 2024:e236

[185] Gomez-Cabellos S, Toorop PE, Fernández-Pascual E, Iannetta PPM, Pritchard HW, Visscher AM. (Epi) genetic control of secondary seed

dormancy depth and germination in Capsella bursa–pastoris. Seed Science Research. 2022;**32**(4):200-221

[186] Cheng X, Wang B, Liu LY, Zhao Z, Ling X, Zhao F, et al. Epigenetic factors altered in dehisced anther correlated to seed dormancy in *Paris polyphylla* var. *Yunnanensis*. Agricultural Sciences. 2019;**10**:1517-1533

[187] Klupczynska EA, Pawłowski TA. Regulation of seed dormancy and germination mechanisms in a changing environment. International Journal of Molecular Science. 2021;**22**:1357

[188] Soltani A, Walter KA, Wiersma AT, Santiago JP, Quiqley M, Chitwood D, et al. The genetics and physiology of seed dormancy, a crucial trait in common bean domestication. BMC Plant Biology. 2021;**21**(1):58

[189] Chiara L, Holness S, De Angelis V, Lepri A, Occhigrossi S, Ruta V, et al. From the outside to the inside: New insights on the main factors that guide seed dormancy and germination. Genes. 2021;**12**(1):52

Chapter 7

Bioprotection by Natural Sources for Sustainable Agriculture

Berna Baş

Abstract

Biocontrol, or protection, involves using living organisms such as microbes, insects, and their by-products to safeguard food security against harmful pathogens and pests in agriculture. Excessive use of synthetic inputs such as fertilizers and pesticides in farming can lead to soil accumulation and crust formation over time. Inorganic chemical pollutants enter the food chain through plant absorption from the soil, resulting in environmental and public health concerns. Compensating for the destructive effects of synthetic pesticides on agroecosystems is challenging. Therefore, it is important to discuss the future of plant health and the restoration of microbial communities in the phyllosphere, endosphere, rhizosphere, and plant growth–promoting microorganisms using natural sources. Local governments should enforce legal restrictions on the frequency and quantity of conventional pesticide use, promote the use of bioagents, and encourage farmers and stakeholders to adopt natural alternatives. This chapter introduces various bio-based materials that can serve as natural alternatives to synthetic chemicals for eradicating or minimizing invasive phytopathogen species through biocontrol in plant protection.

Keywords: biopesticide, biocontrol, bioprotection, natural resources, phytopathogen microorganisms

1. Introduction

The beginning of agriculture goes back to the neolithic ages down to the permanent colonization of human communities. The first agricultural activities began with the cultivation of grains and legumes by the earliest societies. However, people about 10.000 years ago were inexperienced in agricultural activities, and this has caused a decline in agricultural productivity after repetitive planting on the same land for several years with identical crops. Therefore, the farmers were leaving the weakened farmlands to nearby virgin new areas for cultivation. After agricommunities formation, they developed straightforward man-made manifold agricultural techniques such as cultivation methods digging soil with sticks, irrigation channels in the form of ditches, drying in baskets, filling crops in sacks, or storing by piling them underground. The emergence of ancient civilizations started in conjunction with settled agricultural life. Different populations have migrated for accommodation to the regions with nutrient-rich soils where the Euphrates and Tigris valleys, Egypt-Nile valley, Indo-Indus valley, ancient China-Yellow River valley, and Mesopotamian-Sumerian civilizations have laid the foundations of civilizations to become history.

The industrial revolution spark of the eighteenth century also caused agrarian reform. Thanks to breakthroughs in the chemical industry leading up to mechanized production, agricultural productivity has hugely been leapt up leveraged favorable outcomes of synthetic fertilizers and chemical pesticides. However, subsequently, this modern agrirevision has resulted in environmental and public health matters. The major drawback of modernization on classical agriculture has given rise to a wild increase in farming production to fulfill the needs of an incremental population as well as the more profit desire of global trade corporations guiding the agricultural sector. The contemporary agricultural revolution has also radically changed the lifestyle of human societies and built the basis of a more civilized life. But even in modern urban life, the healthy lifestyle of humans is dependent on cultivation farming. In the mid-twentieth century (the 1940s), organic farming interest initially began as a response to the negative consequences of impurities created by the overdense use of synthetic inputs in agricultural activities. Synthetic pesticides are dangerous chemicals that have destructive impacts on agroecosystem and human health and are very difficult to compensate for. Also, heavy metal elements including cobalt, copper, nickel, and zinc that are essential for normal plant physiology are present inside synthetic fertilizers [1], which are one of the major sources of heavy metals within agrisoils. A significant amount of heavy metals are included in synthetic fertilizer mixtures during the manufacturing process. Overuse of these synthetic inputs in farming can lead to gradual accumulation and crusting in the soil. As a result, inorganic chemical pollutants such as fertilizers and pesticides, which are absorbed from the soil by plants, may enter the food chain and become toxic to humans and animals. This can trigger various health problems, including cancer, mutations, genetic disorders, and physiological abnormalities. The toxicological profiles of the ingredients of legal pesticide chemicals are listed on the ATSDR portal [2]. Shortly, the adverse impacts of the agricultural revolution have roughly reminded me of agrarian history. Because of the undesirable side effects of modern agriculture, volunteers included regenerative agriculture organizations, commercial companies, growers, scientists, and trained individuals, as pioneers in organic farming were directed toward the discovery of natural resources substituted to synthetic pesticides for protection, which is a part of regenerative farming.

Since the second half of the twentieth century, human activities have caused significant damage to the environment through the use of chemical pesticides and the increase in greenhouse gases. This has led to a global climate crisis. To ensure the sustainability of agroecosystems for the benefit of mankind, it is important to manage phytopathogens and pest organisms using natural sources. The European Green Deal aims to reduce the use and risk of chemical pesticides by 50% by 2030 and ultimately achieve pesticide-free agriculture in Europe by 2050 [3].

Biopesticides are occasionally called bioprotection and are the rising interest along with the expansion of organic farming practices for agricultural food security. Biocontrol or protection is the use of living materials such as microbes, insects, and their derivative metabolites to maintain food security against detrimental pathogens/pests/abiotic factors in agriculture. Slowly growing the pest biocontrol market has reached the highest growth rate with 13.6% annual growth including new compounds from the last year [4]. There is no resistance problem in pests/microbes against natural sources, because of their shorter shelf life, moderately phytotoxic effects, insufficient bioavailability, instability of natural compounds, and quick decomposition. All these are reasons to reduce the performance of biopesticides. Even if the preference by farmers for biopesticides is behind synthetic chemicals, it has made important progress toward the advantage of biopesticides using recent technologies like nanotechnology and microencapsulation methods. Encapsulation methods create positive incremental influence for the continuing quality of biological, chemical, and physical properties of a bioagent material. Hence encapsulation applications have drawn outstanding attention in industrial applications as in food processing and the pharmaceutical industry. The cost of a new synthetic pesticide to develop is estimated at roughly $ 250 million and is necessary for 10 years, whereas the discovery of natural pesticides is up to 1/10 of the expenditure of chemical pesticides and a period of 3 years is enough [5].

The first biopesticide *Paecilomyces fumosoroseus* is an entomopathogenic fungus labeled insecticide named PreFeRal®. It was for recourse to the European Union Registration Directive in 1994 and received a temporary provision in Belgium in 1997 [6]. BioSafe Systems, a manufacturer of biodegradable products, has developed a biofungicide/bactericide Guarda®. Its active ingredient is thyme oil enriched with additive adjuvant by HOLDit technology to increase efficacy [7]. This powerful adjuvant is designed to enhance the active ingredient's ability to maintain prolonged contact with the target, ensuring maximum effectiveness. Unfortunately, due to the vaporization of thyme oil, the performance of biopesticide without the adjuvant is declined and is mostly wasted. Also, the company announced on their website that it is the first biofungicide/biobactericide registered with Environmental Protection Agency (EPA) [8]. EPA presented a list of the 390 data inputs of biopesticide active ingredients (biochemical and microbial) registered between 1962 and August 31, 2020. There were only seven patented active compounds of biopesticide in 2020; afterward, data entry was interrupted. Yuan [9] reported that China registered nearly 3800 biopesticides with 115 active compounds in 2016.

The purpose of this chapter is to present the current toolbox of bio-based natural resources that can be used to replace agrochemicals in managing diseases caused by plant-damaging microorganisms. This review focuses on biocontrol of plant diseases and excludes discussions on controlling insect pests. The information in this review was gathered from open-access documents available on websites such as SpringerLink, Google Scholar, Web of Science, PubMed, ScienceDirect, Elsevier, and the journal "Pest Management Science." Keywords such as biocontrol, bioprotection, biopesticide, fermentation in bioprotection, supernatant, and their combinations were used to find relevant papers. Due to space limitations, the focus was primarily on recent original research articles published between 2023 and 2024, although a few exceptional older manuscripts were also included. The data in **Table 1** was compiled from original research articles published between 2023 and 2024, with one exception. Archives predating 2023 related to research articles are generally overlooked. This section introduces a variety of bio-based materials that could serve as natural alternatives to synthetic chemicals for eradicating or minimizing invasive phytopathogenic species through biocontrol in plant protection. It also includes examples of diverse action mechanisms from different biological sources.

2. Biopesticide sources

There are no standard definitions for biopesticides, as their intended use can vary depending on the researchers. Biopesticides are derived from natural materials such as animals, plants, bacteria, and certain minerals, as defined by the EPA. For example, canola oil and baking soda are considered biopesticides due to their pesticidal applications [30]. Generally, biopesticides are products of living organisms like plants, animals, microorganisms, and algae. They can be either the organisms themselves, known as bioinoculants, or a part of their constituents.

Bio-based resources for plant protection can be obtained from three main groups, as summarized in **Figure 1**:

1. *Macroscopic organisms (as plants and insect groups—mites, predators, parasitoids, nematodes, other invertebrates) and their components*

    - Plant extracts (majorly secondary metabolites) and essential oils
    - Insects (semiochemicals such as pheromones, allelochemicals)
    - Other invertebrates

2. *Microscopic organisms (bacteria, fungi, viruses) and their components*

    - Microbial inoculants (e.g., bacterization of plant explants as seed, root, plantlet, or soil contaminants as bioinoculant)
    - Microorganism-based elicitors

3. *Fermented microbial and extracts*

    - Valorization of industrial/agricultural wastes by recycling *via* solid/liquid fermentation

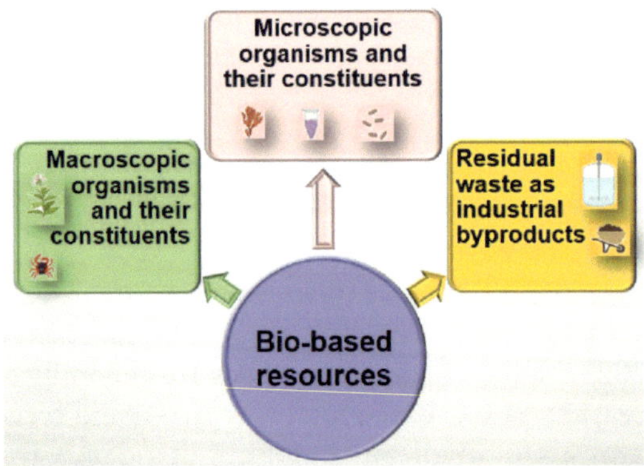

Figure 1.
Bio-based resources originated from macroscopic/microscopic organisms and their ingredients and by valorization of residual waste from industrial by-products.

3. Biopesticide applications

3.1 Treatments of macroscopic/microscopic organisms as bioinoculants and their constitutions as natural biosources

In recent research, yeasts offer novel and powerful alternatives to dangerous synthetic chemical pesticides from preharvest in the field to postharvest in storage for crop disease control. The antimicrobial activity of yeasts is attributed to competing with rival groups for reproduction, nutritional requirements, and colonization of a spatial domain. Nonselective cultivation and simple nutrition requirements of yeasts make ideal organisms for this group of unicellular fungi as biocide reservoirs. The potential effects of *Filobasidium wieringae* and *Pichia kluyveri* yeasts to control the severity of purple spot disease by *Alternaria porri*, which is the highly virulent invader strain on onion leaves, were investigated [31]. *P. kluyveri*, an endophytic yeast isolated from healthy onions, reduced disease severity by approximately 77%, but when both yeasts were used as a cocktail, the total effect rate increased to 85%. The wine industry is important commercially in Portugal but faces challenges with mold infections in vineyards. A study was conducted to determine the effectiveness of using multiple yeasts during wine fermentation on grapes in order to minimize the use of agrochemical products [32]. It was found that the wine yeast *Metschnikowia pulcherrima* inhibited the growth of *Penicillium* before and after harvest, while the yeasts *Lachancea thermotolerans*, *Hanseniaspora uvarum*, and *Starmerella bacillaris* showed strong antagonistic activity against other pathogenic fungi such as *Mucor*, *Aspergillus*, and *Botrytis*, respectively, to prevent fruit decay. The study observed that all yeasts used in the experiment produced diffusible volatile metabolites that suppressed fungal growth. The mechanisms of fungal suppression by yeasts, including toxin and enzyme production, competition, mycoparasitism, vapor compounds, and triggered plant immunity, were studied in detail as antagonistic biocontrol mechanisms against phytopathogens [33].

A remarkable investigation from Biotalys company settled in Belgium has been reported. Researchers from the firm have developed a protein-based biofungicide to protect from the fungal disease on tomatoes, cucurbits, and grapes infected with *Botrytis* and powdery mildew [34]. The bioactive compound they named, AGROBODY (other ECOVA, probably commercial name), was produced from ilama immunized with Folch lower phase lipid extract from *Fusarium oxysporum*. They stated that biological attributions of AGROBODY (such as ready biodegradability, product quality, and shelf life) were superior than the performance of many chemical pesticides and biological references to *Bacillus pumilus* in repeated field trials. So protein-based antibody products to any proteinaceous constituent of phytopathogen organism are an unprobed search field to dig out novel biosources. But the action mechanism is ambiguous.

Miniecology of the phyllosphere harbors diverse microbial communities and offers tremendous application opportunities in agrobiotechnology. Members of the plant phyllosphere are a wide variety of species such as bacteria, yeast, fungi, viruses, nematodes, and protozoans, and definite residents can vary depending on the plant type. *Trichoderma* spp. phyllosphere habitats from the most common species have been used commercially and integrated into the IPM program for years [35]. Moreover, 209 microbial strains from 94 diverse species are used commercially in pest control worldwide [36]. Due to the wide range of hosts, survival of spores, and the production of toxic volatile and nonvolatile organic substances by *Trichoderma harzianum*, it has great potential for ameliorating soil fertility and promoting plant growth in agricultural biotechnology.

After incorporating *Trichoderma harzianum* as a bioinoculant into soil infected with mycelial plugs of *Sclerotinia sclerotium*, the development of sclerotia fungi was not observed after 1 month of incubation [37]. Most probably, fungal *Trichoderma* strains in parasitism are in contact with pathogen fungi, followed by winding around hyphae of pathogen fungi and damage to the pathogen releasing its toxic chemicals and diffusible volatiles acting as biostimulators till overcoming the pathogen [38]. The bacteria *Pseudomonas* spp. and *Achromobacter* spp., isolated from the coffee phyllosphere against *Hemileia vastatrix* yellow rust agent that has caused defoliation in coffee plants in Peru, were evaluated for antagonistic effects to biocontrol in field works [39]. While bacterial strains have prevented the germination of rust uredospores by 81 and 82%, respectively, propiconazole used as a synthetical fungicide has caused 91% inhibition. These bacterial strains showing close effects to chemical synthetic fungicides may be important prospective biocontrol agents for trading purposes in the future. It is assumed that endophyte *Pseudomonas* spp. generates siderophores to acquire ferric iron from the microbiome around plant roots hampering phytopathogen attainment to it [40]. In addition, siderophores are antimicrobials featured in some phytopathogens.

One of the important biosources is bacteriophages of plentiful microorganism groups in nature. Bacteriophages are obligate parasitic viruses; they only infect and replicate in bacteria cells. As a result of killing selectively target bacteria, phages present different approaches in the biocontrol for bacterial disease in phytopathology. Bacteriophages eradicate phytopathogenic bacteria that infect plants by integrating into the bacterial genome and modulating their genes, thereby supporting plant health. The biocontrol mechanism of bacteriophages is based on the regulation of host physiology and the modulation of host genome to fit their requirements. Because of obligate endoparasite, some phages allow their host together to survive modifying host virulence by inactivation of membrane receptors of hosts that may be in contact with other phages. However, when their hosts are in threat, viruses kill their hosts with lytic cycle enzymes. These are general defense mechanisms attributed to viruses [41]. Bacteriophage-invaded phytopathogen bacteria are not infectious to plants, eco-friendly, healthful, and safe tools as biocontrol agents. To be a biocontrol mediator, it is needed to reveal which plant defense mechanisms are also engaged by bacteriophage inside infected plants. The bacteriophage LDT325 was isolated from soil around the root of tea infected with *Pseudomonas syringae* and was examined *in vitro* and *in vivo* for bud blight disease control [42]. They reported lesion size was lessened by 50% in 3 days, but the activity of bacteriophage LTD325 with great potential was excellent for alleviating bacterial disease. Bacteriophages Fifi044 and Fifi318 were freshly isolated from pear orchard soil where fire blight disease by *Erwinia amylovora* occurred. When used in the cocktail including both phages, synergistic interaction has better suppressed bacterial disease than synthetical antibiotic application [43]. But their antibiological mechanisms of are yet unknown.

Mycoviruses are viruses that naturally infect fungi, functioning in a manner similar to bacteriophages. Although the information on their commensal interplays with plants, pathogenic fungi, and other viral factors are usually limited to a small number with decently good enlightened studies, some of the mycoviruses well-known in plant fungi have caused hypovirulence, and some have minimized their host's mycotoxin production. Still, they have the potential to open up new avenues for biocontrol strategies against plant diseases. *Colletotrichum* spp. has a wide host range and causes anthracnose. Mycovirus CaPV1 isolated from this fungus weakens the pathogenicity ability of the fungus damaging its vesicle transport system [44]. In the aforementioned study, the CaPV1 mycovirus hosted by the phytopathogen *Colletotrichum alienum* that invaded *Camellia oleifera* was transfected into the other anthracnose agents *C. fructicola*, *C. spaethianum*, and *C. gloeosporioides* obtained from various plants. It was also observed that the mycovirus caused a decrease in virulence power in those fungi. In addition, mycovirus-associated hypovirulence has disturbed the fungal vesicle transport system by changing transcription.

Moreno-Velandia [45] has used the liquid culture of the bacterium to evaluate the antagonistic activity of the *Bacillus velezensis* Bs006 strain against *Fusarium* wilt in golden berry. After Bs006 was grown in a liquid culture medium, the effects of the bacteria-containing supernatant (SN) and the bacteria-free SN were tested against *Fusarium oxysporum* f. sp. *physalis in vitro* and *in vivo* studies. They noticed that the liquid culture medium with antagonistic bacteria on disease progress by *Fusarium* protected the plants by 37%, while the protective effect of the SN without bacteria was level up to 53%. Also, fengycin purified from this SN has reduced the disease severity by 39%. *Bacillus* spp. species used for biocontrol purposes produce antimicrobial metabolites such as iturin, fengycin, cyclic lipopeptide–containing surfactants, and similar substances. These metabolites impair the cell membrane of phytopathogenic organisms, causing them to swell and lyse, particularly affecting filamentous organisms such as *Fusarium* [46, 47]. The usage of crude SN may be a feasible alternative for a sustainable system against synthetic pesticides over time. The SN is a simple filtrate and is sometimes referred to as fermentation liquid. It is obtained through a chemical extraction process from any biological source by mashing or submerging in water or an organic solvent. As crude extracts, the supernatant solutions have an undeniable wealth of nutrient content about many macromolecule fractions such as proteins, amino acids, carbohydrates, lipids, vitamins, enzymes, minerals, and secondary metabolites [48]. However, the quantity of each chemical fraction in the total content of impure raw material is at a low concentration. The production of supernatant from biomaterial is more efficient, as it requires less time, involves fewer steps, is easier, and is cheaper than purifying complex chemicals in the supernatant. Cell-free SN formulations can be easily customized for large-scale production using simple chemical methods such as centrifugation, filtration, and desiccation, making it a viable option for the pesticide industry without requiring substantial investment.

A phthalide, 3-BPH, from *Angelica sinensis* Chinese herb was *in vitro* tested for *Fusarium graminearum* head blight agent in cereals [23]. In contrast to trading fungicide azoxystrobin, very strong mycotoxin deoxynivalenol (DON) produced by *F. graminearum* was inhibited after exposure to 3-BPH from the herb. DON is a highly toxic metabolite to cause damage to human and livestock well-being and is directly or indirectly included in the food chain through cereal crops contaminated with the *Fusarium* species while growing in a field or while stocked in a warehouse [49]. Mycotoxins are a serious issue due to their occurrence in exported dry foods like spices and grains traded worldwide, so natural eco-friendly biocides can open up a new perspective responsible for minimizing or eliminating noxious secondary metabolites from phytopathogens.

Marine resources are of elevating trends. A microalgae *Amphidinium carterae* produces polyketides, amphidinol-18, which is liable to vigorous biopesticidal activity to battle too many phytopathogens *in vitro* tests [50, 51]. They reported that crude methanolic extracts from *A. carterae* biomass have suppressed fully about 100% of fungal development against *Colletotrichum acutatum*, *Verticillium dahliae*, *Fusarium proliferatum*, *Fusarium cubense*, *Botrytis cinerea*, and *Magnaporthe grisea* [52]. From marine isolates, *Trichoderma asperelloides* T203 and *Trichoderma longibrachiatum* T7407 strains were tried *in vitro*, in pots, and in fields for antagonistic effects to bioprotection against *Magnaporthiopsis maydis*, the late wilt disease agent of maize plants [53]. Researchers notified that strain T203 showed an inhibitory effect separating the hyphae of the pathogen fungus into fragments, and T7404 also functioned as a fungicide and killed the pathogen. While fungal growth was attenuated in field conditions, the biological parameters (fresh biomass, plant length, and yield) for plant improvement have given positive results when sown with fungal mycelium and maize seed. *Trichoderma* species have a broader host range in all life domains including plants, animals, and fungi, and they can adapt biologically to disparate life forms (i.e., aerobic, anaerobic, facultative, and saprophyte) in different ways according to their fluctuating environments in fragile ecologies [54, 55]. In plant disease management, *Trichoderma* species offer several potential advantages: They are broad-spectrum, inexpensive, environmentally safe, and highly effective. They can compete with various microorganisms and are resistant to many chemicals. However, it is important to highlight that, in rare cases, *Trichoderma* can cause suprainfections in humans, such as skin and liver infections. The *Trichoderma* genome has gene clusters with a richness of various lytic enzymes and secondary metabolite [56]. This greatness of biodiversity and biohabitats is undoubtedly linked to environmental adaptability because of versatile genes encoding many functions to be essential for life. Gene expression of microorganisms, microbial interactions, and microbiome structure is shaped by external factors related to biotic and abiotic components such as type of microorganisms, host plant variety, nature of soil, and climatic factors, meanwhile modulated by a degree of synergism. Secondary metabolites of microorganisms facilitate the establishment of interactions between microbial communities and their hosts and are of great value biochemicals with antibiotic, antioxidant, phytotoxic, and mycotoxic activities. Since commercial *Trichoderma* bioproducts in worldwide agriculture are widespread and the most used, there are many types of research and review articles belonging before 2020. However, a comprehensive survey of the chemical diversity, content, and bioactivities of secondary metabolites of the *Trichoderma* genus was reviewed by Ref. [57].

Certain strains of *Trichoderma* are extremophile fungi, for instance, halotolerant *T. longibrachiatum* HL167 and xerotolerant *T. harzianum* 81Y1 [58]. When microbial extremophiles are introduced into the soil as biomodulators to mitigate extreme conditions such as salinity, drought, extreme pH, low nutrient availability, and temperature, the probability of inter- or intraspecific naturally horizontal genes transferring in microbial entities in the microbiome may ascend. These possibilities might permit the origination of new variants with a high degree of adaptability to dynamic environments. Indeed, gaining drastic properties of extremotolerant microorganisms can be crucial to creating innovative new renewable agroecosystems. Even their unique metabolites produced by their genetic systems in extreme conditions to be lethal for others may impede the thrive of soil-borne phytopathogens. **Table 1** is formed to offer more options for possible biopesticide nominees that are not mentioned in this chapter.

Natural biosource	Biopesticide type	Targeted organism	Host	Ref.
Bacillus velezensis P1	Fungicide	*Aspergillus carbonarius*	Grape	[10]
Pseudomonas aeruginosa JQ-41, *Serratia marcescens* S16, *Stenotrophomonas rhizophila* CASMBAUDAL2, *Streptomyces* sp., *Trichoderma* sp., either bacterial consortia or/and microbial consortia	Fungicide	*Ganoderma boninense*	Oil palm	[11]
In vitro synthesis RNAi	Acaricide	*Amphitetranychus viennensis*	Woody ornamental	[12]
Carvacrol commercial form in nanoemulsion	Bactericide	*Xanthomonas axonopodis* pv. *Cyamopsidis*	Cluster bean/ guar	[13]
Plant-derived artificial miRNA	Insecticide	Green peach aphid	Lettuce	[14]
Abietane diterpenoid from *Salvia canariensis*	Fungicide	*Alternaria alternata* *Botrytis cinerea* *Fusarium oxysporum*		[15]
Elicitor-based biopesticide PSP1 (AsES) commercial form (extracellular protease subtilisin from *Sarocladium strictum*)	Biostimulant	*Fusarium graminearum* s.s.	Wheat	[16]
Flavonoids from *Desmodium caudatum*	Antiviral	Tobacco mosaic virus	Tobacco	[17]
A mixture of grape cane extract formulations plus apple extract either alone or together with copper-based agents	Fungicide	*Plasmopara viticola*	Grapevine	[18]
Bioinoculants from desert plant endophytes (*Bacillus subtilis, B. velezensis, B. tequilensis, Stenotrophomonas maltophilia, Klebsiella pneumoniae*)	Antifungal	*Fusarium verticillioides*	Maize	[19]
Oleoresin from chili pepper pod extract	Fungicide	*Botrytis cinerea* *Guignardia bidwellii* *Plasmopara viticola*	Grapevine	[20]
Mycotoxin citrinin commercial	Herbicide	Ageratina adenophora	Weed	[21]
Lycorine from Lycoris spp.	Fungicide	*Magnaporthe oryzae*	*Rice*	[22]
3-Butylidenephthalide from *Angelica sinensis* roots	Fungicide	*Fusarium graminearum*		[23]
Microencapsulated *Yamadazyma mexicana* LPa14 as inoculant	Fungicide	*Colletotrichum gloeosporioides*	Avocado	[24]
Bioinoculant *Chryseobacterium cucumeris* PcE1	Bactericide	*Pectobacterium carotovorum* subsp. *Carotovorum*	Chinese cabbage	[25]
Diphenyl ethers from endophytic *Rhexocercosporidium* sp. Dzf14	Antibacterial	*Clavibacter michiganensis*, *Bacillus subtilis*		[26]
The supernatant of *Bacillus wiedmannii* isolate ZT	Herbicide	*Lolium temulentum*	Ryegrass-*Lolium temulentum*	[27]

Natural biosource	Biopesticide type	Targeted organism	Host	Ref.
Bipolaris yamadae HXDC-1-2	Mycoherbicide	*Echinochloa crus-galli*	Weed-*Microstegium vimineum*	[28]
Compost (plant waste + poultry manure)	Fungicide	*Fusarium oxysporum* f. sp. *albedinis*	Date palm	[29]

Table 1.
Some biopesticide candidates from natural materials published in 2023–2024.

3.2 Seed treatments *via* natural sources

The recent surge in seed-coating research reveals a powerful opportunity to greatly curtail pathogen effects in the rhizosphere, paving the way for healthier, more resilient crops. This innovative solution could transform agricultural practices and enhance food security. Today, global seed coating market is in upward movement with the meeting of innovative coating technologies, controlled release coatings, and biopolymer coating films [59–61]. Seed coating technologies play a crucial role in safeguarding seeds from pathogens, pests, weeds, and soil salinity. By utilizing these advanced coatings, we can significantly enhance germination quality and support the development of robust seedlings, making them far superior to untreated seeds [59].

An antagonistic yeast *Williopsis saturnus* varieties (*Ws* var. *saturnus* and *Ws* var. *mrakii*) and their produced lethal toxins have retarded the growth of some pathogens [60–62]. When alive *Williopsis saturnus* var. *saturnus* in a solution with and without edible coating material that is whey protein concentrate was sprayed on peanut surface, it stopped both the advancement of *Aspergillus flavus* and aflatoxin generation of fungi in three months duration [63]. A triad mixture of cotton seeds, antagonistic *Streptomyces globisporus* (alive or inactive powder preparation), and pathogen *Verticillium dahliae* by seed coating has ameliorated plant growth index and lessened disease severity in pods [64]. Foliar vegetation, leaf size, and root vitality have increased by about 50–70% despite the attack of the pathogen *V. dahliae*, and disease index dropped off to ~66% on seedlings. Covering with inactive spores (dried heat at 170°C 2 h) of *S. globisporus* promoted plant development, and the paradoxical result was defined as some metabolites related to plant growth being highly resistant to heat.

While bacterization (like antagonist bacteria) of *Pisum sativum* seeds has amplified the germination rate, the prevalence of *Rhizoctonia solani* damping-off symptoms was diminished [65]. However, coating rice seeds with Zn without arbuscular mycorrhizae fungi was more beneficial than the control group on germination, and soil application of Zn alone was better for germination than the seed coating method [66]. Similarly in a study, soil inoculation with *Pseudomonas* sp. could donate better nodulation and growth than seed inoculation [67]; nevertheless, seed coating with plant-alive beneficial microorganisms (BM) could better improve seed germination and protect in case of biotic agents like pathogens, abiotic factors such as salinity, and agrochemicals residues [68]. Although the shelf life of alive BM or their bioactive ingredients is a priority in commercialization, encapsulation technologies, as described in the ensuing section, such as micro- and nanoencapsulation, can sort out these problems by enabling uniformity of active compounds. Since seed coating is a kind of simple encapsulation system, the camouflage of a bioactive entity by relevant covering material seems very protective to allow keeping their biological properties safe. If cotton seeds are coated with the proper dosage of living antagonists or similar beneficial organisms, this coating can also foster the enlargement of cotton plantlets with strong roots and be very protective against pathogens through active metabolites produced by the alive microorganism spores [64]. Hence, when using living bioinoculants or bioactive substances for seed coating, the germination quality of the trade-off coated seeds versus uncoated seeds should be well-considered in order to compensate for potential damage from exposure to soil-borne pathogens, diseases, and abiotic stresses.

Seed applications with natural biocides can be applied in formulations such as gas, liquid, solid, and foam/gel formulations depending on the physical contact of the biomaterial with the seed, according to the closed storage area or open field practices. In prolonged storage duration of kernels, natural resource trials to preclude seed germination and microbial contamination can conserve the viability of seeds under optimal conditions as well as the disallowance of aflatoxin production, especially in grain cereals. Seed coating with carrier material is still a prevalent method to keep seed viability and protection during long-term storage safe. The biological coating material is a microorganism as bioinoculant from symbionts in plant microbiome and one of their active ingredients.

3.3 Encapsulation treatment of biosources

Recently, nano- and microencapsulation technologies have become increasingly important in the fields of food, pharmacology, and agriculture. Particularly in agri-farming, it has uncountable prominent benefits for the maintenance of agricultural commodities from the planting stage to the preparation of products for market before and after harvest. Due to the elasticity of the biological features of natural resources, for instance, the bioactivity of plant-based agents in field applications deteriorates in contrast to chemical pesticides. They have shorter shelf life, weaker stability, and lower bioavailability. Nanotechnology addresses these issues by designing packaging at the nanoscale to enhance the effectiveness of active compounds [69]. The main principle of nanotechnology is to deliver the active substance in a nano-sized transporter material to the target [70]. Encapsulation is a highly effective coating process that protects active substances, solid, or liquid, by surrounding them. This protective envelope works at both macro- and micro-levels, ensuring the stability and effectiveness of the encapsulated materials. Core material condensed in a capsule prevents direct chemical and physical connections until it reaches the target, maintaining its functionality and quality and thus increasing efficacy and prolonging shelf life [71]. Both nanoencapsulation and encapsulation of the active ingredient increase the bioavailability of biomaterial, protect nontarget organisms from toxic effects, maximize the usefulness of a smaller amount of active ingredient, minimize waste of the active compound, and extend the shelf life of the active ingredient.

Artemisia herba-alba essential oil in microcapsule exerted fairly well *in vitro* antifungal activity to *Botrytis cinerea*, inhibiting about 38 with 0.1% concentration compared to 12% inhibition obtained with free essential oil [72]. The active ingredient of *Oregano vulgare*, carvacrol in commercial form, was tested for antibacterial impact to bean bacterial blight agent *Xanthomonas axonopodis* pv. *cyamopsidis* using nanotechnology [13]. While nanoemulsified carvacrol depending on the dosage has blocked bacterial growth completely *in vitro* petri plates, in field experiments, disease severity was reduced to 14%, demonstrating effects similar to those of streptomycin sulfate, a well-established positive control. In commercial avocado orchards, micro-encapsulated *Yamadazyma mexicana* yeast is a candidate to be a biofungicide to be displaced to synthetic fungicide Hidrocob 77 to control anthracnose disease caused by *Colletotrichum* species [24]. Encapsulated *Y. mexicana* application in field preharvest and storage postharvest to avocado repressed almost 100% severity of anthracnose.

While the natural antimicrobial peptide from *Streptomyces natalensis*, natamycin, in nanofilm, increased the germination ratio of peanut seeds, fungal development and aflatoxin biosynthesis decreased during the annual storage period [73]. Seeds of barley and maize naturally infected with *Aspergillus flavus*, which biosynthesizes aflatoxin, were mantled with a preparation of talc powder mixed with *Trichoderma viride* from a culture collection [74]. As a result, this formulation prevented the progress of *A. flavus by* about 31% and diminished aflatoxin production. Savitharani [75] indicated biopesticidal potential essential oil (EO) of *Anisomeles indica* leaf to bacterial blight *Xanthomonas* strains on husked paddy/rice seeds in pods. Disease severity of plantlets from rice seeds treated with nanoemulsified EO was as less as 6.66%, while the protective potential of free EO and synthetic pesticides were nearly 13.33 and 18.33%, respectively. Phosphate-solubilizing *Pseudomonas* isolates procured from the roots of healthy tomatoes were combined with natural rock phosphate, and then this mixture was used in the seed coating process [76]. After treatment, encapsulated bacteria improved plant growth parameters and induced systemic resistance in plants to bacterial cancer. Pomegranate infected with *Penicillium implicatum* was dipped into the solution obtained by converting potato peel waste into polymers using lactobacillus strains. The fungal progression was prohibited for 14 consecutive days, and bruised pomegranate was redeemed with approximately 5–15% recovery in the fruit [77]. These nonsynthetic polymers are utilized in the food processing industry for manifold purposes in encapsulation technologies as an edible encapsulation matrix. It appears it can be safely feasible to film material for other types of grains, also postharvest fruits in storehouses, and vegetables in greengrocer markets.

However, even though nanopesticides are conditioned and designed in laboratories and are given the desired results, it is impossible to control natural conditions in the field. When transforming from laboratory setting to scale-up factory during the commercialization phase of nanomaterials, additional researches will be required to optimize the stability of various parameters. These parameters include the ongoing biopesticide delivery system, biocompatibility, bioavailability, and biodegradation of the nanomaterials.

3.4 Bioresources fermentation

Some lactic acid bacteria can ferment carbohydrates converted to polymers like starch to produce lactic acid [78, 79]. Lactobacillus strains are abundant in soil and on above-ground decomposing plant residues and they acidify altering pH around their habitat exporting numerous typical antimicrobial acidic metabolites synthesized through solid-state fermentation (SSF) [80]. Thus, this microbiome-mediated shield system naturally hampers soil-borne phytopathogenic organisms to infect plant roots.

Some researchers engage assorted composts to debilitate the activity of soil-borne pathogenic organisms [29]. The antifungal influence of compost consisting of date palm wastes and poultry manures has been compared to commercial composts. Researchers reported the natural compost had repressed completely by 100% growth of *Fusarium oxysporum* f. sp. *albedinis*. Agroindustrial waste valorization *via* SSF and submerged liquid fermentation (SmL) technologies has been examined as a potential renewable resource to fortify thriving symbionts. In actuality, recycling leftover agrowaste can be a valuable gain in donating circular economy. Symbionts *Trichoderma harzianum* and entomopathogenic fungus *Beauveria bassiana* are both antifungal microorganisms to phytopathogenic fungi [81, 82]. In the work, debris of husks and fibers of rice, wheat stalks, peels of orange and potato, whiskey draff, and soybean fibers were fermented in laboratory volume vessels and successfully blended in a culture medium to propagate this symbiont. Research has shown that the performance of fungal growth varies based on several factors, including the type of fermentation, the quality of raw materials, and the nutrient content. These elements are important factors in determining the fungus's resistance to external conditions [81]. In SSF technology, heat, humidity, biodegradability properties of the organic material, pH, C/N ratio, and porous structure of the substrate are important for the growth of fungus on the solid substrates [81], while nitrogen source in organic waste influences the resilience to virulence, tolerance of heat and UV-B in SmF system has the key task to amend biological trait of blastospores of *B. bassiana* [82]. Through different scenarios, these results can be replaced with disposed of debris like household and industrial wastes as new biopesticide sources with lower operational cost and materials. Similarly, as fermentation filtrate, supernatant of *Streptomyces virginiae* XDS1–5 that is an antagonist strain from soil rhizosphere against peach brown rot by *Monilinia fructicola* showed inhibition achievement of 80% *in vitro* assay and 66% *in vivo* assay [83].

As with other agricultural noxious organisms, tolerance to synthetic herbicides in weeds has also troubled persistently the last three decades. Hence, there is a need for new herbicide formulations based on natural substances and new herbs. Microbial metabolites are more favorable to control weeds than alive bioinoculants for long periods of stability and impact [4]. Endophytic fungus *Mycoleptodiscus indicus* from *Conyza* sp. is cultivated for estimation of bioherbicidal activity in solid or liquid state fermentation in three different culture media [84]. The fungal toxin in the mixture of sugarcane bagasse broth has promised to be potentially a potent bioherbicide impairing cucumber plantlets used as indicator plants and hindering their seed germination. This experiment was exhibited to have dissimilar chemical compositions of each of the three different media contents. These very informative results indicated that the content of any biosource may diversify the type, density, and quantity of active components bypassing the normal chemical reactions in biosynthesis pathways contingent on cultural conditions. As a result, registration of biopesticide products should be regional in local norms rather than EPA.

4. Conclusion

In today's global climate crisis, we should discuss the future of plant health and rehabilitation of microbial community terrains—phyllosphere, endosphere, rhizosphere residing of symbionts, and plant growth–promoting microorganisms—by natural sources. We should discuss a holistic approach to consolidating regenerative agricultural practices that address national agrifood security and tackle agricultural issues using indigenous biosources. The variability of microbial communities in its life spheres appears under the influence of edaphic factors linked to geographic locations. During the coevolution of plant-pathogen interactions, plants evolve to eliminate or minimize the effect of elicitors and effectors secreted by microorganisms, while pathogens also evolve greatly diversifying the number of elicitors/effectors to suppress or completely defeat the plant defenses. Likewise, saprophytes and symbionts shared identical niches in the pathosystem and would develop multiple mechanisms to struggle with other challengers, to meet the needs of food supply and reproduction, and to occupy space as a cost of survival. A bioinoculant that is optimal within the first isolated location, previously registered to EPA, might shift activity in a habitat of another type of terrestrial ecosystem that is determined by changes in climatic and edaphic factors. Accordingly, rather than EPA confirmation, it might be advised in the registration of new biocontrol resources that it could be aimed at the strategy to manage renewable agriculture systems in coverage of national programs with the collaboration of regional plant protection organizations. This synergistic interaction between foundations will encourage the use of biosources by farmers and stakeholders. Another important issue is that relationships between universities, research institutions, and commercial corporations are not strong in collaboration. After a scientist spends time and announces his or her works with academic publications, he or she does not take care of the work and abandons the results. Commercial laboratories conduct entirely free investigations in their small independent laboratories. The establishment of solidarity alongside academic and commercial entrepreneurship will enhance the use of natural resources through increased biopesticide inputs, providing farmers with greater choices to adopt sustainable innovations.

Despite the diversity of bioagents, the reasons they come from chemical pesticides can be as follows:

- Inadequate furnished experts to demonstrate the use and properties of biomaterial for farmers.
- Unconsciousness of farmers about the superior qualities of biomaterials over chemicals.
- Inconsistency of biological attributes of biomaterial (such as half-life and shelf life) in environment and flexibility in performance.
- Insufficient endeavor to commercialize academic research results.
- Intricate registration processes.
- The absence of bio–pest control units within the Ministry of Agriculture may be caused by inappropriate legal regulations and quality control.
- Governments should impose legal restrictions on the periodicity and quota of exerted conventional pesticides to provide the use of bioagents and to encourage farmers and stakeholders.

Author details

Berna Baş
Gaziantep University, Gaziantep, Turkey

*Address all correspondence to: bas@gantep.edu.tr

References

[1] Li Q. Imran: Mitigation strategies for heavy metal toxicity and its negative effects on soil and plants. International Journal of Phytoremediation. 2024;**26**(9):1439-1452. DOI: 10.1080/15226514.2024.2327611

[2] ARDSR Agency for Toxic Substances and Disease Registry [Internet]. 2024. Available from: https://wwwn.cdc.gov/TSP/substances/ToxHealthReferences.aspx [Accessed: July 5, 2024]

[3] European Commission [Internet]. 2024. Available from: https://food.ec.europa.eu/plants/pesticides/sustainable-use-pesticides_en [Accessed: July 5, 2024]

[4] Marrone PG. Status of the biopesticide market and prospects for new bioherbicides. Pest Management Science. 2024;**80**(1):81-86. DOI: 10.1002/ps.7403

[5] AgriBusiness Global [Internet]. 2024. Available from: https://www.agribusinessglobal.com/biopesticides/biorational-processes-in-biopesticide-discovery-and-development/ [Accessed: July 5, 2024]

[6] Neale M. The regulation of natural products as crop-protection agents. Pest Management Science: formerly Pesticide Science. 2000;**56**(8):677-680

[7] BioSafe Systems [Internet]. 2024. Available from: https://biosafesystems.com/product/guarda/ [Accessed: July 5, 2024]

[8] VGN Vegetables Growers News [Internet]. 2024. Available from: https://vegetablegrowersnews.com/news/biosafe-systems-introduces-guarda-a-new-thyme-oil-antimicrobial [Accessed: July 5, 2024]

[9] Yuan S, Wang Y, Shi L, Zhou P. Current situation and prospects of biological pesticides related standards in China. Chinese Journal of Biological Control. 2018;**34**(1):1-7. DOI: 10.16409/j.cnki.2095-039x2018.01.001

[10] Silveira RD, Veras FF, Hernandes KC, Bach E, Passaglia LMP, Zini CA, et al. Genomic analysis reveals genes that encode the synthesis of volatile compounds by a *Bacillus velezensis*-based biofungicide used in the treatment of grapes to control *Aspergillus carbonarius*. International Journal of Food Microbiology. 2024;**415**:110644. DOI: 10.1016/j.ijfoodmicro.2024.110644

[11] Rupaedah B, Prasetyo AE, Hidayat F, Nia Asiani N, Abdul Wahid A, Nurlaila Lutfia A. Evaluation of microbial biocontrol agents for *Ganoderma boninense* management in oil palm nurseries. Journal of the Saudia Society of Agricultural Sciences. 2024;**23**(3):236-244. DOI: 10.1016/j.jssas.2023.12.001

[12] Yang J, Zhang Y, Zhao J, Gao Y, Liu Z, Zhang P, et al. Target gene selection for RNAi-based biopesticides against the hawthorn spider mite, *Amphitetranychus viennensis* (Acari: Tetranychidae). Pest Management Science. 2023;**79**(7):2482-2492. DOI: 10.1002/ps.7437

[13] Choudhary P, Bhanjana G, Kumar S, Dilbaghi N. Development and evaluation of eco-friendly carvacrol nanoemulsion as a sustainable biopesticide against bacterial leaf blight of cluster bean. Pest Management Science. 2024;**80**(2):452-462. DOI: 10.1002/ps.7776

[14] Wang Y, Li X, Zhu C, Yi S, Zhang Y, Hong Z. Plant-derived artificial miRNA effectively reduced the proliferation of aphid (Aphidoidea) through

spray-induced gene silencing. Pest Management Science. 2024;**80**(9):4322-4332. DOI: 10.1002/ps.8138

[15] López-Cabeza R, Rodríguez-Sabina S, Reyes CP, Expósito DG, Giménez C, Jiménez IA, et al. Bio-guided isolation of aromatic abietane diterpenoids from *Salvia canariensis* as biopesticides in the control of phytopathogenic fungi. Pest Management Science. 2024;**80**(4):2199-2207. DOI: 10.1002/ps.7958

[16] Martínez M, Arata A, Dinolfo MI, Lázaro L, Welin B, Stenglein S. Evaluation of PSP1 biostimulant on *Fusarium graminearum*-wheat pathosystem: Impact on disease parameters, grain yield, and grain quality. Pest Management Science. 2024;**80**(7):3578-3589. DOI: 10.1002/ps.8062

[17] Li YK, Xiong RF, Wu QY, Yao S, Qu X, Huang ZH, et al. C-alkylated flavonoids from the whole plants of *Desmodium caudatum* and their anti-TMV activity. Pest Management Science. 2023;**79**(10):3721-3730. DOI: 10.1002/ps.7589

[18] Besrukow P, Will F, Dussling S, Berkelmann-Löhnertz B, Schweiggert R. Additive and synergistic antifungal effects of copper and phenolic extracts from grape cane and apples. Pest Management Science. 2023;**79**(9):3334-3341. DOI: 10.1002/ps.7519

[19] Dinango VN, Dhouib H, Wakam LN, Kouokap LK, Youmbi DY, Eke P, et al. Bacterial endophytes inhabiting desert plants provide protection against seed rot caused by *Fusarium verticillioides* and promote growth in maize. Pest Management Science. 2024;**80**(3):1206-1218. DOI: 10.1002/ps.7850

[20] Vuerich M, Petrussa E, Filippi A, Cluzet S, Fonayet JV, Sepulcri A, et al. Antifungal activity of chili pepper extract with potential for the control of some major pathogens in grapevine. Pest Management Science. 2023;**79**(7):2503-2516. DOI: 10.1002/ps.7435

[21] Yang Q, Guo Y, Wang H, Luo Z, Chen Y, Jiang M, et al. Action of the fungal compound citrinin, a bioherbicide candidate, on photosystem II. Pest Management Science. 2024;**80**(1):133-148. DOI: 10.1002/ps.7513

[22] Qiao S, Yao J, Wang Q, Li L, Wang B, Feng X, et al. Antifungal effects of amaryllidaceous alkaloids from bulbs of *Lycoris* spp. against *Magnaporthe oryzae*. Pest Management Science. 2023;**79**(7):2423-2432. DOI: 10.1002/ps.7420

[23] Ding YY, Sun Y, Luo XF, Zhang SY, Wang R, Yang ZG, et al. Anti-phytopathogenic activity and the mechanisms of phthalides from *Angelica sinensis* (Oliv.) Diels. Pest Management Science. 2023;**79**(6):2135-2146. DOI: 10.1002/ps.7389

[24] González-Gutiérrez KN, Ragazzo-Sánchez JA, Calderón-Santoyo M. Field and postharvest application of microencapsulated Yamadazyma mexicana LPa14: Anthracnose control and effect on postharvest quality in avocado (*Persea americana* mill. cv. Hass). Pest Management Science. 2024;**80**(7):3459-3469. DOI: 10.1002/ps.8052

[25] Lin L, Tao M, He WM, Wu QH, Huang HK, Murero AK, et al. Identification of non-canonical antagonistic bacteria via interspecies contact-dependent killing. Pest Management Science. 2024;**80**(8):3997-4005. DOI: 10.1002/ps.8103

[26] Gu G, Hou X, Xue M, Pan X, Dong J, Yang Y, et al. Diphenyl ethers from

endophytic fungus *Rhexocercosporidium* sp. Dzf14 and their antibacterial activity by affecting homeostasis of cell membranes. Pest Management Science. 2024;**80**(6):2658-2667. DOI: 10.1002/ps.7972

[27] Eigharlou M, Hashemi Z, Mohammadi A, Khelghatibana F, Nami Y, Sadeghi A. Herbicidal proteins from *Bacillus wiedmannii* isolate ZT selectively inhibit ryegrass (*Lolium temulentum* L.). Pest Management Science. 2024;**80**(7):3478-3490. DOI: 10.1002/ps.8053

[28] Tan M, Zhang Y, Zhang Y, Vurro M, Qiang S. Effects of *Bipolaris yamadae* strain HXDC-1-2 as a bioherbicide against *Echinochloa crus-galli* in rice and dry fields. Pest Management Science. 2024;**80**(8):3786-3794. DOI: 10.1002/ps.8081

[29] Mohamed OZ, Yassine B, El Hassan A, Abdellah H, Rachid B. Evaluation of compost quality and bioprotection potential against *Fusarium* wilt of date palm. Waste Management. 2020;**113**:12-19. DOI: 10.1016/j.wasman.2020.05.035

[30] EPA, United States Environmental Protection Agency [Internet]. 2023. Available from: https://www.epa.gov/ingredients-used-pesticide-products/what-are-biopesticides [Accessed: June 10, 2024]

[31] Abo-Elyousr KAM, Imran M, Sallam NM, Abdel Aal AM, Assiri ME, Abdel-Rahim IR. Sustainable biocontrol of purple blotch disease in *Allium cepa* L. by biocontrol yeasts, *Pichia kluyveri* and *Filobasidium wieringae*. Egyptian Journal of Biological Pest Control. 2024;**34**:11. DOI: 10.1186/s41938-024-00776-6

[32] Esteves M, Lage P, Sousa J, Centeno F, de Fátima TM, Tenreiro R, et al. Biocontrol potential of wine yeasts against four grape phytopathogenic fungi disclosed by time-course monitoring of inhibitory activities. Frontiers in Microbiology. 2023;**14**:1146065. DOI: 10.3389/fmicb.2023.1146065

[33] Freimoser FM, Rueda-Mejia MP, Tilocca B, Migheli Q. Biocontrol yeasts: Mechanisms and applications. World Journal of Microbiology and Biotechnology. 2019;**35**:154. DOI: 10.1007/s11274-019-2728-4

[34] Boutton C, Leybaert K, Geerinck J, Van Daele I, Hanssens J, Reyes C, et al. EVOCA™, the first biofungicide developed by Biotalys' AGROBODY foundry™ platform. Journal of Plant Diseases and Protection. 2024;**1**:9. DOI: 10.1007/s41348-024-00880-7

[35] Waghunde RR, Shelke RM, Sabalpara AN. Trichoderme: A significant fungus for agriculture and environment. African Journal of Agricultural Research. 2016;**11**(22):1952-1965. DOI: 10.5897/AJAR2015.10584

[36] van Lenteren JC, Bolckmans K, Köhl J, Ravensberg WJ, Urbaneja A. Biological control using invertebrates and microorganisms: Plenty of new opportunities. BioControl. 2018;**63**:39-59. DOI: 10.1007/s10526-017-9801-4

[37] Swathy K, Nisha V, Vivekanandhan P. Biological control effect of *Trichoderma harzianum* (Hypocreales: Hypocreaceae) against phytopathogens. Environmental Quality Management. 2024;**34**:e22227. DOI: 10.1002/tqem.22227

[38] Alizadeh M, Qaderi S, Roshanroo M, Karimzadeh S, Fazli M, Saeedi M, et al. Contouring multifaceted biological activities and applications of *Trichoderma* spp. for managing plant health. Journal of Crop Health.

2024;**76**:561-597. DOI: 10.1007/s10343-024-00976-y

[39] Ogata-Gutiérrez K, Chumpitaz-Segovia C, Lirio-Paredes J, Zúñiga-Dávila D. Antifungal activity of *Phyllospheric bacteria* isolated from *Coffea arabica* against *Hemileia vastatrix*. Microorganisms. 2024;**12**(3):582. DOI: 10.3390/microorganisms12030582

[40] Ibrahim E, Nasser R, Hafeez R, Ogunyemi SO, Abdallah Y, Khattak AA, et al. Biocontrol efficacy of endophyte *Pseudomonas poae* to alleviate *Fusarium* seedling blight by refining the morpho-physiological attributes of wheats. Plants. 2023;**12**(12):2277. DOI: 10.3390/plants12122277

[41] Dicks LM, Vermeulen W. Bacteriophage-host interactions and the therapeutic potential of bacteriophages. Viruses. 2024;**16**(3):478. DOI: 10.3390/v16030478

[42] Liu L, Wang B, Huang A, Zhang H, Li Y, Wang L. Biological characteristics of the bacteriophage LDT325 and its potential application against the plant pathogen *Pseudomonas syringae*. Frontiers in Microbiology. 2024;**15**:1370332. DOI: 10.3389/fmicb.2024.1370332

[43] Kim B, Lee SY, Park J, Song S, Kim KP, Roh E. Bacteriophage cocktail comprising Fifi044 and Fifi318 for biocontrol of *Erwinia amylovora*. The Plant Pathology Journal. 2024;**40**(2):160-170. DOI: 10.5423/PPJ.OA.01.2024.0005

[44] Zhu JZ, Qiu ZL, Gao BD, Li XG, Zhong J. A novel partitivirus conferring hypovirulence by affecting vesicle transport in the fungus *Colletotrichum*. MBio. 2024;**15**(2):e02530-e02523. DOI: 10.1128/mbio.02530-23

[45] Moreno-Velandia CA, Ongena M, Cotes AM. Effects of Fengycins and Iturins on *Fusarium oxysporum* f. sp. *Physali* and root colonization by *Bacillus velezensis* Bs006 protect golden berry against vascular wilt. Phytopathology. 2021;**111**(12):2227-2237. DOI: 10.1094/PHYTO-01-21-0001-R

[46] Han Q, Wu F, Wang X, Qi H, Shi L, Ren A, et al. The bacterial lipopeptide iturins induce *Verticillium dahliae* cell death by affecting fungal signaling pathways and mediate plant defence responses involved in pathogen-associated molecular pattern-triggered immunity. Environmental Microbiology. 2015;**17**:1166-1188. DOI: 10.1111/1462-2920.12538

[47] Fan B, Blom J, Klenk HP, Borriss R. *Bacillus amyloliquefaciens*, *Bacillus velezensis*, and *Bacillus siamensis* form an "operational group *B. amyloliquefaciens*" within the *B. subtilis* species complex. Frontiers in Microbiology. 2017;**8**:22. DOI: 10.3389/fmicb.2017.00022

[48] Diab MK, Mead HM, Ahmed Khedr MM, Abu-Elsaoud AM, El-Shatoury SA. Actinomycetes are a natural resource for sustainable pest control and safeguarding agriculture. Archives of Microbiology. 2024;**206**(6):268. DOI: 10.1007/s00203-024-03975-9

[49] Zhang Y, Ouyang B, Zhang W, Guang C, Xu W, Mu W. Deoxynivalenol: Occurrence, toxicity, and degradation. Food Control. 2023;**155**:110027. DOI: 10.1016/j.foodcont.2023.110027

[50] Thomas Y, Thiebeauld O. U.S. Patent No. 11,793,197. Washington, DC: U.S. Patent and Trademark Office; 2023

[51] ImmuneRise A Natural Marine Biopesticide [Internet]. 2024. Available from: https://www.immunrisebiocontrol.fr/en/biopesticide/ [Accessed: July 5, 2024]

[52] Molina-Miras A, Abreu AC, Rosales LL, Cerón-García MC, Sánchez-Mirón A, Fernández I, et al. A step forward in sustainable pesticide production from *Amphidinium carterae* biomass via photobioreactor cultivation with urea as a nitrogen source. Bioresource Technology. 2023;**387**:129643. DOI: 10.1016/j.biotech.2023.129643

[53] Degani O, Dor S. *Trichoderma* biological control to protect sensitive maize hybrids against late wilt disease in the field. Journal of Fungi. 2021;**7**(4):315. DOI: 10.3390/jof7040315

[54] Schuster A, Schmoll M. Biology and biotechnology of *Trichoderma*. Applied Microbiology and Biotechnology. 2010;**87**:787-799. DOI: 10.1007/s00253-010-2632-1

[55] Druzhinina I, Seidl-Seiboth V, Herare-Estrella A, Horwitz BA, Kenerley CM, Monte E, et al. *Trichoderma*: The genomics of opportunistic success. Nature Reviews Microbiology. 2011;**9**(10):749-759. DOI: 10.1038/nrmicro2637

[56] Scott K, Konkel Z, Gluck-Thaler E, Valero David GE, Simmt CF, Grootmyers D, et al. Endophyte genomes support greater metabolic gene cluster diversity compared with non-endophytes in *Trichoderma*. PLoS One. 2023;**18**(12):e0289280. DOI: 10.1371/journal.pone.0289280

[57] Zhang JL, Tang WL, Huang QR, Li YZ, Wei ML, Jiang LL, et al. *Trichoderma*: A treasure house of structurally diverse secondary metabolites with medicinal importance. Frontiers in Microbiology. 2021;**12**:723828. DOI: 10.3389/fmicb.2021.723828

[58] Zenteno-Alegria CO, Yarzabal Rodrigues LA, Ciancas Jimenez J, Alvarez Gutierrez PE, Gunde-Cimerman N, Batista-Garcia RA. Fungi beyond limits: The agricultural promise of extremophiles. Microbial Biotechnology. 2024;**17**(3):e14439. DOI: 10.1111/1751.14439

[59] Seed Coating Market [Internet]. 2024. Available from https://www.marketsandmarkets.com/Market-Reports/seed-coating-materials-market-149045530.html [Accessed: August 1, 2024]

[60] Guyard C, Séguy N, Cailliez JC, Drobecq H, Polonelli L, Dei-Cas E, et al. Characterization of a *Williopsis saturnus* var. mrakii high molecular weight secreted killer toxin with broad-spectrum antimicrobial activity. Journal of Antimicrobial Chemotherapy. 2002;**49**(6):961-971. DOI: 10.1093/jac/dkf040

[61] Liu SQ, Tsao M. Biocontrol of spoilage yeasts and moulds by *Williopsis saturnus* var. *saturnus* in yoghurt. Nutrition and Food Science. 2010;**40**:166-175. DOI: 10.1108/00346651011029192

[62] Karabulut G, Cagri-Mehmetoglu A. Antifungal, mechanical, and physical properties of edible film containing *Williopsis saturnus* var. *saturnus* antagonistic yeast. Journal of Food Science. 2018;**83**:763-769. DOI: 10.1111/1750-3841.14062

[63] Ulutasdemir T, Cagri-Mehmetoglu A. Effects of edible coating containing *Williopsis saturnus* var. *saturnus* on fungal growth and aflatoxin production by *Aspergillus flavus* in peanuts. Journal of Food Safety. 2019;**39**:e12698. DOI: 10.111/jfs./12698

[64] Chen Q, Bai S, Zhang T, Duan C, Zhao J, Xue Q, et al. Effects of seed-coating preparations of living

Streptomyces globisporus on plant growth promotion and disease control aganist *Verticillium* wilt in cotton. Sustainability. 2021;**13**(11):6001. DOI: 10.3390/su13116001

[65] Barberio A, Palmieri D, Barone G, Del Grosso C, Lima G, De Curtis F. Selection and characterization of beneficial bacteria and seed bacterization of leguminoses as a biocontrol strategy against *Rhizoctonia solani* damping-off. In: Abstracts of Presentations at the XXIV Congress of the Italian Phytopathological Society (Journal of Plant Pathology). Vol. 100. Ancona, Italy: Journal of Plant Pathology; 2018. pp. 613-653. DOI: 10.1007/s42161-018-0130-y

[66] Mehmood H, Ali MA, Hussain S, Baig KS, Farooq U, Ajmal M, et al. Synchronization of *Arbuscular mycorrhizae* fungi inoculation with different zinc application methods for improvement in BASMATI rice growth and yield in alkaline calcareous soil. Journal of King Saud University-Science. 2022;**34**(5):102053. DOI: 10.1016/j.jksus.2022.102053

[67] Bhattacharjya S, Chandra R. Effect of inoculation methods of *Mesorhizobium ciceri* and PGPR in chickpea (*Cicer areietinum* L.) on symbiotic traits, yields, nutrient uptake and soil properties. Legume Research-An International Journal. 2013;**36**(4):331-337

[68] Rocha I, Ma Y, Souza-Alonso P, Vosátka M, Freitas H, Oliveira RS. Seed coating: A tool for delivering beneficial microbes to agricultural crops. Frontiers in Plant Science. 2019;**10**:1357. DOI: 10.3389/fpls.2019.01357

[69] Lammari N, Rabti H, Louaer O, Elaissari A, Meniai AH. Encapsulation of bioactive compounds: Role of nanotechnology. In: Sefat F, Farzi G, Mozafari M, editors. Principles of Biomaterials Encapsulation. 1st ed. Woodhead Publishing: Elsevier; 2023. pp. 39-65. DOI: 10.1016/B978-0-323-85947-9.00007-8

[70] Malik S, Muhammad K, Waheed Y. Nanotechnology: A revolution in modern industry. Molecules (Basel, Switzerland). 2023;**28**(2):661. DOI: 10.3390/molecules28020661

[71] Beheshtizadeh N, Vaez A, Siminzar P, Azami M, Farzin A. Encapsulation: Shell and core. In: Sefat F, Farzi G, Mozafari M, editors. Principles of Biomaterials Encapsulation. Woodhead Publishing: Elsevier; 2023. pp. 115-148. DOI: 10.1016/B978-0-12-824345-9.00016-7

[72] Ez-Zoubi A, Ez-Zoubi Y, Bentata F, El-Mrabet A, Ben Tahir C, Labhilili M, et al. Preparation and characterization of a biopesticide based on *Artemisia herba-*alba essential oil encapsulated with succinic acid-modified beta-cyclodextrin. Journal of Chemistry. 2023;**1**:8. DOI: 10.1155/2023/3830819

[73] Wang J, Wu S. Natamycin loaded onto chitosan-based coatings for the quality preservation of peanut seed from fungal contamination. Journal of Plant Disease and Protection. 2023;**130**:899-903. DOI: 10.1007/s41348-023-00734-8

[74] El-Dawy EG, Gherbawy YA, Hussein MA. Characterization of *Aspergillus* section *Flavi* associated with stored grains. Mycotoxin Research. 2024;**40**(1):187-202. DOI: 10.1007/s12550-023-00514-1

[75] Savitharani M, Kumar V, Manjunatha T, Ruksana F, Vishala E, Prathap HM. Biopesticide potential of *Anisomeles indica* Kuntze leaf essential oil and its nanoemulsion aganist bacterial blight-*Xanthomonas* strains. Journal of Natural Pesticide Research.

2024;**8**:100075. DOI: 10.1016/j.napere.2024.100075

[76] Bakki M, Banane M, Marhane O, Esmaeel Q, Hatimi A, Barka EA, et al. Phosphate solubilizing *Pseudomonas* and *Bacillus* combined with rock phosphates promoting tomato growth and reducing bacterial cancer disease. Frontiers in Microbiology. 2024;**15**:1289466. DOI: 10.3389/fmicb.2024.1289466

[77] Ghosh M, Chouhan D, Kamra A, Sharma V. Sustainable utilization of potato industry waste for antifungal biopolymer production by *Lactobacillus helveticus* and its application on pomegranates (*Punica granatum* L.). Circular Economy and Sustainability. 2021;**1**:1297-1312. DOI: 10.1007/s43615-021-00120-2

[78] Raman J, Kim JS, Choi KR, Eun H, Yang D, Ko YJ, et al. Application of lactic acid bacteria (LAB) in sustainable agriculture: Advantages and limitations. International Journal of Molecular Sciences. 2022;**23**(14):7784. DOI: 10.3390/ijms23147784

[79] Oliveira PM, Zannini E, Arendt EK. Cereal fungal infection, mycotoxins, and lactic acid bacteria mediated bioprotection: From crop farming to cereal products. Food Microbiology. 2014;**37**:78-95. DOI: 10.1016/j.fm.2013.06.003

[80] Krishnan SV, Nampoothiri KM, Suresh A, Linh NT, Balakumaran PA, Pocsi I, et al. *Fusarium* biocontrol: Antagonism and mycotoxin elimination by lactic acid bacteria. Frontiers in Microbiology. 2024;**14**:1260166. DOI: 10.3389/fmicb.2023.1260166

[81] Sala A, Vittone S, Barrena R, Sanches A, Artola A. Scanning agro-industrial wastes as substrates for fungal biopesticide production: Use of *Beauveria bassiana* and *Trichoderma harzianum* in solid-state fermentation. Journal of Environmental Management. 2021;**295**:113113. DOI: 10.1016/j.jenvman.2021.113113

[82] Lima VH, Matugawa AT, Mascarin GM, Fernandes EKK. Complex nitrogen sources from agro-industrial byproducts: Impact on production, multi-stress tolerance, virulence, and quality of *Beauveria bassiana* blastospores. Microbiology Spectrum. 2024;**12**(6):e0404023. DOI: 10.1128/spectrum.04040-23

[83] Chen M, Jia F, Chen S, Zheng Y, Hu Y, Liu W, et al. *Streptomyces virginae* XDS1-5, antagonistic actinomycete, as a biocontrol to peach brown rot caused by *Monilinia fructicola*. Journal of the Science of Food and Agriculture. 2024;**104**(12):7514-7523. DOI: 10.1002/jsfa.13573

[84] Portela VO, da Rosa OR, Souza M, de Oliveira AL, Dalcol II, Jacques RJS. Chemical composition and bioherbicidal activity of metabolites of *Mycoleptodiscus indicus* produced in three culture media. Biocatalysis and Agricultural Biotechnology. 2024;**58**:103147. DOI: 10.1016/jbcab.2024.103147

Chapter 8

Role of Microorganisms in Seed Germination

Faquir C. Garg

Abstract

Quality of the seed, the backbone of modern agriculture, is an important factor in the successful establishment and performance of any crop. Two indicators of seed quality are uniform seed germination and seedling vigour. To enhance germination, many types of treatments, including beneficial microbes belonging to arbuscular mycorrhizal fungi, *Trichoderma* spp., rhizobia and other bacteria, have been tried on seeds before sowing *via* coating or bio-priming treatments and increase in seed germination of different crops including cereal crops, oil seeds and vegetables, have been reported. The role of endophytes and seed-borne microorganisms on seed germination and the mechanism of action of microorganisms in seed germination have also been discussed.

Keywords: PGPB, endophytes, seed germination, seed microbiome, auxins

1. Introduction

Seed germination is a very complex process that involves many biochemical, physiological and morphological changes in seeds. There are three requirements for a seed to germinate *viz.* (i) the seed must be viable, which means the embryo should be alive and able to germinate; (ii) the seed should not be dormant, i.e., there should be no chemical barrier to germination and (iii) environmental conditions such as humidity, temperature, air (O_2) and light must be available in sufficient quantity. If all these conditions are met, the inactive embryo in the seed will commence the germination process. To break the exogenous dormancy of seeds, optimal temperature and humidity are a must. In addition, some factors such as plant growth regulators can affect the rate and uniformity of mature seed germination [1].

Seed quality, in terms of uniform germination and high seedling vigour, contributes to successful crop establishment and thus crop performance [2]. When seedlings emerge quickly and vigorously, they are more likely to capture nutrients, tolerate biotic and abiotic stress, compete with weeds and of course cope with adverse environmental conditions [3]. Thus, seed quality is a primary objective of the agroindustry. Several different types of seed treatment technologies are now available and used to improve seed quality. Among these treatments, pre-sowing seed treatment is of great interest due to its effectiveness and environmental benefits. In general, various seed treatment technologies are used to increase seed germination, including chemical and biological treatment. Since seeds come in contact with microorganisms

 IntechOpen

present in soil and can be useful in seed germination, it is interesting to consider the effectiveness of beneficial microorganisms, present in soil or applied deliberately, on seeds of various crops for enhancing germination.

According to market survey reports, global biological seed treatment including the use of biofertilisers, biopesticides and biostimulants [4] is expected to touch USD 1.7 billion by 2025, and that of only seed treatment with microorganisms and natural substances applied as biostimulants may be around USD 338 million by this time. Pre-sowing seed treatments with beneficial microorganisms have relatively low application costs, as they require a single treatment [5–7]. These treatments include seed coating and/or seed priming comprising of application of a thin layer of external material onto the seed surface, without affecting the shape, size, or weight of the seed [8, 9]. Several different priming techniques are used to solve the germination problems under different environmental conditions. Among these biopriming, i.e., the application of microorganisms to seeds is important [10]. Seed bio-priming is known to improve seedling emergence. Under adverse environmental conditions, biopriming results in the rapid emergence of seedlings [11]. If beneficial microorganisms are added during bio-priming, they proliferate on the seed surface and further help in germination, thus it has practical agronomic importance in crop production [12]. In biopriming, seeds are primed with different endophytic microbes that easily integrate into seeds and help improve the biochemical and germination parameters [13]. Seed treatment does not induce any change within the seed and the active ingredient applied provides an advantage for the induction of germination and enhances seedling growth. Thus, the application of beneficial microorganisms to the seeds to increase their germination, seedling vigour and biomass, as well as the capability to overcome abiotic stress both during and after seedling emergence has been evaluated in many crops including cereal grains, oil seed crops and vegetables [14–17]. The role of endophytes and seed-borne microorganisms on seed germination in wildflowers and grasses has also been explored and discussed in several studies [18, 19].

2. Beneficial effects of microorganisms

Seed germination in open fields is never achieved 100% and, in some cases, more than 50% of the seeds used fail to germinate, resulting in the failure of crops and economic loss to the agribusiness. Microorganisms are inoculated on seed surfaces as biofertilisers, biostimulants or bioprotectants often to protect the seeds from insect pests and plant pathogens [20]. How these microorganisms influence seed germination, a pre-requirement for the establishment and growth of plants have been studied by several researchers in different crops (**Table 1**).

Application of beneficial microorganisms to the seed directly has two significant advantages, (i) use of the lesser amount of microbial inoculum per plant and (ii) instantaneous contact of microbes with roots after germination and early developmental stages [37, 38]. The microorganisms mainly used for seed inoculation include fungi like *Trichoderma* spp., arbuscular mycorrhizal (AM) fungi and bacteria known as plant-growth-promoting bacteria (PGPB) and bacterial symbionts like *Rhizobium*, *Bradyrhizobium*, *Azorhizobium*, *Mesorhizobium* etc. [39]. These microbes have been applied on seed surfaces of cereals, oil seeds, vegetables, pulse crops and fibre and forage crops [40] to assess their effect on seed germination and other growth parameters. In recent years, a number of studies have been conducted

Crop	Microorganism	Method/growth condition	Germination	Source
Bacteria				
Wheat	*Pseudomonas balearica*	Seed coating	Germination 100%	[16]
Maize	*Pseudomonas putida*, *P. fluorescens, Azospirillum lipoferum*, alone or in combination	Seed coating; growth chamber	Germination improved by +16%; Vigour index +75%	[21]
	Mixta theicola	Seed soaking; growth chamber	Germination increased by +38%, Seedling vigour index +117%,	[22]
Rice	*Bacillus* sp.	Seed coating; growth chamber	Germination improved	[23]
	Paenibacillus yonginensis	Seed soaking; growth chamber	Germination improved by 4%	[24]
Quinoa	*Bacillus licheniformis* *Enterobacter asburiae*	Seed coating	Germination 116% Germination 132%	[25]
Soybean	*Bacillus velezensis*	Seed coating	Germination increased from 55 to 64%	[26]
Sunflower	*Enterobacter*, *Bacillus* sp., *Paraburkholderia phytofirmans*	Seed priming; growth chamber	Germination and vigour index improved	[15]
	Pseudomonas fluorescens	Seed priming; growth chamber	Germination and vigour index improved	[27]
Carrot	*S. marcescenes* or *Pseudomonas fluorescenes* or *P. putida* or *Klebsiella aerogenes* or *Bacillus cereus*	Seed treatment; Greenhouse	Germination improved from 115 to 157%	[28]
Tomato	*Pseudomonas* spp. S3	Seed treatment or Soil drenching; Green house	Germination 111%	[14]
Cucumber	*Acinetobacter radioresisstens* + *Pseudomonas paralactis* + *Bacillus cereus*	Seed priming; Open field	Germination improved	[29]

Crop	Microorganism	Method/growth condition	Germination	Source
Cyanobacteria				
Maize	*Spirulina platensis*	Seed priming; Growth chamber; Cd stress	Germination increased by +63%,	[30]
Fungi				
Wheat	*Trichoderma harzianum*	Seed coating; Growth chamber	Germination increased by +35%, Vigour index +120%	[31]
Yeast				
Wheat	*Meyerozyma guilliermondii*	Seed coating; Growth chamber	Germination increased by +97%,	[32]
Bacteria + Fungi				
Tomato	*Trichoderma harzanium + P. fluorescens*	Seed coating	Germination 48%	[33]
Canola	*Bacillus subtilis*, *Macrophomina phaseolina*, alone or in combination	Seed priming; Growth chamber; Salt stress	Germination	[34]
	Bacillus subtilis + Trichoderma harzianum	Salt stress	Germination	[35]
Cyanobacteria + Fungi				
Maize	*Anabaena torulosa + Nostoc carneum + N. piscinale + A. doliolum* or *A. torulosa + Trichoderma viride*	Seed coating; Greenhouse	Germination improved by +16%; Germination-related enzymes improved α-amylase by 10%, Invertase by 13%	[36]

Table 1.
Effect of microorganisms on seed germination.

to determine the role of microorganisms in the germination of different types of seeds and the results reported have been encouraging.

2.1 Cereal crops

*2.1.1 Wheat [*Triticum aestivum *L.]*

One of the most important cereal crops grown world over and used to prepare various domestic and bakery products is wheat. Colla et al. [41] applied a consortium of endophytic microorganisms consisting of *Rhizoglomus intraradices* BEG72, *Funneliformis mosseae* and *Trichoderma atroviride* MUCL 45632 to evaluate their ability to promote emergence and growth of seedlings. After 17 days of sowing, a significant effect of the application of microbial inoculum was observed on the germination and growth parameters of seedlings.

Kthiri et al. [31] applied different strains of *Trichoderma harzianum* to the outer surface of seeds of wheat and obtained higher values for germination, seedling growth and other growth parameters in comparison to untreated. In another study, Kthiri et al. [32] conducted an experiment in pots under controlled conditions where they applied *Meyerozyma guilliermondii*, a yeast on Durum-wheat seeds cv. Karim. Seed treatment with yeast promoted germination, which increased from 47% (untreated seeds) to 97%.

To examine the effect of microbes on germination and seedling growth of *T. aestivum* cv. Salavat Yulaev under adverse environmental conditions, Lastochkina et al. [42] applied *Bacillus subtilis* strain 10-4 on wheat seeds. They observed an increase in germination and plant growth after 6 days of sowing. PGPB such as *Proteus mirabilis* R2, *Pseudomonas balearica* RF-2 and *Cronobacter sakazakii* RF-4 improved germination of wheat seeds under normal as well as water-stress conditions. Germination and promptness indices were found to be at the highest values, 100 and 68% respectively, for seed treatment with *P. balearica* [16]. Significant improvement in the rate of wheat seed germination on coating with *Bacillus pumilus* MA9 and *Virgibacillus halodenitrificans* MA14, attributed to secretion of auxin, solubilisation of inorganic phosphate, and ACC deaminase activity has also been reported by Brahim et al. [43].

*2.1.2 Maize [*Zea mays *L.]*

Maize grown under a wide range of soil and agroclimatic conditions is a very important crop both for human consumption as well as animal feed [44]. Several studies have focused on applying beneficial microorganisms to the seed to induce uniform germination and better seedling growth. An indigenous strain of *T. harzianum*, when used as a coating agent on maize seeds in an open-field experiment in Kenya, proved to be effective [45]. The inoculated seeds showed a higher germination percentage and seedling growth under *T. harzianum*-coating treatment. Sharma et al. [36] compared the effect of a cyanobacterial consortium (BF1-4) comprising *Anabaena torulosa*, *A. dolionum*, *Nostoc carneum* and *N. piscinale* or cyanobacterium *Anabaena torulosa* as the matrix and *Trichoderma viride* as the partner (An-Tr biofilm), applied through bio-priming treatments on two maize cultivars. Germination and germination-related enzymes (α-amylase and invertase activity) were detected after 96 h of sowing. Seed inoculation with microorganisms (BF1-4 or An-Tr biofilm) increased germination percentage and enzyme activities for both the cultivars of maize. However, the highest increase in germination, by 16% was noted in one of the cultivars, HKI323PV, inoculated with An-Tr biofilm. Biopriming maize-seed using the cyanobacterium *Spirulina platensis* accelerated germination in both control as well as cadmium (Cd) -toxicity conditions. A maximum increase of 16% was reported in the absence of Cd contamination [30].

Noumavo et al. [21] applied PGPB like *Pseudomonas putida*, *P. fluorescens* and *Azospirillum lipoferum*, alone or in combination, and found an increase in seed germination of maize. The best results for germination percentage of 16% and growth parameters were observed with the combination of *P. putida* and *P. fluorescens*. Earlier, higher germination and seedling vigour have been reported in young maize plantlets grown under field conditions after seed treatment with endophytes such as *P. putida*, *P. fluorescens*, *A. liofarum* and *A. brasilense* [46, 47].

Recently, an experiment was conducted with the aim of testing the effect of the endophytic bacterium *Mixta theicola* isolated from the roots of *Solenostemma argel*, a wild herb in maize [22]. Inoculation resulted in a significant enhancement of seed germination of about 38%, root elongation, and seedling vigour index of 15-day-old plantlets.

*2.1.3 Rice [*Oryza sativa *L.]*

Rice, the most widely consumed grain in the world, is generally cultivated under wetland conditions. However, presently, dry-direct seeding is preferred as it needs less water and reduces labour requirements. Seed coating of rice seeds using *Bacillus* sp. KS-54 was found to be effective in enhancing germination in both controlled and field conditions compared to noncoated seeds [23]. The coating treatment not only enhanced final germination but also lowered the mean germination time. Under field conditions, it increased germination and germination index values, which indicate rapid and synchronised germination. Coating with *Paenibacillus yonginensis* together with SiO_2 was found to increase germination and average plant length [24].

*2.1.4 Quinoa [*Chenopodium quinoa*]*

Quinoa, indigenous to the Andean region of South America, is an edible seed. The rate of quinoa seed germination increased by 116 and 132% after the application of *Bacillus licheniformis* QA1 and *Enterobacter asburiae* QF11, respectively [25]. These strains showed high phosphorus solubilisation activity; however, *E. asburiae* QF11 exhibited overproduction of auxin, and *B. licheniformis* QA1 produced ammonia and siderophores.

2.2 Oil seed crops

*2.2.1 Soybean [*Glycine max *L.]*

Soybean, an important oilseed crop, is grown worldwide. Soybean plants are naturally associated with arbuscular mycorrhizal fungi and *Bradyrhizobium japonicum* [48]. When *Bacillus velezensis* strain CMRP 4490 was applied to soybean seeds as a coating, it resulted in an increase in seed germination from 55.5 to 64% as compared to the control [26].

*2.2.2 Canola [*Brassica napus *L.]*

Canola has become one of the world's most important oilseed crops. With the aim to promote canola-seedling growth under field conditions, biopriming treatments with

Bacillus subtilis or *Macrophomina phaseolina*, or a combination of both, enhanced germination parameters of canola seeds, even under high salinity conditions [34]. Similarly, *B. subtilis* and *T. harzianum* have been reported to increase germination as well as root length and seedling vigour index when applied as a coating to canola seeds [35].

*2.2.3 Sunflower [*Helianthus annuus *L.]*

Besides soybean and canola, sunflower is another dynamic crop cultivated for vegetable oil. Application of different strains of *P. fluorescens* as coating agents resulted in improvements in seed germination and the vigour index of sunflower seedlings over the untreated control [27]. Den [15] bio-primed seeds of different hybrid cultivars of sunflower with selected endophytic bacterial species: *Enterobacter* sp. (FD-17), *Burkholderia phytofirmans* (PsJN) and *Bacillus* sp. (KS-54). In cultivar FH620, 100% germination and the maximum percentage of mean germination time were observed in seeds treated with bacterial agents. Seeds treated with *Enterobacter* (FD-17) exhibited higher vigour indices compared to unprimed seeds. *Enterobacter* (FD-17) gave the maximum germination percentage and mean germination time in cultivar FH615, followed by *Bacillus* sp. (KS-54).

In seed cultivar FH620, treatment with *Enterobacter* (FD-17) also showed a high germination percentage in comparison to other bacterial strains and unprimed seeds as well. FH620 showed an optimum germination rate of 40% after biopriming with *Enterobacter* (FD-17) compared to 30% in untreated seeds. Thus, the endophyte *Enterobacter* (FD17) showed more effective results on the germination parameters of sunflowers.

2.3 Horticultural crops

*2.3.1 Tomato [*Solanum lycopersicum *L.]*

Tomato is a very important vegetable crop grown worldwide in open fields, greenhouses and net houses. Mastouri et al. [33] evaluated the effect of seed treatment with *Trichoderma* sp. to maintain good performance of seedlings under abiotic stress conditions. Treatment with the fungus *Trichoderma* sp. under osmotic, salt, or suboptimal temperature conditions resulted in faster and more uniform germination in comparison to control. A significantly higher germination rate (more than 48%) and a lower mean germination time of tomato seeds after coating with *T. harzianum* and *P. fluorescens,* either singly or in combination, has also been reported by Srivastava et al. [49]. It was further observed that combinations of inoculants were more effective than single-isolate treatments. A relative germination increases of 111% compared to the control was observed in tomato seeds with the application of *Pseudomonas* sp. S3 [14].

*2.3.2 Carrot [*Daucus carota *L.]*

Carrot seed germination is poor ranging from 50 to 85%, causing a huge economic loss to the farmers. Thus, increasing the germination rate can lead to higher yield and lesser seed loss [50]. Beneficial microorganisms can be successfully applied to carrot seeds during priming [51]. The seeds are hydrated through the controlled addition of water which helps in the start of the physiological process of germination before planting. It ensures that the entire batch of seeds is at the same stage of the process of germination resulting in rapid and uniform emergence of planted seed than unprimed seed [52]. The beneficial microorganisms can be added to the water used to hydrate the seed during drum priming which proliferate to high numbers on the seed surface [53]. An experiment was conducted in greenhouse using carrot seeds bio-primed by drum bio-priming with selected beneficial microorganisms such as *Clonostachys rosea* IK726, *Pseudomonas chlororaphis* MA342, *P. fluorescens* CHA0, *T. harzianum* T22 and *T. viride* S17a and improved emergence in carrot seed was observed. Using *C. rosea* IK726 resulted in additional improvement in emergence time in carrots [54]. Likewise, the ability of 10 selected microbial isolates to promote the germination of carrot seed was investigated by Fiodor et al. [28] who reported variation in seed germination among tested bacteria. The seed germination improved by 156.88 ± 2.35% compared to the control with the application of *S. marcescens* AF8I1. In addition, carrot seeds inoculated with *P. fluorescens* AF8I4, *P. putida* AF1I1, *K. aerogenes* AF3II1 and *B. cereus* AF8II13 also showed increased relative seed germination.

*2.3.3 Onion [*Allium cepa *L.]*

Since seed bio-priming is known to improve seedling emergence, onion seeds were primed using microorganisms such as *Clonostachys rosea* IK726, *P. chlororaphis* MA342, *P. fluorescens* CHA0, *T. harzianum* T22 and *T. viride* S17a in a glasshouse experiment and an improvement in the emergence of onion seed was observed [54]. The treatment with *C. rosea* further reduced germination time but the results were not consistent and a negative effect on emergence was observed in another experiment which has been attributed to the proliferation of an unidentified indigenous microorganism during priming. However, the combined data of 3 years indicated a positive effect of bio-priming on germination of onion seeds.

*2.3.4 Cucumber [*Cucumis sativus*]*

Considering that PGPB improves yield, Perez-Garcia et al. [29] evaluated the effect of PGPB on the germination and phytocompound production in cucumber seedlings. Cucumber seed germination was significantly promoted by application of *Acinetobacter radioresistens* (KBENdo3P1), *Pseudomonas paralactis* (KBENdo6P7) and *Bacillus cereus* (KBENdo4P6). It has also been reported that the use of *T. harzianum* as a coating agent results in higher seedling emergence compared to untreated cucumber seeds [55].

*2.3.5 Pumpkin [*Cucurbita pepo*]*

In pumpkin, it has also been established that when the seeds of pumpkin were inoculated with a commercial formulation of *T. harzianum*, the germination and growth of pumpkin seedlings under saline stress conditions improved [56]. Thus, seed treatment with *T. harzianum* leads to improvements in biotic, abiotic and physiological stresses in germinating seeds and seedlings.

*2.3.6 Radish [*Raphanus sativus *L.]*

The effect of bio-priming radish seeds with five bacterial strains, *Agrobacterium rubi* A 16, *Burkholderia gladii* BA 7, *Pseudomonas putida* BA 8, *Bacillus subtilis* BA 142 and *Bacillus megaterium* M 3 on the germination under different saline (NaCl) conditions was examined using three radish cultivars, 'Antep', 'Beyaz' and 'Siyah' [57]. Bio-priming with bacteria significantly improved the percentage of seed germination under saline conditions. Germination percentage was observed to be higher in bio-priming with *Burkholderia gladii* BA 7, *P. putida* BA 8, *B. subtilis* BA 142 and *A. rubi* A 16 in comparison to control and *B. megaterium* M 3. The seeds of cultivars 'Beyaz' and 'Siyah' did not germinate in the control and treatment with *B. megaterium* M 3 but there was a significant increase [$P < 0.01$] in germination percentage in other bio-priming treatments in the presence of 20 g l^{-1}. NaCl which was found to be statistically significant ($P < 0.01$). However, the effect of bio-priming with PGPR on percent germination varied with bacterial strain and cultivar.

*2.3.7 Cumin [*Cuminum cyminum *L.]*

Cumin, one of the most important spices and medicinal plants, is grown in many countries, such as Iran, Egypt, India and several other Asian countries. One of the problems with cultivation of cumin is its small seed, weak vigour and thus poor emergence. Because of the difficulty faced in the cultivation of cumin, the pelleting technique is used to reduce variation in size and increase the accuracy of sowing. An experiment consisting of six combined levels of treatments, such as seed inoculation with strain P.F2, CHA0 *P. fluorescence* bacteria, strain T36, T39 of *T. harzianum* fungi, coating and drought stress was conducted [58]. The highest seedling emergence percentage was obtained by inoculating coated seeds with the CHA0 strain of *P. flourescence*. Further, biopriming the coated seed with T36 fungi or CHA0 bacteria resulted in an improved seed germination and seedling establishment under stress conditions.

2.4 Wild plants

2.4.1 Flowers

Seed treatment with native microbes significantly increases the rate of seed germination, plant biomass, seed yield and resistance to abiotic stress in agricultural crops [33, 40]. Barrera et al. [18] reported that treatment with native soil microbial wash improved the germination of two native wildflower species, Mexican Hat (*Ratibida columnifera* Nutt.) and Cowpens Daisy (*Verbesina encelioides* Cav.) used for the restoration of habitat in semi-arid subtropical climate. Germination of *R. columnifera* increased significantly higher than that of *V. encelioides* ($P < 0.0001$). The germination percentage of *R. columnifera* was 75%, while that of *V. encelioides* was 30% only. In addition, *R. columnifera* seeds germinated faster in comparison to *V. encelioides.* While the seeds of *R. columnifera* seeds started germinating on 4th day of sowing, the seeds of *V. encelioides* seeds germinated only on 6th day.

2.4.2 Cryptocarya densiflora L.

Cryptocarya densiflora, a member of the Lauraceae family is a woody plant of high economic value. It contains more than 350 varieties that are extensively distributed throughout the globe. Almost all parts of this plant can be utilised for building materials or as a source of substances of medicinal value [59, 60]. For the cultivation of *C. densiflora* plants, seedlings are collected from around the mother plant because the seed coat of this plant is very hard which prevents its commercial propagation. Alternatively, the seeds are soaked in warm water or treated with acid to soften the hard seed coat which accelerates the germination process [61]. Sugiharto et al. [19] evaluated the effects of endophytic *Aspergillus niger* and microbial consortia on the germination and growth of *C. densiflora* seedlings under greenhouse conditions. The results obtained indicate that the inoculation of seeds of *C. densiflora* with *A. niger* and consortia formula affected the germination ability of *C. densiflora* seeds which was found to be different in sterile and non-sterile media. The seed germination was induced varying from 33.3% in sterile medium and 25% in non-sterile medium.

3. Role of seed microbiome in germination

The seed microbiome consists of microbes colonising internal seed tissues, termed seed endophytes, and microbes present on seed surfaces called seed epiphytes. The two groups differ in that endophytes emerge from internal seed tissues without being affected by the external environment, whereas epiphytes originate exclusively from the surface of the seed or surrounding environment. This division, however, seems arbitrary because epiphytes may become endophytes and vice versa [62]. Seeds do harbour microorganisms, both endophytes as well as epiphytes, but only a small fraction of microbes colonise the seeds in comparison to other plant parts [63, 64]. The fact is that seed microbiome is an important determinant and driver of seed health and seed germination and thus the quality of seed [65, 66]. There are many reports on the promotion of seed germination by the seed microbiome, particularly under unfavourable conditions, such as those found at high altitudes [67–69]. Beneficial microbes inhabiting seeds have been found to play a profound role in mediating seed dormancy and preparing a suitable environment for seed germination by managing hormone levels [69–72]. Thus, microorganisms help in lessening stressful conditions by moderating seed hormone levels of gibberellic acid and cytokinin [65], assisting in seed germination.

Although mutualistic associations of plants with microbes have been reported to enhance their fitness [73, 74], few studies have assessed the role and potential application of the seed microbiome in seed germination and seedling growth. In a study with tall fescue seeds, those harvested before attaining physiological maturity were found to inhabit a lower number of endophytic fungi. These seeds with low infection were later found to germinate poorly, affecting seedling vigour [75], indicating the role of endophytes in seed germination.

Endophytes transmitted *via* seeds of *Phragmites australis* can increase seed germination and seedling growth of invasive Phragmites [76]. *Phragmites australis* ssp. *australis* (Commonly known as Reed), is an invasive wetland grass of European origin that has spread throughout North America. The effect of fungal endophyte inoculations on plant performance as a potential control method for invasive Phragmites has been investigated. It was hypothesised that different endophytes, which were isolated from invasive Phragmites, would have variable effects on Phragmites growth. Fungal endophytes of seeds of invasive, non-native Phragmites were isolated to determine if fungal symbiosis could contribute to invasiveness through their effects on seed germination and seedling growth. One-third of surface sterilised seeds yielded endophyte isolates belonging to taxa like *Alternaria* sp. representing 54%, *Phoma* sp. (21%) and *Penicillium corylophilum* (12%). Overall germination of seeds harbouring an isolate was significantly higher (36%) than seeds not harbouring an endophyte (20%). *Penicillium* has been reported as strongly associated with increased germination of seeds.

Enterobacter cloacae, a well-known plant endophyte generally found in seeds enhanced seed germination and seedling growth [77]. The gastrointestinal tract of birds harbours many facultative anaerobes [78], which are known to be associated with plants, enhance plant health [79, 80] and protect germinating seeds and seedlings from fungal pathogen infections [81]. These bacteria could colonise seeds during gut passage and probably are responsible for enhanced seed germination [82] possibly due to the breaking of dormancy or the removal of germination inhibitors.

Adverse effects of the seed microbiome on seed germination have also been observed, such as in the seeds of the invasive grass *Lolium rigidum* in the wheat fields of the Western Australian grain belt. *Lolium rigidum* shows poor germination due to dormancy during the crop-growing season, rendering it difficult to remove the invasive grass. This dormancy has been attributed to abscisic acid and cytokinin levels mediated by endogenous seed endophytes [83].

These studies do indicate that endophytes play an important role in enhancing seed germination. The role of epiphytic microorganisms has rarely been discussed in seed microbiome studies. Whether the seed microbes that amplify during germination and are transmitted to seedlings are derived from epiphytic or endophytic microbiomes is not well understood and may be a subject for future research.

4. Adverse effects of microorganisms on seed germination

The effect of microorganisms on seed germination varies with the microbial strain applied and the prevailing environmental conditions. Only certain strains are able to promote seed germination. For this reason, quite a few numbers of attempts on the application of microorganisms on seed to improve germination have failed. *Cuscuta campestris* seeds when treated with *Bacillus* sp. had no significant effect on seed germination in comparison to the control [84]. Soybean plants are naturally found to be associated with arbuscular mycorrhizal fungi and rhizobia [48]. When assessed for the effect of soybean seed coating with different isolates of *Trichoderma virens* on the germination of seeds and development of soybean seedlings, five of the seven strains tested inhibited germination, and none of the strains showed improvement in the final germination percentage compared to uncoated seeds [85].

There are reports on the inhibitory effect of the treatment of seeds with microorganisms on germination rate and plant height in several crops. In one of the studies, lettuce seeds were coated with a consortium of *Trichoderma* spp., *Beauveria bassiana*, *Metarhizium anisopliae* and arbuscular mycorrhizal fungi resulting in the reduction of germination [86]. Similar results have been reported by coating sweet pepper seeds with *T. viride*, *T. polysporhum*, *T. stromaticum*, *B. bassiana*, *M. anisopliae* and arbuscular mycorrhizal fungi [40]. *Trichoderma* spp., are known to produce and secrete a wide range of extracellular hydrolytic enzymes capable of degrading plant cell walls, attacking the seed tegument and damaging it, thus leading to a reduction in seed germination and growth of seedlings [87, 88].

The inhibition of germination of seeds of barley and wheat by fungi has been reported and attributed to the affinity of microorganisms for oxygen which compete for the same with the seed [89, 90]. Inhibition of seed germination after inoculation with *A. chroococcum* grown on a medium containing nitrate was inhibited [91]. Whereas the cells, but not the culture filtrate, inhibited germination which appears to be due to competition between viable bacteria and seed for available oxygen. The cultures grown on the nitrate-containing medium were inhibitory only when the inoculum contained much higher numbers of viable cells compared to inoculum cultured on nitrogen-free medium. These studies suggest a limit on the number of viable cells that can be inoculated on each seed to prevent any adverse effect on its germination.

5. Mechanism of action of microorganisms

How the inoculation of microorganisms induces germination is not well understood. Microbial isolates that exhibited a positive effect on seed germination have been characterised for their ability to solubilise phosphates, produce various metabolites like siderophores and hydrocyanic acid, synthesise antibiotics, enzymes, and phytohormones such as auxin, cytokinin and gibberellic acid [92–94]. Most of these isolates showed only one of the investigated properties. Findings suggest that there is no relationship between the presence of biocontrol abilities, iron chelation by siderophores or antibiotic production with seed germination. The modes of action of microorganisms in the germination of seeds vary with the microorganisms. Modes that have been reported for different microorganisms are the production of auxins, metabolites and degradative enzymes.

5.1 Auxins

Soaking of seeds in preparations of IAA-producing bacteria has been reported to influence seed germination positively. Improvement in germination in maize seeds soaked in a bacterial suspension of the IAA producer *Mixta theicola* SAR by up to 27% has been reported by Hagaggi et al. [22]. Carrot seed germination is stimulated by IAA production by microorganisms [28]. Microorganisms vary in their ability to produce IAA. While *S. marcescens* AF8I1 showed high IAA production, *B. cereus* AF8II13 did not synthesise this hormone. Some other isolates like *S. plymuthica* EDC15, *B. ambifaria* AF8II10 and *S. plymuthica* EEPC5 produced IAA ranging from 15.23 to 51.27 μg mL^{-1} and showed a positive effect on seed germination. The increase in germination of carrot seeds by application of these microbes was 8.22, 10.22 and 10.32%, respectively.

Yeast isolates, *Meyerozyma guilliermondii* which increased wheat seed germination from 40 to 97% was found to produce indole-3-acetic acid [32]. Significantly improved wheat seed germination by application of IAA-producing strains of *Bacillus pumilus* MA9 and *Virgibacillus halodenitrificans* MA14 has also been reported by Brahim et al. [43]. *Pseudomonas* sp. S3 strain which induced seed germination up to 111% in tomatoes was positive for the production of IAA, ACC deaminase, and siderophores and solubilised phosphate and zinc [14]. Soil microbes are known to improve seed germination rate in maize also *via* the production of plant growth hormones [95]. However, isolates of *Bacillus* spp. which produced IAA when applied to maize seeds did not show any effect on germination rate as compared to the control [96]. Although indole acetic acid-producing microorganisms can promote seed germination but it is not a universal fact. These results indicate a great potential of some of these isolates as plant growth-promoting rhizobacteria and as biostimulants for seed biopriming.

5.2 Metabolic products

Metabolic products of *A. chroococcum* appear to be involved in the stimulation of seed germination. Although the presence of cells of *A. chroococcum* increased seed germination, the stimulatory effect of metabolites produced during culturing of the inoculum without the necessity for colonisation of microorganisms on their seedling indicates the role of microbial metabolites. However, the stimulatory effect was not consistent and failed sometimes demanding further investigations. The application of *Bacillus pumilus* MA9 and *Virgibacillus halodenitrificans* MA14 significantly improved wheat germination rate which has been assigned to the solubilisation of inorganic phosphate, and ACC deaminase activity [43]. The isolate *B. cereus* AF8II13 which possesses the ability to solubilise phosphate has been found to exert a positive effect on carrot seed germination in the absence of IAA synthesis [28].

5.3 Enzyme production

The hydrolytic enzymes produced by microorganisms have been reported to be helpful in the germination of hard-coated seeds, such as that of *C. densiflora*. The seed coat of *C. densiflora* is hard and composed of lignin and cellulose. Bilkay et al. [97] found that *A. niger*, a known endophyte, helps in softening the skin of seeds of *C. densiflora*, thus breaking its dormancy. The endophyte *A. niger* produces ligninolytic and cellulase enzymes that make the seed skin soft [98, 99] and are responsible for the induction of seed germination in *C. densiflora*.

6. Conclusion

The primary objective of the agricultural industry is improvement in seed quality. To enhance seed quality, several different types of available technologies are being used. One such technology is the use of beneficial microorganisms for seed treatment in various crops involving direct seed sowing. Since quickly emerging healthy seedlings can immediately utilise available resources, tolerate biotic and abiotic stresses and other adverse environmental conditions, rapid seed germination and seedling vigour contribute to successful crop establishment and performance. The role of beneficial microbes like arbuscular mycorrhizal fungi, *Trichoderma* spp., rhizobia and other bacteria in seed germination and the emergence of seedlings have been evaluated by numerous workers and have been found to exert a positive effect. Application of beneficial microorganisms directly to the seed helps in the reduction of the amount of microbial inoculum per plant, saves labour (economic advantage) and ensures early contact between microbes and rootlets (agronomic benefit). The germination of crops like wheat, maize, rice, soybean, canola, sunflower, tomato, carrot, cucumber, pumpkin, etc. increased significantly by the application of beneficial microorganisms. Although there are many reports on the effect of microorganisms on the growth and yield of different crops, investigations on the influence of microorganisms on seed germination are scant and need to be strengthened.

Acknowledgements

The moral support and platform provided by the Chairman, Gaurav Bharati Siksha Sansthan, Dehradun, and The President and Vice-chancellor, Sardar Bhagwan Singh University, Dehradun is highly acknowledged. The author also acknowledges the authors of the research papers used in this compilation.

Author details

Faquir C. Garg
Sardar Bhagwan Singh University, Balawala, Dehradun, Uttarakhand, India

*Address all correspondence to: fcgarg1@rediffmail.com

References

[1] Begum N, Hasanuzzaman M, Li Y, Akhtar K, Zhang C, Zhao T. Seed germination behaviour, growth, physiology and antioxidant metabolism of four contrasting cultivars under combined drought and salinity in soybean. Antioxidants. 2022;**11**:1-23. DOI: 10.3390/antiox11030498

[2] Damalas CA, Koutroubas SD, Fotiadis S. Hydro-priming effects on seed germination and field performance of faba bean in spring sowing. Agriculture. 2019;**9**:201. DOI: 10.3390/agriculture9090201

[3] Osburn RM, Schroth M. Effect of osmopriming sugar beet seed on germination rate and incidence of *Pythium ultimum* damping-off. Plant Disease. 1989;**73**:21-24. DOI: 10.1094/PD-73-0021

[4] Markets and markets- biological seed treatment market by type (microbials and botanicals), crop (corn, wheat, soybean, cotton, sunflower, and vegetable crops), function (seed protection and seed enhancement), and region—global forecast to 2025. Available from: https://www.marketsandmarkets.com/Market-Reports/biological-seed-treatment-market-162422288.html [Accessed: November 2, 2021]

[5] Rouphael Y, Cardarelli M, Bonini P, Colla G. Synergistic action of a microbial-based biostimulant and a plant derived-protein hydrolysate enhances lettuce tolerance to alkalinity and salinity. Frontiers in Plant Science. 2017;**8**:131. DOI: 10.3389/fpls.2017.00131

[6] Amirkhani M, Mayton HS, Netravali AN, Taylor AG. A seed coating delivery system for bio-based biostimulants to enhance plant growth. Sustainability. 2019;**11**:5304. DOI: 10.3390/su11195304

[7] Qiu YQ, Amirkhani M, Mayton H, Chen Z, Taylor AG. Biostimulant seed coating treatments to improve cover crop germination and seedling growth. Agronomy. 2020;**10**:154. DOI: 10.3390/agronomy10020154

[8] Halmer P. Commercial seed treatment technology. In: Black M, Bewley JD, editors. Seed Technology and its Biological Basis. Sheffield. Sheffield, UK: Academic Press; 2000. pp. 257-286

[9] Halmer P. Seed technology and seed enhancement. Acta Horticulturae. 2008;**771**:17-26. DOI: 10.17660/ActaHortic.2008.771.1

[10] Zaheer A, Munir AS, Amjad H, ud Din S. Investigation of antioxidant enzymes and biochemical changes in the wheat seeds (freed) induced by different pre-sowing treatments. World Applied Sciences Journal. 2012;**1**:31-36

[11] Rao MSL, Kulkarni SI, Lingaraju SI, Nadaf HL. Bio priming of seeds: A potential tool in the integrated management of *Alternaria* blight of sunflower. Helia. 2009;**50**:107-114

[12] McDonald MB. Seed priming. In: Seed Technology and its Biological Basis. Sheffield: Sheffield Academic Press; 2000. pp. 287-325

[13] Warwate SI, Kandoliya UK, Bhadja NV, Golakiya BA. The effect of seed priming with plant growth promoting rhizobacteria (PGPR) on growth of coriander (*Coriandrum sativum* L.) seedling. International Journal of Current Microbiology and Applied Sciences. 2017;**6**:1926-1934

[14] Pandey S, Gupta S. Evaluation of *Pseudomonas* sp. for its multifarious plant growth promoting potential and its ability to alleviate biotic and abiotic stress in tomato (*Solanum lycopersicum*) plants. Scientific Reports. 2020;**10**:20951. DOI: 10.1038/s41598-020-77850-0

[15] Den NZU, Bukhari SA, Iftikhar T, Mustafa G. Biochemical and phenolic acid profiling of sunflower hybrid varieties' seeds treated with different bio-priming agents. Pakistan Journal of Botany. 2021;**53**:981-989. DOI: 10.30848/PJB2021-3(4)

[16] Zia R, Nawaz MS, Yousaf S, Amin I, Hakim S, Mirza MS, et al. Seed inoculation of desert-plant growth-promoting rhizobacteria induce biochemical alterations and develop resistance against water stress in wheat. Physiologia Plantarum. 2021;**172**:990-1006. DOI: 10.1111/ppl.13362

[17] Cardarelli M, Woo SC, Rouphael Y, Colla G. Seed treatment with microorganisms can have a biostimulant effect by influencing germination and seedling growth of crops. Plants (Basel). 2022;**11**(3):259. DOI: 10.3390/plants11030259

[18] Barrera D, Luera J, Lavallee K, Soti P. Ecological Processes. 2021;**10**:19-26. Article Number 19

[19] Sugiharto A, Latifah D, Rakhmawati SU, Subouo YB, Harto. Study of the effect of microbial addition in seed germination and seedling growth of *Cryptocarya densiflora* L. In: Second International Conference on Tropical Wetland Biodiversity and Conservation. IOP Conf. Series: Earth and Environmental Science. Vol. 976. Bristol, England: IOP Publishing Ltd; 2022. p. 012055

[20] Zahir ZA, Arshad M, Frankenberger WT. Plant growth promoting rhizobacteria: Applications and perspectives in agriculture. Advances in Agronomy. 2004;**81**:98-169

[21] Noumavo PA, Kochoni E, Didagbé YO, Adjanohoun A, Allagbé M, Sikirou R, et al. Effect of different plant growth promoting *Rhizobacteria* on maize seed germination and seedling development. American Journal of Plant Sciences. 2013;**4**:1013-1021. DOI: 10.4236/ajps.2013.45125

[22] Hagaggi NSA, Mohamed AAA. Enhancement of *Zea mays* (L.) growth performance using indole acetic acid producing endophyte *Mixta theicola* isolated from *Solenostemma argel* (Hayne). South African Journal of Botany. 2020;**134**:64-67. DOI: 10.1016/j.sajb.2020.02.034

[23] Javed T, Afzal I, Mauro RP. Seed coating in direct seeded rice: An innovative and sustainable approach to enhance grain yield and weed management under submerged conditions. Sustainability. 2021;**13**:2190. DOI: 10.3390/su13042190

[24] Choi ES, Sukweenadhi J, Kim YJ, Jung KH, Koh SC, Hoang VA, et al. The effects of rice seed dressing with *Paenibacillus yonginensis* and silicon on crop development on South Korea's reclaimed tidal land. Field Crops Research. 2016;**188**:121-132. DOI: 10.1016/j.fcr.2016.01.005

[25] Mahdi I, Fahsi N, Hafidi M, Allaoui A, Biskri L. Plant growth enhancement using rhizospheric halotolerant phosphate solubilizing bacterium *Bacillus licheniformis* QA1 and *Enterobacter asburiae* QFf11 isolated from *Chenopodium quinoa* Willd. Microorganisms. 2020;**8**:1-21. DOI: 10.3390/microorganisms.8060948

[26] Teixeira GM, Mosela M, Abreu Nicoletto ML, Ribeiro RA, Hungria M,

Youssef K, et al. Genomic insights into the antifungal activity and plant growth-promoting ability in *Bacillus velezensis* CMRP 4490. Frontiers in Microbiology. 2021;**11**:618415. DOI: 10.3389/fmicb.2020.618415

[27] Moeinzadeh A, Sharif-Zadeh F, Ahmadzadeh M, Tajabadi FH. Biopriming of sunflower (*Helianthus annuus* L.) seed with *Pseudomonas fluorescens* for improvement of seed invigoration and seedling growth. Australian Journal of Crop Science. 2010;**4**:564-570

[28] Fiodor A, Ajijah N, Dziewit L, Kumar P. Biopriming of seed with plant growth-promoting bacteria for improved germination and seedling growth. Frontiers in Microbiology. 2023;**14**:1142966. DOI: 10.3389/fmicb.2023.1142966

[29] Perez-Garcia L, Saenz-Mata J, Fortis-Hernandez M, Navarro-Munoz CE, Palacio-Rodriguez R, Preciado-Ragel P. Plant-growth-promoting rhizobacteria improve germination and bioactive compounds in cucumber seedlings. Agronomy. 2023;**13**(2):315

[30] Seifikalhor M, Hassani SB, Aliniaeifard S. Seed priming by cyanobacteria (*Spirulina platensis*) and salep gum enhances tolerance of maize plant against cadmium toxicity. Journal of Plant Growth Regulation. 2020;**39**:1009-1021. DOI: 10.1007/s00344-019-10038

[31] Kthiri Z, Ben Jabeur M, Machraoui M, Gargouri S, Hiba K, Hamada W. Coating seeds with *Trichoderma* strains promotes plant growth and enhance the systemic resistance against *Fusarium* crown rot in durum wheat. Egyptian Journal of Biological Pest Control. 2020;**30**:139. DOI: 10.1186/s41938-020-00338-6

[32] Kthiri Z, Ben Jabeur M, Chairi F, López-Cristoffanini C, López-Carbonell M, Serret MD, et al. Exploring the potential of *Meyerozyma guilliermondii* on physiological performances and defense response against *Fusarium* crown rot on durum wheat. Pathogens. 2021;**10**:52. DOI: 10.3390/pathogens10010052

[33] Mastouri F, Björkman T, Harman GE. Seed treatment with *Trichoderma harzianum* alleviates biotic, abiotic, and physiological stresses in germinating seeds and seedlings. Phytopathology. 2010;**100**:1213-1221. DOI: 10.1094/PHYTO-03-10-0091

[34] Mousavi M, Omidi H. Seed priming with bio-priming improves stand establishment, seed germination and salinity tolerance in canola cultivar (Hayola 401). Iranian Journal of Plant Physiology. 2019;**9**:2807-2817

[35] Somagh HA, Mousavi SM, Omidi H, Mohammadian E, Hemmati M. Canola seed germination and seedling growth in response to saline condition and bio-priming. Iranian Journal of Plant Physiology. 2017;**7**:2149-2156

[36] Sharma V, Prasanna R, Hossain F, Muthusamy V, Nain L, Das S, et al. Priming maize seeds with cyanobacteria enhances seed vigour and plant growth in elite maize inbreds. Biotech. 2020;**10**:154. DOI: 10.1007/s13205-020-2141-6

[37] Philippot L, Raaijmakers JM, Lemanceau P, van der Putten WH. Going back to the roots: The microbial ecology of the rhizosphere. Nature Reviews. Microbiology. 2013;**11**:789-799. DOI: 10.1038/nrmicro3109

[38] Oliveira RS, Rocha I, Ma Y, Vosátka M, Freitas H. Seed coating with arbuscular mycorrhizal fungi as

an ecotechnological approach for sustainable agricultural production of common wheat (*Triticum aestivum* L.). Journal of Toxicology & Environmental Health Part A: Current Issues. 2016;**79**:329-337. DOI: 10.1080/15287394.2016.1153448

[39] Paparella S, Araujo SS, Rossi G, Wijayasinghe M, Carbonera D, Balestrazzi A. Seed priming: State of the art and new perspectives. Plant Cell Reports. 2015;**34**:1281-1293. DOI: 10.1007/s00299-015-1784-y

[40] Rocha I, Ma Y, Souza-Alonso P, Vosátka M, Freitas H, Oliveira RS. Seed coating: A tool for delivering beneficial microbes to agricultural crops. Frontiers in Plant Science. 2019;**10**:1357. DOI: 10.3389/fpls.2019.01357

[41] Colla G, Rouphael Y, Bonini P, Cardarelli M. Coating seeds with endophytic fungi enhances growth, nutrient uptake, yield and grain quality of winter wheat. International Journal of Plant Production. 2015;**9**:171-190

[42] Lastochkina O, Garshina D, Ivanov S, Yuldashev R, Khafizova R, Allagulova C, et al. Seed priming with endophytic *Bacillus subtilis* modulates physiological responses of two different *Triticum aestivum* L. cultivars under drought stress. Plants. 2020;**9**:1810. DOI: 10.3390/plants9121810

[43] Brahim AH, Ali MB, Daoud L, Jlidi M, Akremi I, Hmani H, et al. Biopriming of durum wheat seeds with endophytic diazotrophic bacteria enhances tolerance to *Fusarium* head blight and salinity. Microorganisms. 2022;**10**:970. DOI: 10.3390/microorganisms10050970

[44] Wu F, Guclu H. Global maize trade and food security: Implications from a social network model. Risk Analysis. 2013;**33**:2168-2178. DOI: 10.1111/risa.12064

[45] Okoth SA, Otadoh JA, Ochanda JO. Improved seedling emergence and growth of maize and beans by *Trichoderma harziunum*. Tropical and Subtropical Agroecosystems. 2011;**13**:65-67

[46] Gholami A, Shahsavani S, Nezarat S. The effect of plant growth promoting rhizobacteria (PGPR) on germination, seedling growth and yield of maize. World Academy of Science, Engineering and Technology. 2009;**37**:19-24

[47] Rudolph N, Labuschagne N, Aveling TAS. The effect of plant growth promoting rhizobacteria on seed germination and seedling growth of maize. Seed Science and Technology. 2015;**43**:507-518. DOI: 10.15258/sst.2015.43.3.04

[48] Pagano C, Miransari M. Abiotic and Biotic Stresses in Soybean Production. The Importance of Soybean Production Worldwide. Cambridge, MA, USA: Academic Press; 2016. pp. 1-26

[49] Srivastava R, Khalid A, Singh US, Sharma AK. Evaluation of arbuscular mycorrhizal fungus, fluorescent *Pseudomonas* and *Trichoderma harzianum* formulation against *Fusarium oxysporum* f. sp. *Lycopersici* for the management of tomato wilt. Biological Control. 2010;**53**:24-31. DOI: 10.1016/j.biocontrol.2009.11.012

[50] Makhaye G, Aremu AO, Gerrano AS, Tesfay S, Du Plooy CP, Amoo SO. Biopriming with seaweed extract and microbial-based commercial biostimulants influences seed germination of five *Abelmoschus esculentus* genotypes. Planning Theory. 2021;**10**:1327. DOI: 10.3390/plants10071327

[51] Bennett AJ, Whipps JM. Dual application of beneficial microorganisms to seed during drum priming. Applied Soil Ecology. 2008;**38**:83-89. DOI: DOI.org/10.1016/j.apsoil.2007.08.001

[52] Rowse H. Drum priming: A non-osmotic method of priming seeds. Seed Science and Technology. 1996;**24**:281-294

[53] Wright B, Rowse H, Whipps JM. Microbial population dynamics on seeds during drum and steeping priming. Plant and Soil. 2003;**255**:631-640. Available from: https://www.jstor.org/stable/24123983

[54] Bennett AJ, Mead A, Whipps JM. Performance of carrot and onion seed primed with beneficial microorganisms in glasshouse and field trials. Biological Control. 2009;**51**:417-426. DOI: 10.1016/j.biocontrol.2009.08.001

[55] Pill WG, Collins CM, Goldberger B, Gregory N. Responses of non-primed or primed seeds of 'Marketmore 76' cucumber (*Cucumis sativus* L.) slurry coated with *Trichoderma* species to planting in growth media infested with *Pythium aphanidermatum*. Scientia Horticulturae. 2009;**121**:54-62. DOI: 10.1016/j.scienta.2009.01.004

[56] Yildirim E, Taylor AG, Spittler TD. Ameliorative effects of biological treatments on growth of squash plants under salt stress. Horticulture Technology. 2006;**111**:1-6

[57] Kaymak HC, Güvenc İ, Yarali F, Dönmez MF. The effects of bio-priming with PGPR on germination of radish (*Raphanus sativus* L.) seeds under saline conditions. Turkish Journal of Agriculture and Forestry. 2009;**33**:173-179. DOI: 10.3906/tar-0806-30

[58] Piri R, Moradi A, Balouchi H, Salehi A. Improvement of cumin (*Cuminum cyminum*) seed performance under drought stress by seed coating and biopriming. Scientia Horticulturae. 2019;**257**:108667. DOI: 10.1016/j.scienta.2019.108667

[59] Aislabie J, Deslippe JR. Soil microbes and their contribution to soil services. In: Dymond JR, editor. Ecosystem Services in New Zealand. Lincoln, New Zealand: Manaaki Whenua Press; 2013. pp. 143-161

[60] Braga RM, Dourado MN, Araújo WL. Microbial interactions: Ecology in a molecular perspective. Brazilian Journal of Microbiology. 2016;**47**(Suppl. 1):86-98

[61] Sharma KK, Singh US, Sharma P, Kumar A, Sharma L. Seed treatment for sustainable agriculture – A review. Journal of Applied and Natural Science. 2015;7:521-539

[62] Barret M, Briand M, Bonneau S, Préveaux A, Valière S, Bouchez O, et al. Emergence shapes the structure of the seed microbiota. Applied and Environmental Microbiology. 2015;**81**(4):1257-1266

[63] Nelson EB. The seed microbiome: Origins, interactions, and impacts. Plant and Soil. 2018;**422**(1):7-34

[64] Tarquinio F, Attlan O, Vanderklift MA, Berry O, Bissett A. Distinct endophytic bacterial communities inhabiting seagrass seeds. Frontiers in Microbiology. 2021;**12**:703014

[65] Chee-Sanford JC, Williams MM, Davis AS, Sims GK. Do microorganisms influence seed-bank dynamics? Weed Science. 2006;**54**(3):575-587

[66] Rodrí-guez CE, Antonielli L, Mitter B, Trognitz F, Sessitsch A. Heritability and functional importance of the *Setaria viridis* bacterial seed

microbiome. Phytobiomes Journal. 2020;**4**(1):40-52

[67] Rodríguez RJ, Mitter B, Sessitsch A, Compant S. Commentary: Seed bacterial inhabitants and their roots of colonization. Plant and Soil. 2018;**422**:1-6

[68] Wassermann B, Cernava T, Müller H, Berg C, Berg G. Seeds of native alpine plants host unique microbial communities embedded in cross-kingdom networks. Microbiome. 2019;**7**(1):1-12

[69] Dalling JW, Davis AS, Arnold AE, Sarmiento C, Zalamea PC. Extending plant defense theory to seeds. Annual Review of Ecology, Evolution, and Systematics. 2020;**51**:123-141

[70] Berg G, Koberl M, Rybakova D, Müller H, Grosch R, Smalla K. Plant microbial diversity is suggested as the key to future biocontrol and health trends. FEMS Microbiology Ecology. 2017;**93**(5): 1-9. DOI: 10.1093/femsec/fix050

[71] Sarmiento C, Zalamea PC, Dalling JW, Davis AS, Stump SM, U'Ren JM, et al. Soilborne fungi have host affinity and host-specific effects on seed germination and survival in a lowland tropical forest. PNAS. 2017;**114**(43):11458-11463

[72] Fuerst EP, James MS, Pollard AT, Okubara PA. Defense enzyme responses in dormant wild oat and wheat caryopses challenged with a seed decay pathogen. Frontiers in Plant Science. 2018;**8**:2259

[73] Compant S, Samad A, Faist H, Sessitsch AA. Review on the plant microbiome: Ecology, functions, and emerging trends in microbial application. Journal of Advanced Research. 2019;**19**:29-37

[74] Bashir I, War AF, Rafiq I, Reshi ZA, Rashid I, Shouche YS. Phyllosphere microbiome: Diversity and functions. Microbiological Research. 2022;**254**:12688

[75] Hill NS, Bouton JH, Hiatt EE, Kittle B. Seed maturity, germination, and endophyte relationships in tall fescue. Crop Science. 2005;**45**(3):859-863

[76] Shearin ZR, Filipek M, Desai R, Bickford WA, Kowalski KP, Clay K. Fungal endophytes from seeds of invasive, non-native *Phragmites australis* and their potential role in germination and seedling growth. Plant and Soil. 2018;**422**(1):183-194

[77] Santoyo G, Moreno-Hagelsieb G, Orozco-Mosqueda MD, Glick BR. Plant growth-promoting bacterial endophytes. Microbiological Research. 2016;**183**:92-99

[78] Waite DW. Taylor MW exploring the avian gut microbiota: Current trends and future directions. Frontiers in Microbiology. 2015;**6**:673-685. DOI: 10.3389/fmicb.2015.00673

[79] Mendes R, Garbeva P, Raaijmakers JM. The rhizosphere microbiome: Significance of plant beneficial, plant pathogenic, and human pathogenic microorganisms. FEMS Microbiology Reviews. 2013;**37**:634-663

[80] Berg G, Grube M, Schloter M, Smalla K. The plant microbiome and its importance for plant and human health. Frontiers in Microbiology. 2014;**5**:491-492. DOI: 10.3389/Fmicb.2014.00491

[81] Nelson EB. Microbial dynamics and interactions in the spermosphere. Annual Review of Phytopathology. 2004;**42**:271-309

[82] Traveset A. Effect of seed passage through vertebrate frugivores' guts on germination: A review. Perspectives in Plant Ecology. 1998;**1**:151-190

[83] Goggin DE, Emery RN, Kurepin LV, Powles SB. A potential role for endogenous microflora in dormancy release, cytokinin metabolism and the response to fluridone in *Lolium rigidum* seeds. Annals of Botany. 2015;**115**(2):293-301

[84] Saric-Krsmanovic M, Bozic D, Radivojevic L, Gajic-Umiljendic J, Santric L, Vrbnicanin S. Effects of plant growth promoting rhizobacteria (PGPR) and cover crops on seed germination and early establishment of field dodder (*Cuscuta campestris* Yunk.). Pesticidi i Fitomedicina. 2017;**32**:105-111. DOI: 10.2298/pif1702105s

[85] Yusnawan E, Inayati A, Baliadi Y. Effect of soybean seed treatment with *Trichoderma virens* on its growth and total phenolic content. AIP Conference Proceedings. 2019;**2120**:020003. DOI: 10.1063/1.5115604

[86] Diniz KA, Oliveira JA, Guimarães RM, De Carvalho MLM, Machado JDC. Incorporation of microorganism, amino acids, micronutrients and growth regulators in lettuce seed through the coating technique. Revista Brasileira de Sementes. 2006;**28**:37-43. DOI: 10.1590/S0101-31222006000300006

[87] Do Vale LHF, Gómez-Mendoza DP, Kim MS, Pandey A, Ricart CAO, Filho EX, et al. Secretome analysis of the fungus *Trichoderma harzianum* grown on cellulose. Proteomics. 2012;**12**:2716-2728. DOI: 10.1002/pmic.201200063

[88] Almeida DA, Crivelente Horta MA, Ferreira Filho JA, Faraj Murada N, Pereirade SA. The synergistic actions of hydrolytic genes reveal the mechanism of *Trichoderma harzianum* for cellulose degradation. Journal of Biotechnology. 2021;**334**:1-10. DOI: 10.1016/j.jbiotec.2021.05.001

[89] Griffin DM. Fungi attacking seeds in dry seed-beds. In: Proceedings of the Linnean Society of New South Wales. Vol. 91. Sydney: Gibbs, Shallard & Co.; 1966. pp. 84-89

[90] Lynch J, Pryn SJ. Interaction between a soil fungus and barley seed. Journal of General Microbiology. 1977;**103**:193-196

[91] Harper SHT, Lynch JM. Effects of *Azotobacter chroococcum* on barley seed germination and seedling development. Journal of General Microbiology. 1979;**112**:45-51

[92] Kazaz S, Erba S, Baydar H. Breaking seed dormancy in oil rose (*Rosa damascena* mill.) by microbial inoculation. African Journal of Biotechnology. 2010;**9**:6503-6508

[93] Perez-Garcia LA, Castañeda-Ramírez JC, Lafuente-Rincón D. Successful and unexpected potential of entomopathogen *Bacillus thuringiensis* as plant growth promoting rhizobacterium. Exitoso e inesperado potencial del entomopatógeno *Bacillus thuringiensis* Como rizobacteria promotora del crecimiento de las plantas. Journal of Natural and Agricultural Sciences. 2019;**6**:40-46

[94] Swarnalakshmi K, Yadav V, Tyagi D, Dhar DW, Kannepalli A, Kumar S. Significance of plant growth promoting rhizobacteria in grain legumes. Growth Promotion and Crop Production. Plants. 2020;**9**:1596

[95] Wu YN, Feng YL, Pare PW, Chen YL, Xu R, Wu S, et al. Beneficial soil microbe promote seed germination, plant growth and photosynthesis in herbal crop *Codonopsis pilosula*. Crop & Pasture Science. 2016;**67**(1):91-98

[96] Lwin KM, Myont MM, Tar T, Moe W. Isolation of plant hormone

(Indole-3-acetic acid - IAA) producing rhizobacteria and study on their effects on maize seedling. Engineering Journal. 2012;**16**(5):137-144. DOI: 10.4186/ej.2012.16.5.137

[97] Bilkay IS, Karakoç Ş, Aksöz N. Indole-3-acetic acid and gibberellic acid production in *Aspergillus Niger*. Turkish Journal of Biology. 2010;**34**:313-318. DOI: 10.3906/biy-0812-15

[98] Pingili M, Marla SR, Raparla R, Vanga S. Isolation and screening of lignocellulosic degrading fungi from degraded fruit litter. International Journal of Current Microbiology and Applied Sciences. 2017;**6**(12):2200-2206

[99] Alberto M, Vali EC, Ayala M, Luis J, Mallol F. Cellulolytic and ligninolytic potential of new strains of fungi for the conversion of fibrous substrates. Biotechnology Research and Innovation Journal. 2019;**3**(1):177-186

Chapter 9

Improvement of Barley Seed Performance and Seedling Growth by Plant Growth-Promoting Bacteria

Masoumeh Asadi Aghbolaghi, Ramin Piri, Bahman Fazeli-Nasab, Zahra Shakeri Chaleshtori, Beata Dedicova, Riyaz Sayyed and Andrea Mastinu

Abstract

Biostimulants are widely acknowledged as an effective method to alleviate drought-induced stress in various plant species, particularly during the crucial germination stage. The goal of this study was to investigate how plant growth-promoting bacteria (*Azospirillum lipoferum* and *Azotobacter chroococcum*) affect specific biochemical characteristics of barley under different levels of drought stress (0, -6, and -12 bar). The study employed a factorial design with three replications using a randomized complete block design. Our results show that the interaction effect of inoculation treatments and stress levels on emergence percentage, mean emergence time, seedling dry weight, and ascorbate peroxidase (APX) activity level is significant at a 1% probability level. Additionally, the interaction effect on the emergence index, proline content, and catalase (CAT) activity level is substantial at a 5% probability level. There are highest emergence percentage (89.33%), emergence index (4.08), and seedling dry weight (0.923 g) and the lowest emergence rate (6.47 days), proline content (0.153 mg/g fresh leaf tissue), and CAT activity (29 mg/g fresh leaf tissue) in *Azospirillum*-inoculated seeds. The highest APX activity (19.30 mg/g fresh leaf tissue) was obtained under non-stress conditions. Overall, inoculation with *Azospirillum* bacteria has an influential role in modulating the destructive effects of drought stress on the evaluated traits.

Keywords: barley, biochemical indices, bio-agents, early growth, osmotic potential

1. Introduction

Barley (*Hordeum vulgare* L.) is an annual plant from the Gramineae family, and its cultivation probably began in Ethiopia and Southeast Asia. It is a well-suited grain production area with favorable climatic conditions, fertile soils with a pH between 7 and 8, and high water-holding capacity. This species is more drought tolerant than wheat, so it can produce the highest yields in areas where water restricts grain production. Like other plants in the germination stage, barley is susceptible to drought stress, which can diminish growth and seedling characteristics in this crop [1, 2].

Drought stress is one of the most significant and everyday environmental stressors that has constrained agricultural production and the efficiency of using semiarid areas [3]. Drought stress reduces the germination index, seedling length, weight, and emergence uniformity, affecting seedling establishment [4–6]. Environmental factors, such as drought stress, can disturb the germination process by reducing the absorption of water by the seed, influencing the supply of seed storage resources, or interfering with the involvement of structural components and protein formation in the emerging embryo [7].

Applying some methods, such as seed bioremediation as a cost-effective, easy, and environmentally friendly treatment, can effectively counter drought stress's effects during germination and improve emergence index and seedling emergence [8]. As a substitute for chemical seed treatments, bio-priming or biological pretreatment uses advantageous microbes, and plant compatibility and stability can be improved by biological agents [8]. As a substitute for chemical seed treatments, bio-priming or biological pretreatment uses advantageous microbes, and plant compatibility and stability can be improved by biological agents [9].

Drought is the most hazardous environmental element that lowers agricultural productivity worldwide. Due to its severity and duration variations, drought may negatively impact agricultural output more than any other abiotic stress. By the end of the twenty-first century, droughts are anticipated to occur more frequently globally and in already dry places due to the continually changing climate, which tends to worsen. According to predictions, there will also likely be more than 8.6 billion people on the earth by 2030. To feed the world's population, agricultural output must increase globally by 70% by 2050 [10–13]. Food security would be a significant issue because more than 65% of people would only depend on agriculture [14]. As a result, it is crucial and essential to stabilize agricultural productivity and provide food security. To accomplish this aim, it is necessary to look into the processes behind crop drought tolerance [10–13].

Abiotic and biotic stressors are the two main categories of plant stress. Abiotic stressors that negatively impact crop and other plant growth, development, and production include salt, drought, water logging, mineral toxicity, and high temperatures. The genetic, molecular, biochemical, and physiological impacts of salt and drought conditions are highly comparable. Both of these affect nutrient availability, absorption, and transport in plants, which affects nutrient uptake [14].

Regional to global ranges of severe drought may impact terrestrial ecosystems, and the severity and length of the drought significantly impact plant production and ecosystem health. Understanding how plants adapt to drought is essential [15]. Plants have developed various complex and successful techniques to adjust to dry circumstances. Among them, maintaining the high water potential in tissues through osmotic adjustment and the generation of abscisic acid and dehydrins has given plants the ability to withstand drought [10].

Plant growth-promoting rhizobacteria (PGPR) can improve plant growth and drought stress tolerance by controlling physiological and biochemical traits [16]. A high-value food crop like barley would benefit from PGPR inoculation as it would encourage the development of stems, roots, and leaves and reduce salt and drought stress in the plant's numerous organs.

According to the definition, the improvement of seed quality refers to seed treatment using compounds and processes that improve characteristics, such as maximum physical purity and the ability and speed of germination, which help the seed to create plants that achieve quantitative yield. Moreover, they make the quality of the product possible; they help. Improving seed quality is considered one of the most important ways to achieve sustainable agricultural systems, and adding plant growth-promoting bacteria (PGPBs) to seed biofortification is one of the newest ways to improve seed quality. So now, various physiological and biological seed priming methods are gradually replacing chemical treatments [17–21]. Improvement of seed health due to the role of these bacteria as pest and plant pathogen inhibitors, as well as their growth-stimulating effect on seed germination and seedling structure through the production of growth-enhancing hormones in the environment of seed and seedling root, is among the most important of these mechanisms [22–25].

Bacterial priming with *Pseudomonas fluorescens* has improved fennel seedlings' germination percentage and vigor index under drought stress conditions [26]. While the dry weight of shoots and roots per Marigold's pot was decreased under drought stress, *Pseudomonas* and *Azotobacter* treatment has improved these indicators [27]. Plant growth-promoting bacteria (PGPBs) have promoted the germination rate and improved corn seedling growth [28]. Inoculation of barley seeds with PGPBs increases the length and weight of barley roots [29]. PGPBs-mediated tolerance to oxidative stress and improved activity of antioxidant enzymes, such as CAT, have been reported in plants under drought stress conditions [30, 31]. Moreover, biological priming of cumin seeds with PGPBs increased the activity of ascorbate peroxidase (APX) and catalase (CAT) enzymes. It could improve germination and emergence under drought stress conditions [32].

Following the environmental pollution crisis, especially soil and water resources, which have contaminated human food resources and threatened human communities' health, many efforts have been made to find appropriate solutions to eliminate contaminants, promote soil quality, and enhance the efficiency of agricultural products. Biofertilizers, especially in nutrient-poor soils, are crucial in sustainable agricultural systems for boosting output rates and preserving soil quality [33, 34]. Many researchers have proved the importance of biological treatments in adapting to the environment. Their role in mitigating the destructive effects of stress at the most critical stage of plant growth, namely the germination stage, can influence many plant species. *Azospirillum lipoferum* and *Azotobacter chroococcum* play significant roles as plant growth-promoting bacteria (PGPBs). *Azospirillum lipoferum* enhances plant growth through various mechanisms, including nitrogen fixation, phytohormone production, and promoting tolerance to abiotic stresses. Studies have shown that *Azospirillum lipoferum* can improve maize growth under saline conditions by enhancing physiological activities [35]. On the other hand, *Azotobacter chroococcum* is another beneficial bacterium that contributes to plant growth promotion. It is involved in nitrogen fixation and can enhance the growth of plants by improving nutrient availability in the soil [36]. *Azospirillum lipoferum* and *Azotobacter chroococcum* are valuable components of sustainable agricultural practices because they support plant growth and productivity through different mechanisms [35].

The increase in the activity of antioxidant enzymes, such as superoxide dismutase (SOD), peroxidase, and CAT of *Catharanthus roseus* seeds and seedlings, as a result of priming with diazotrophic bacteria, such as *Azospirillum* and *Azotobacter*, has led to the improvement of germination and germination index [37]. Inoculation of peanut seeds with two strains of *Pseudomonas fluorescens* has increased seed germination percentage by 15–30% [38]. A significant increase in the germination percentage of lettuce seeds inoculated with *Azospirillum* compared to non-inoculated seeds has also been shown [39]. A substantial increase in the germination percentage of biologically primed radish seeds with *Pseudomonas putida* has been reported [40].

Based on these publications and observations, our research aimed to study the effect of biological priming of *Azospirillum lipoferum* and *Azotobacter chroococcum* on germination components, seedling development, and some biochemical characteristics of barley plants under drought stress.

2. Materials and methods

2.1 Plant material

Barley seeds were obtained from Pakan-Bazr Company, Isfahan, Iran. To assess the effect of PGPBs on germination, seed growth indices, and biochemical changes of barley cv. Valfajr, under drought stress, a factorial experiment was conducted in a randomized complete block design with three replications in the Laboratory of Agronomy at the Faculty of Agriculture, Shahrekord University, Iran. The studied treatments involved three seed inoculation treatments with bacteria (*Azospirillum lipoferum* strain OF (NCIMB 11861), *Azotobacter chroococcum* strain W5 (GenBank accession no. MT299751), control, and three levels of drought treatment (0, −6, and − 12 bar) in three replications. These bacteria are cultivated under controlled conditions to ensure their viability and effectiveness as plant growth-promoting bacteria. For bacterial inoculation, the seeds were sterilized with 2.5% sodium hypochlorite for 1 minute and washed three times with distilled water. After mixing the seeds with gum arabic, 7 grams of each bacterium (containing 10^8 CFU/ml (colony-forming units) per milliliter) per kilogram (kg) of seeds was added for bacterial inoculation. To reach the inoculum density of 10^8 CFU/ml, this absorption density (bacterial suspension) reaches 0.5 in a wavelength of 600 nm by a spectrophotometer. Totally, 25 seeds from each seed treatment were planted in sand-filled pots. Each replicate for each treatment consisted of five pots. After sowing, each pot was irrigated with the specified drought treatment level. Drought treatments were applied from the beginning of the germination stage. Industrial Grade Polyethylene Glycol 8000 (PEG8000) was used to apply drought stress (Formula (1)) [41].

$$\Psi S = -\left(1.8\times10^{-2}\right)C-\left(1.8\times10^{-4}\right)C^2+\left(2.67\times10^{-4}\right)CT+\left(8.39\times10^{-7}\right)C^2T \quad (1)$$

Here, ΨS is the osmotic pressure in mega-pascal (MPa), *C* is the polyethylene glycol (*PEG*) consumption in grams per kilogram of water, and *T* is the germination medium temperature in degrees Celsius.

The experimental period lasted for 25 days. Then, emergence percentage (EP), emergence index (EI), seed vigor, rate, seedling dry weight (SDW), seedling height, number of leaves, proline content, and the activity of CAT and APX enzymes were measured. To measure the seedlings' dry weight, the plants were harvested entirely and subsequently dried in an oven at 72°C for 72 hours. Then, they were weighed using a scale with 0.0001 milligram (mg) readability.

2.2 Measurement of seedling biochemical characteristics

Fresh leaves of 25-day-old plants were sampled, and the enzyme activity was measured: Leaf samples (1 g) were frozen in liquid nitrogen and then ground in 10 mL of extraction buffer. The extraction buffer was prepared by combining 0.1 M phosphate buffer (pH 7.5) with 0.5 mM ethylenediaminetetraacetic acid (EDTA) to estimate CAT, proline, and APX. The extract was filtered through a four-layered cheesecloth and then centrifuged at 15000 × g for 20 minutes at 4°C. The enzyme activity was assessed using the supernatant. A temperature of 4°C was maintained throughout the preparation of the enzyme extract.

Catalase activity was estimated by measuring the decrease in hydrogen peroxide (H_2O_2) concentration [42]. To prepare the reaction mixture, 0.5 mL of H_2O_2 (75 mM), 1.5 mL of phosphate buffer (0.1 M, pH 7), and 50 μL of diluted enzyme extract were mixed. The absorbance at 240 nm was monitored for 1 minute to observe the decrease in H_2O_2 concentration. The catalase enzyme activity was determined based on the amount of H_2O_2 that was decomposed.

Ascorbate peroxidase activity was determined by measuring the decrease in optical density caused by ascorbate at 290 nm, following the method described by Nakano and Asada [43]. A reaction mixture of 3 mL was prepared by combining 50 mM potassium phosphate buffer (pH 7.0), 0.1 mL H_2O_2 (0.1 mM), EDTA (0.1 mM), ascorbate (0.5 mM), enzyme extract (0.1 mL), and water. The decrease in absorbance was measured using a spectrophotometer, and the activity was determined by comparing the results to a standard curve based on the reduction of ascorbate concentration.

Proline activity was measured using a spectrometer with an absorbance rate of 520 nm. The concentration of proline in milligrams per gram (mg/g) of fresh leaf tissue was determined using a standard curve [44].

To measure the protein content, 1 ml of Bradford solution was completely mixed with 100 μl of enzyme extract, and the solution's absorbance rate was recorded at 595 nm using a spectrometer. A standard curve determined proline content as mg/g of fresh leaf tissue [45].

2.3 Statistical analysis

Analysis of variance (ANOVA Statistix 10 software) was used. The mean comparison was conducted utilizing the Duncan test at 1 and 5% probability levels. The graphs were created using the Excel software.

3. Results

3.1 ANOVA test

The results demonstrated that the main effects of seed bacterial inoculation and stress levels on all measured traits are significant at the 1% probability level. Moreover, the interaction effect of inoculation treatments and stress levels was significant on emergence percentage, mean emergence time (MET), seedling dry weight, and APX activity at the 1% probability level and on emergence index, proline content, and CAT activity at the 5% probability level (**Table 1**).

S.O.V. (Sources of variation)	Seedling emergence	Emergence index	Mean emergence time	Seedling dry weight	Seedling length	Proline content	CAT activity	APX activity
Biological treatment (A)	570.110**	1.393**	3.380**	0.028**	40.11**	0.0005**	30.04**	17.16**
Osmotic potential (B)	12419**	33.35**	99.09**	1.577**	1825**	0.0164**	355. 5**	177.5**
A*B	152.11*	0.316*	1.208**	0.006**	1.89^{ns}	0.0024**	13.26**	5.250**
Error	47.037	0.109	0.083	0.001	0.94	0.0007	4.28	1.261
C.V. (Coefficient of variation) (%)	13.07	16.87	2.89	7.61	5.09	20.03	10.07	8.5

*ns*non-significant and significant at 5% probability levels, respectively.*
***non-significant and significant at 1% probability levels, respectively.*

Table 1.
shows the analysis of variance (mean square) for the impact of seed inoculation and drought on barley's measured traits.

3.2 Emergence percentage and emergence index

A comparison of the mean interaction of barley seed bacterial inoculation and drought stress on germination percentage and germination index showed that with increasing drought stress, there is a significant decrease in germination percentage. Using PGPBs increased the emergence index so that the highest emergence percentage, with an average of 89.33%, and the highest emergence index, with an average of 4.08, were allocated to seed inoculation with *Azospirillum* bacteria under non-stress conditions. The lowest emergence percentage (11.5%) and emergence index (0.17) were related to non-inoculated treatment with a − 12 bar drought stress level (**Figures 1** and **2**).

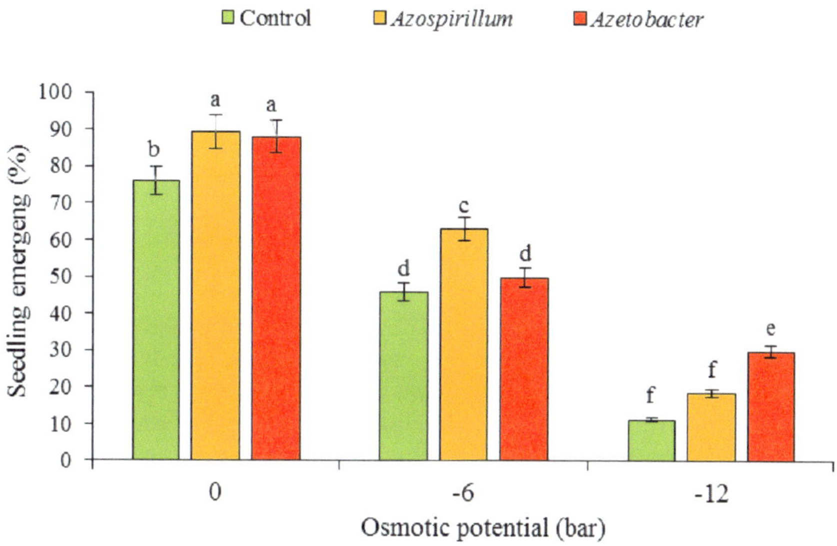

Figure 1.
The effect of inoculation treatments on emergence percentage of barley seedlings under drought stress conditions. Common letters in each column indicate insignificance level.

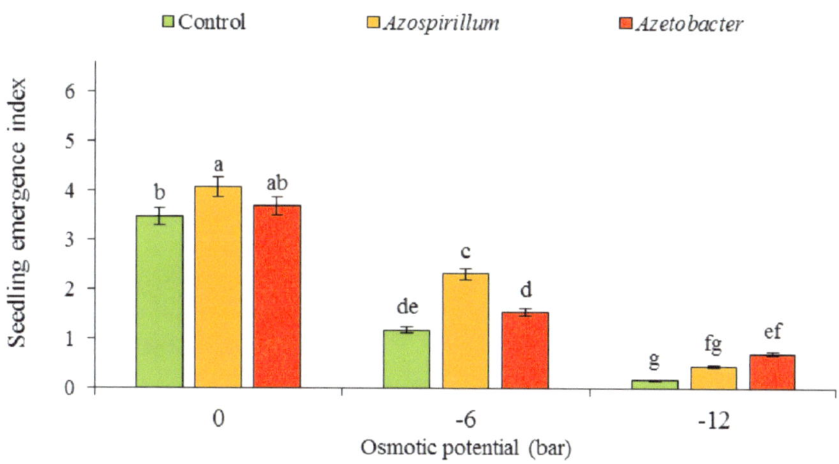

Figure 2.
The effect of inoculation treatments on emergence index of barley seedlings under drought stress conditions. Common letters in each column indicate insignificance level.

3.3 Mean emergence time (MET)

A comparison of the mean interaction of barley seed bacterial inoculation and drought stress on MET showed that, with increasing the level of osmotic potential due to drought stress, MEt also increases significantly. The use of PGPBs could reduce this parameter under stress and optimal conditions so that the lowest MET (6.47 days) was related to non-stress *Azospirillum* inoculation treatment, and the highest MET (13.44 days) was obtained in non-inoculation treatment with –12 bar stress level (**Figure 3**).

3.4 Seedling dry weight

With increasing drought stress, drought stress intensity significantly affected seedling dry weight, so a decreasing trend was observed with increasing stress levels. Bacterial inoculation treatments had a positive effect as well. The highest seedling dry weight (0.92 g) belonged to *Azospirillum* inoculation treatment under non-stress conditions, and the lowest seedling dry weight (0.1 g) was observed in non-inoculation treatment with a – 12 bar stress level (**Figure 4**).

3.5 Seedling length

A comparison of the mean indicated that bacterial seed injection causes an increase in barley seedling length so that the highest seedling length (21.22 cm) was related to seed inoculation with *Azospirillum,* and the lowest (17 cm) was allocated to control treatment (no inoculation). Moreover, with increasing stress levels, barley seedling length decreased significantly so that among the stress levels, the highest seedling length (25.44 cm) was observed in non-stress conditions (control), and the lowest seedling length (9.66 cm) was observed at –12 bar osmotic potential level (**Figure 5A-B**).

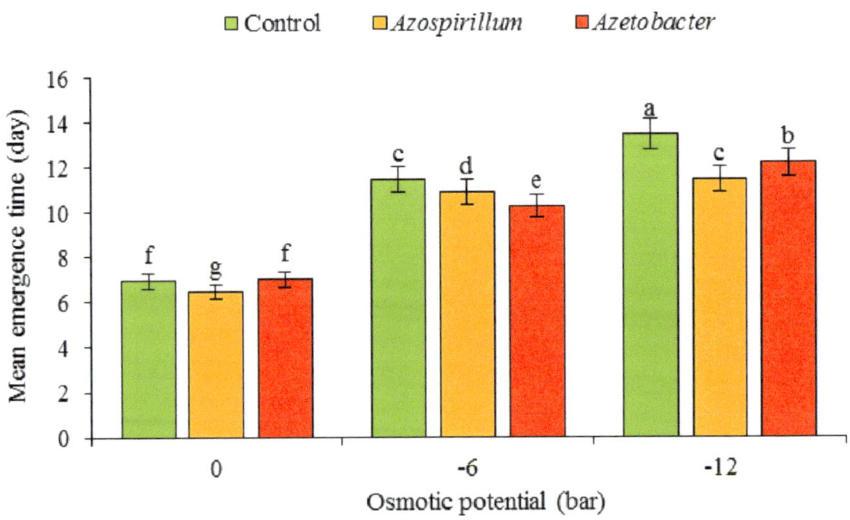

Figure 3.
The effect of inoculation treatments on MET of barley seedlings under drought stress conditions. Common letters in each column indicate insignificance level.

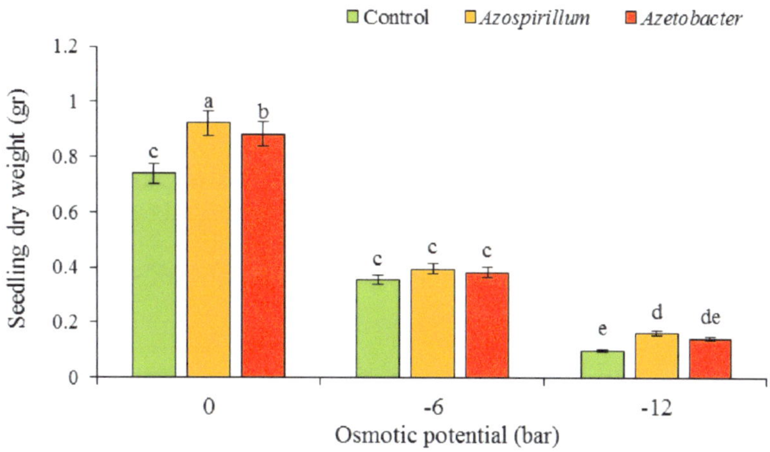

Figure 4.
The effect of inoculation treatments on seedling dry weight of barley under drought stress conditions. Common letters in each column indicate insignificance level.

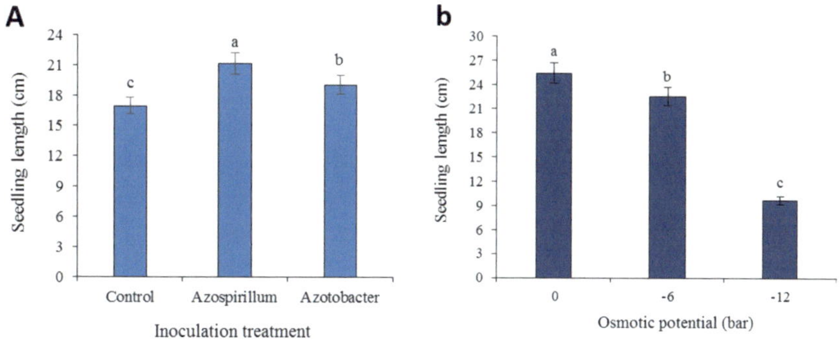

Figure 5.
The effect of inoculation treatments (a) and osmotic potential (b) on barley seedling length.

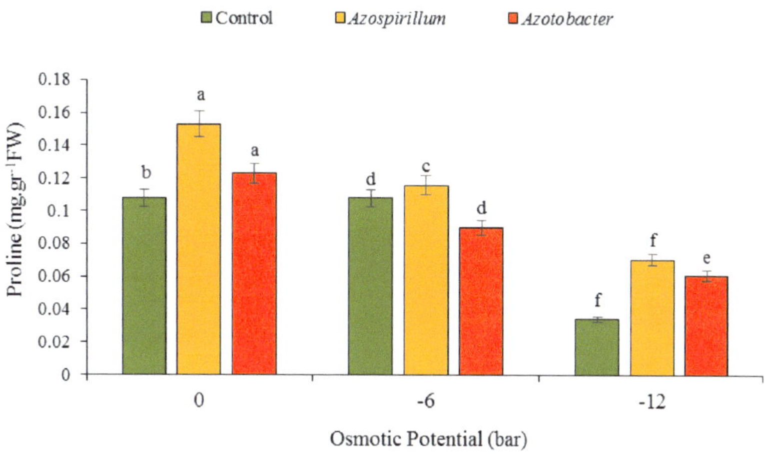

Figure 6.
The effect of inoculation treatments on the proline content of barley seedlings under drought-stress conditions. Common letters in each column indicate insignificance level.

3.6 Proline content

A comparison of the mean interaction of barley seed bacterial inoculation and drought stress on proline content and APX and CAT activity showed that with increasing drought stress, proline content and CAT and APX activity decrease significantly. The use of PGPBs boosted proline levels and APX and CAT activity so that the highest proline content (0.153 mg/g of fresh tissue) was related to *Azospirillum* treatment under non-stress conditions and the most minor proline content (0.034 mg/g of fresh tissue) was observed in non-inoculation treatment with −12 bar stress level (**Figure 6**).

3.7 Catalase (CAT) and ascorbate peroxidase (APX) enzyme activity

The activity of CAT and APX enzymes was also affected by the treatments used in this study. There was a decrease in the activity of these enzymes with an increase in osmotic potential. The highest CAT (29 mg protein/min) and APX (19.3 mg protein/min) activity were observed in the *Azospirillum* and *Azotobacter* treatments under non-stress conditions. The lowest CAT (12.5 mg protein/min) and APX (7.91 mg protein/min) activity were observed in non-inoculation treatment at −12 bar osmotic potential level (**Figure 7A-B**).

3.8 Correlation between the studied traits

The correlation results showed a significant difference between Emergence percentage (EP), Emergence index (EI), Mean emergence time (MET), Seedling dry weight (SDW), Proline, CAT, and APX traits. The lowest correlation level (−0.9621) was observed between MET and Seedling dry weight (SDW) traits, and the highest correlation level (0.9547) was observed between EI and SDW traits (**Figure 8**).

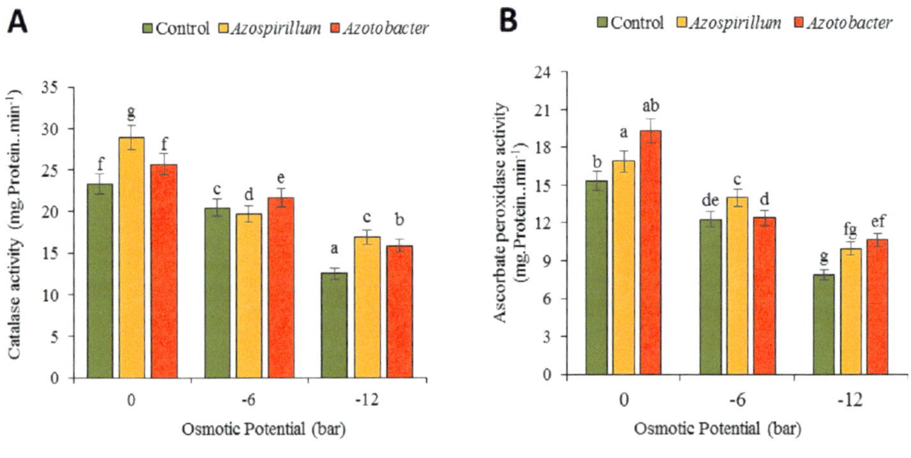

Figure 7.
The effect of inoculation treatments on CAT (A) and APX (B) activity of barley seedlings under drought stress conditions. Common letters in each column indicate insignificance level.

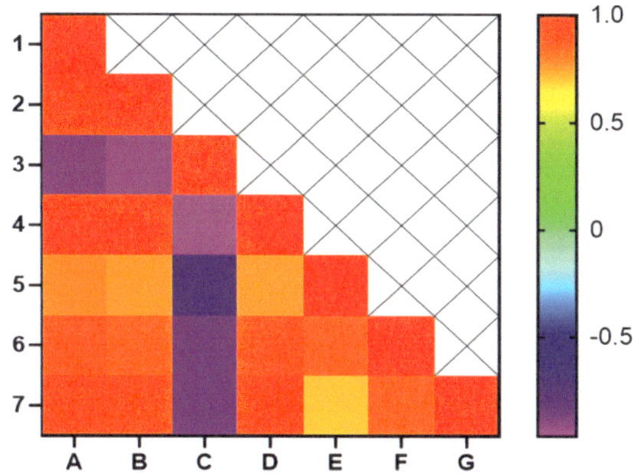

Figure 8.
The correlation between Emergence percentage (1, A), Emergence index (2, B), Mean emergence time (3, C), Seedling dry weight (4, D), Proline (5, E), CAT (6, F), and APX (7, G) (Heat MAP).

4. Discussion

Adverse effects of drought stress on the percentage and germination rate of fennel plants have been reported [26]. According to this study, *Azospirillum* and *Azotobacter* PGPBs have increased alfalfa's germination and emergence percentages under salinity stress conditions [46]. Our results showed a significant decrease in germination percentage and germination index with increasing drought stress, though using PGPBs could increase the emergence index.

As drought stress increased, the mean germination time increased. Still, using different seed pretreatments decreased the mean germination time, indicating faster seed germination in a shorter period [47, 48]. Our results showed a significant increase in barley emergence time with increasing the osmotic potential level due to drought stress. However, using PGPBs could reduce this parameter under pressure and optimal conditions.

Inoculating barley seeds with PGPBs has increased seedling dry weight and length [29]. Similarly, our result showed a significant decrease in seedling dry weight with increasing drought stress.

Seedling length decreases significantly with increasing drought stress levels [47]. The use of bacterial treatments has caused an increase in corn seedling length [49]. The present study found that inoculating seeds with bacteria increases the size of barley seedlings.

Environmental stresses have reduced plant growth and germination indices; nevertheless, using various seed pretreatments, including priming and bio-priming, has improved these indices. In addition, PGPBs promote plant growth indices in multiple ways, such as the production and secretion of the hormones such as auxin, gibberellin, and cytokinin, as indicated in these studies [50, 51], which are similar to our results.

Drought stress, as one of the biological stresses, can diminish the germination and emergence of various plants. With increasing drought stress, germination rate and percentage are significantly reduced [52, 53]. In this study, barley seedling emergence and growth indices decreased considerably with increasing osmotic potential levels. Decreased stress-induced germination and emergence may be associated with reduced seed water uptake. Suppose the water uptake by the seed is impaired or slow. In that case, the germination metabolic activities take place slowly, resulting in a more extended period for the radicle to emerge from the seed, thus reducing the emergence rate [54, 55].

The reduction or non-transfer of nutrients from seed storage tissues to the embryo is one of the reasons for decreased seedling length under drought stress conditions [56, 57]. Diminished germination percentage and seedling dry weight can be due to a drop in seed consumed material weight and rate of seed storage material [52]. PGPBs could improve morphological traits under stress conditions through the activity of biochemicals such as oxidant and proline enzymes [58, 59]. The CAT enzyme converts this molecule into water and oxygen in a reaction using ascorbate as the hydrogen donor. Moreover, it reduces the production of toxic substances such as oxygen free radicals and hydrogen peroxide and activates plant defense mechanisms against stress conditions [60, 61].

Based on the provided sources, it has been shown that *Azospirillum lipoferum* and *Azotobacter chroococcum* can increase proline content in barley under certain conditions. Inoculating maize with these bacteria has also improved physiological activities and promoted growth, especially under saline conditions. Although the sources do not explicitly mention proline in barley, the overall enhancement of physiological activities and development in plants like maize suggests that these bacteria could also increase proline content in barley. Proline accumulation is a typical response of plants to various stresses, including salinity, and the positive effects of these bacteria on plant growth under stressful conditions indicate a potential role in increasing proline content in barley, too [35].

According to the research, *Azospirillum lipoferum* and *Azotobacter chroococcum* do not directly boost catalase activity in barley. However, they positively affect plant growth and nutrient absorption, particularly when exposed to stress conditions such as cadmium exposure. The study discovered that *Azospirillum lipoferum* significantly increased root length and biomass in cadmium-treated barley, enhancing plant nutrient content. Although these bacteria did not directly influence catalase levels, their presence improved plant growth and reduced the toxicity effects of cadmium on barley plants [36, 62].

The effects of *Azospirillum lipoferum* and *Azotobacter chroococcum* on barley plants have been studied based on the sources provided. The research shows explicitly that *Azospirillum lipoferum* positively impacts barley seedlings. It increases the nutrient content in the roots and shoots and promotes root length and biomass growth, even in the presence of cadmium chloride ($CdCl_2$). However, the sources do not explicitly mention the direct effect of these bacteria on the APX enzyme in barley. Therefore, based on the available information, no direct evidence suggests that *Azospirillum lipoferum* or *Azotobacter chroococcum* increase the APX enzyme in barley [36].

Plant growth-promoting bacteria (PGPBs) promote plant growth through direct mechanisms such as nitrogen fixation, element dissolution, and the synthesis of plant growth regulators, as well as indirect means such as phytotoxicity reduction and removal and competition with plant pathogens [63]. PGPBs can promote plant growth by developing root structures and producing plant hormones (including auxin, cytokinin, and gibberellin) [64]. Reports of auxin and gibberellin production in the soil rhizosphere have significantly increased root growth in different plant species [65]. *Azotobacter* has a high potential for producing auxin and decreasing ethylene levels, which considerably promotes the growth and formation of root structures. PGPBs produce the cytokinin hormone, stimulating cell division and improving root growth and development [58].

One of the effects of PGPBs is that they increase the solubility and availability of elements, such as potassium, phosphorus, and iron, in the root environment and actively participate in nitrogen fixation and uptake [66]. These bacteria increase the plant's mineral uptake from the soil by helping to increase root area, develop capillary roots, and improve ion uptake [58].

This research was significant in the correlation between MET and SDW traits. Distilled water allowed okra seeds to germinate faster, with a mean germination time of about 42 hours. However, when rinsed with a wastewater treatment plant, the germination rate was relatively lower, and the mean germination time was longer. Their results on mean germination time are comparable to those reported by Okçu [67], who showed that the mean germination time of seed increases with salinity [68]. In research, the correlation between the germination percentage and the indicators of germination speed, dry weight, and seedling length was significant [69]. Simple correlation coefficients between seed germination ability and some related traits have shown a substantial correlation between MET and Seedling dry weight traits. Also, in this research, the correlation of characteristics like primary shoot and root dry weight and seedling vigor index was significant with the MET trait [70].

Studies have demonstrated that *Azospirillum lipoferum* and *Azotobacter chroococcum* positively influence barley plant growth, particularly in root and shoot development. *Azospirillum lipoferum*, a species within the *Azospirillum* genus, secretes phytohormones that induce changes in root architecture, encouraging the growth of adventitious roots and root hairs, which are advantageous for the plant. Moreover, *Azospirillum* bacteria have been proven to enhance crop yields in various plants, such as wheat, corn, rice, and sugar cane, by using nitrogen fixation and the production of growth regulators [36]. On the other hand, *Azotobacter chroococcum*, another beneficial bacterium, is known as a biofertilizer that enhances plant growth by improving the mobility and absorption of nutrients and plant growth hormones [71]. While the provided sources do not explicitly mention the specific impact of *Azotobacter chroococcum* on barley plants, its known role in promoting plant growth suggests that it could contribute to the development of both the root and shoot parts in barley. Therefore, considering the available information, it is reasonable to conclude that both *Azospirillum lipoferum* and *Azotobacter chroococcum* can potentially enhance the growth of both the root and shoot parts in barley plants [36, 71].

This study suggests that *Azotobacter* (*Azotobacter* spp.) and *Azospirillum* (*Azospirillum* spp.), as the most critical PGPBs, have modulated the destructive effects of stress by increasing the activity of antioxidant enzymes such as CAT and APX. They also improved the initial growth of barley seedlings by significantly producing growth-promoting hormones, particularly auxins, gibberellins, and cytokinins.

5. Conclusion

The results of the present study revealed that drought stress-induced osmotic potential diminishes all morphological and biochemical traits in barley plants. The use of PGPBs in this study could partially affect these plant characteristics and improve them under stress and optimal conditions. All indicators showed a significant decreasing trend with increasing osmotic potential. *Azospirillum* inoculation treatment was the best treatment for drought stress, which partially mitigated the destructive effects of increased osmotic potential, enhancing drought stress tolerance in barley plants. From this point of view, applying the *Azospirillum* of PGPBs to other plants under various stress conditions should be addressed.

Conflict of interest

The authors declare that they have no conflict of interest.

Author details

Masoumeh Asadi Aghbolaghi[1*], Ramin Piri[1], Bahman Fazeli-Nasab[2], Zahra Shakeri Chaleshtori[3], Beata Dedicova[4*], Riyaz Sayyed[5] and Andrea Mastinu[6]

1 Department of Agronomy and Plant Breeding, University of Tehran, Tehran, Iran

2 Department of Agronomy and Plant Breeding, Agriculture Institute, Research Institute of Zabol, Zabol, Iran

3 Faculty of Agriculture, Department of Agronomy and Plant Breeding, Shahrekord University, Shahrekord, Iran

4 Department of Plant Breeding, Swedish University of Agricultural Sciences (SLU), Alnarp, Sweden

5 Department of Microbiology, PSGVP Mandal's Shri S. I. Patil Arts, G B Patel Science, & STKVS Commerce College, Shahada, Maharashtra, India

6 Department of Molecular and Translational Medicine, Division of Pharmacology, University of Brescia, Brescia, Italy

*Address all correspondence to: asadi.m66@gmail.com and beata.dedicova@slu.se

References

[1] Saldaña-Robles N, Damián-Ascencio C, Gutiérrez-Chávez A, Zanor G, Guerra-Sánchez R, Herrera-Díaz I, et al. Spatio-temporal groundwater arsenic distribution in Central Mexico: Implications in accumulation of arsenic in barley (*Hordeum vulgare* L.) agrosystem. Environmental Science and Pollution Research. 2021;**28**(9):11333-11347. DOI: 10.1007/s11356-020-11317-7

[2] Ghorbanpour M, Mohammadi H, Kariman K. Nanosilicon-based recovery of barley (*Hordeum vulgare*) plants subjected to drought stress. Environmental Science: Nano. 2020;**7**(2):443-461. DOI: 10.1039/C9EN00973F

[3] Shamloo-Dashtpagerdi R, Lindlöf A, Aliakbari M, Pirasteh-Anosheh H. Plausible association between drought stress tolerance of barley (*Hordeum vulgare* L.) and programmed cell death via MC1 and TSN1 genes. Physiologia Plantarum. 2020;**170**(1):46-59. DOI: 10.1111/ppl.13102

[4] Kaczmarek M, Fedorowicz-Strońska O, Głowacka K, Waśkiewicz A, Sadowski J. CaCl2 treatment improves drought stress tolerance in barley (*Hordeum vulgare* L.). Acta Physiologiae Plantarum. 2017;**39**(1):41. DOI: 10.1007/s11738-016-2336-y

[5] Springer TL, Mornhinweg DW. Seed germination and early seedling growth of barley at negative water potentials. Agronomy. 2019;**9**(11):671. DOI: 10.3390/agronomy9110671

[6] Shahmoradi S, Chaichi MR, Mozafari J, Mazaheri D, Sharifzadeh F. Effects of drought stress on seed dormancy in *Hordeum spontaneum*. Journal of Plant Research (Iranian Journal of Biology). 2016;**28**(5):1028-1042 Available from: https://www.dorl.net/dor/20.1001.1.23832592.1394.28.5.11.4

[7] Voigt EL, Almeida TD, Chagas RM, Ponte LFA, Viégas RA, Silveira JAG. Source–sink regulation of cotyledonary reserve mobilization during cashew (*Anacardium occidentale*) seedling establishment under NaCl salinity. Journal of Plant Physiology. 2009;**166**(1):80-89. DOI: 10.1016/j.jplph.2008.02.008

[8] Piri R, Moradi A, Balouchi H. Improvement of salinity stress in cumin (*Cuminum cyminum*) seedling by inoculation with *Rhizobacteria*. Indian Journal of Agricultural Sciences. 2020;**90**(2):371-375

[9] Bennett AJ, Whipps JM. Dual application of beneficial microorganisms to seed during drum priming. Applied Soil Ecology. 2008;**38**(1):83-89. DOI: 10.1016/j.apsoil.2007.08.001

[10] Cai K, Gao H, Wu X, Zhang S, Han Z, Chen X, et al. The ability to regulate transmembrane potassium transport in root is critical for drought tolerance in barley. International Journal of Molecular Sciences. 2019;**20**(17):4111. DOI: 10.3390/ijms20174111

[11] Farooq M, Wahid A, Kobayashi N, Fujita D, Basra SMA. Plant drought stress: Effects, mechanisms, and management. In: Lichtfouse E, Navarrete M, Debaeke P, Véronique S, Alberola C, editors. Sustainable Agriculture. Dordrecht: Springer Netherlands; 2009. pp. 153-188. DOI: 10.1007/978-90-481-2666-8_12

[12] Boyer JS. Plant productivity and environment. Science.

1982;**218**(4571):443-448. DOI: 10.1126/science.218.4571.443

[13] Skowron E, Trojak M. Effect of exogenously-applied abscisic acid, putrescine, and hydrogen peroxide on drought tolerance of barley. Biologia. 2021;**76**(2):453-468. DOI: 10.2478/s11756-020- 00644-2

[14] Zaib S, Zubair A, Abbas S, Hussain J, Ahmad I, Shakeel SN. Plant growth-promoting Rhizobacteria (PGPR) reduce adverse effects of salinity and drought stresses by regulating nutritional profile of barley. Applied and Environmental Soil Science. 2023;**2023**:1-20. Article ID 7261784. DOI: 10.1155/2023/7261784

[15] Feng X, Liu W, Qiu CW, Zeng F, Wang Y, Zhang G, et al. HvAKT2 and HvHAK1 confer drought tolerance in barley through enhanced leaf mesophyll H+ homeostasis. Plant Biotechnology Journal. 2020;**18**(8):1683-1696. DOI: 10.1111/pbi.13332

[16] Slimani A, Raklami A, Oufdou K, Meddich A. Isolation and characterization of PGPR and their Potenzial for drought alleviation in barley plants. Gesunde Pflanzen. 2023;**75**(2):377-391. DOI: 10.1007/s10343-022-00709-z

[17] Atta BM, Saleem M, Abro S, Rizwan M, Sarwar G, Farooq A. Enhancement of germination and yield of cotton through optical seed priming: Lab. And diverse environment studies. PLoS One. 2023;**18**(7):e0288255. DOI: 10.1371/journal.pone.0288255

[18] de Andrade LA, Santos CHB, Frezarin ET, Sales LR, Rigobelo EC. Plant growth-promoting rhizobacteria for sustainable agricultural production. Microorganisms. 2023;**11**(4):1-16. DOI: 10.3390/microorganisms11041088

[19] Mashabela MD, Tugizimana F, Steenkamp PA, Piater LA, Dubery IA, Terefe T, et al. Metabolomic evaluation of PGPR defense priming in wheat (*Triticum aestivum* L.) cultivars infected with Puccinia striiformis f. sp. tritici (stripe rust), front. Plant Science. 2023;**14**:1103413. DOI: 10.3389/fpls.2023.1103413

[20] Shah A, Subramanian S, Smith DL. Seed priming with Devosia sp. cell-free supernatant (CFS) and citrus bioflavonoids enhance canola and soybean seed germination. Molecules. 2022;**27**(11):1-15. DOI: 10.3390/molecules27113410

[21] Shelar A, Singh AV, Dietrich P, Maharjan RS, Thissen A, Didwal PN, et al. Emerging cold plasma treatment and machine learning prospects for seed priming: A step towards sustainable food production. RSC Advances. 2022;**12**(17):10467-10488. DOI: 10.1039/d2ra00809b

[22] Bomle DV, Kiran A, Kumar JK, Nagaraj LS, Pradeep CK, Ansari MA, et al. Plants saline environment in perception with rhizosphere bacteria containing 1-Aminocyclopropane-1-carboxylate deaminase. International Journal of Molecular Sciences. 2021;**22**(21):1-35. DOI: 10.3390/ijms222111461

[23] Farooq T, Akram MN, Hameed A, Ahmed T, Hameed A. Nanopriming-mediated memory imprints reduce salt toxicity in wheat seedlings by modulating physiochemical attributes. BMC Plant Biology. 2022;**22**(1):540. DOI: 10.1186/s12870-022-03912-2

[24] Li Y, Liang Y, Liu M, Zhang Q, Wang Z, Fan J, et al. Genome-wide association studies provide insights into the genetic architecture of seed germination traits in maize. Frontiers

in Plant Science. 2022;**13**:930438. DOI: 10.3389/fpls.2022.930438

[25] Louis N, Dhankher OP, Puthur JT. Seed priming can enhance and retain stress tolerance in ensuing generations by inducing epigenetic changes and trans-generational memory. Physiologia Plantarum. 2023;**175**(2):e13881. DOI: 10.1111/ppl.13881

[26] Hoseini Moghaddam M, Salehi A, Rezaei R. The effect of various biological treatments on germination and some seedling indices of fennel (*Foeniculum vulgare* L.) under drought stress. Iranian Journal of Seed Science and Technology. 2019;**7**(2):59-74. DOI: 10.22034/ijsst.2019.110904.1100

[27] Sahib Hasan M, Selahvarzi Y, Nabati J, Aziz M. Effects of drought stress and bio-fertilizers on some growth, photosynthetic pigments, morphophysiological and biochemical traits of *Calendula officinalis*. Journal of Plant Process and Function. 2020;**9**(36):135-152. Available from: http://www.dorl.net/dor/20.1001.1.23222727.1399.9.36.2.5

[28] Gholami A, Shahsavani S, Nezarat S. The effect of plant growth promoting rhizobacteria (PGPR) on germination, seedling growth and yield of maize. World Academy of Science, Engineering and Technology. 2009;**49**(1):19-24

[29] Çakmakçı R, Erat M, Erdoğan Ü, Dönmez MF. The influence of plant growth–promoting rhizobacteria on growth and enzyme activities in wheat and spinach plants. Journal of Plant Nutrition and Soil Science. 2007;**170**(2):288-295. DOI: 10.1002/jpln.200625105

[30] Poorabtehaj M, Habibi D, Paknejad F, Fazeli F, Davoodi Fard M. Investigation of the effect of growth-promoting bacteria and foliar application of salicylic acid and amino acids on the activity of antioxidant enzymes under drought stress conditions in barley (*Hordeum vulgare* L.). Agronomy and Plant Breeding. 2012;**8**(2):147-160

[31] Sairam R, Srivastava G. Changes in antioxidant activity in sub-cellular fractions of tolerant and susceptible wheat genotypes in response to long-term salt stress. Plant Science. 2002;**162**(6):897-904. DOI: 10.1016/S0168-9452(02)00037-7

[32] Piri R, Moradi A, Balouchi H, Salehi A. Improvement of cumin (*Cuminum cyminum*) seed performance under drought stress by seed coating and biopriming. Scientia Horticulturae. 2019;**257**:108667. DOI: 10.1016/j.scienta.2019.108667

[33] Schwarzenbach RP, Egli T, Hofstetter TB, Von Gunten U, Wehrli B. Global water pollution and human health. Annual Review of Environment and Resources. 2010;**35**:109-136. DOI: 10.1146/annurev-environ-100809-125342

[34] Madhav S, Ahamad A, Singh AK, Kushawaha J, Chauhan JS, Sharma S, et al. Water Pollutants: Sources and Impact on the Environment and Human Health, Sensors in Water Pollutants Monitoring: Role of Material. Springer Nature Singapore PTE Ltd; 2020. pp. 43-62. DOI: 10.1007/978-981-15-0671-0_4

[35] Abdel Latef AAH, Abu Alhmad MF, Kordrostami M, Abo-Baker A-BA-E, Zakir A. Inoculation with *Azospirillum lipoferum* or *Azotobacter chroococcum* reinforces maize growth by improving physiological activities under saline conditions. Journal of Plant Growth

Regulation. 2020;**39**(3):1293-1306. DOI: 10.1007/s00344-020-10065-9

[36] Cruz-Hernández MA, Mendoza-Herrera A, Bocanegra-García V, Rivera G. *Azospirillum* spp. from plant growth-promoting bacteria to their use in bioremediation. Microorganisms. 2022;**10**(5):1-13. DOI: 10.3390%2Fmicroorganisms10051057

[37] Karthikeyan B, Jaleel CA, Gopi R, Deiveekasundaram M. Alterations in seedling vigor and antioxidant enzyme activities in Catharanthus roseus under seed priming with native diazotrophs. Journal of Zhejiang University Science B. 2007;**8**:453-457. DOI: 10.1631/jzus.2007.B0453

[38] Shweta B, Maheshwari DK, Dubey RC, Arora DS, Bajpai VK, Kang SC. Beneficial effects of fluorescent pseudomonads on seed germination, growth promotion, and suppression of charcoal rot in groundnut (*Arachis hypogea* L.). Journal of Microbiology and Biotechnology. 2008;**18**(9):1578-1583. Available from: https://www.pubmed.ncbi.nlm.nih.gov/18852515/

[39] Barassi C, Ayrault G, Creus C, Sueldo R, Sobrero M. Seed inoculation with Azospirillum mitigates NaCl effects on lettuce. Scientia Horticulturae. 2006;**109**(1):8-14. DOI: 10.1016/j.scienta.2006.02.025

[40] Kaymak HÇ, Güvenç İ, Yarali F, Dönmez MF. The effects of bio-priming with PGPR on germination of radish (*Raphanus sativus* L.) seeds under saline conditions. Turkish Journal of Agriculture and Forestry. 2009;**33**(2):173-179. DOI: 10.3906/tar-0806-30

[41] Michel BE, Kaufmann MR. The osmotic potential of polyethylene glycol 6000. Plant Physiology. 1973;**51**(5):914-916. DOI: 10.1104/pp.51.5.914

[42] Aebi H. [13] Catalase in vitro. Methods in Enzymology. 1984;**105**:121-126. DOI: 10.1016/S0076-6879(84)05016-3

[43] Nakano Y, Asada K. Spinach chloroplasts scavenge hydrogen peroxide on illumination. Plant and Cell Physiology. 1980;**21**(8):1295-1307. DOI: 10.1093/oxfordjournals.pcp.a076128

[44] Bates LS, Waldren RP, Teare I. Rapid determination of free proline for water-stress studies. Plant and Soil. 1973;**39**(1):205-207. DOI: 10.1007/BF00018060

[45] Bradford MM. A rapid and sensitive method for the quantitation of microgram quantities of protein utilizing the principle of protein-dye binding. Analytical Biochemistry. 1976;**72**(1-2):248-254

[46] Younesi O, Poustini K, Chaichi MR, Pourbabaie AA. Effect of growth promoting rhizobacteria on germination and early growth of two alfalfa cultivars under salinity stress condition. Journal of Crops Improvement. 2013;**14**(2):83-97. DOI: 10.22059/jci.2013.29503

[47] Ansari O, Azadi M, Sharif-Zadeh F, Younesi E. Effect of hormone priming on germination characteristics and enzyme activity of mountain rye (*Secale montanum*) seeds under drought stress conditions. Journal of Stress Physiology & Biochemistry. 2013;**9**(3):61-71

[48] Costa A, Paiva E, Torres S, Souza M, Pereira K, Leite M, et al. Osmoprotection in *Salvia hispanica* L. seeds under water stress attenuators. Brazilian Journal of Biology. 2021;**82**(e233547):1-9. DOI: 10.1590/1519-6984.233547

[49] Efthimiadou A, Katsenios N, Chanioti S, Giannoglou M, Djordjevic N,

Katsaros G. Effect of foliar and soil application of plant growth promoting bacteria on growth, physiology, yield and seed quality of maize under Mediterranean conditions. Scientific Reports. 2020;**10**(1):1-11. DOI: 10.1038/s41598-020-78034-6

[50] Shukla PS, Agarwal PK, Jha B. Improved salinity tolerance of *Arachis hypogaea* (L.) by the interaction of halotolerant plant-growth-promoting rhizobacteria. Journal of Plant Growth Regulation. 2012;**31**(2):195-206. DOI: 10.1007/s00344-011-9231-y

[51] Egamberdieva D, Wirth S, Bellingrath-Kimura SD, Mishra J, Arora NK. Salt-tolerant plant growth promoting rhizobacteria for enhancing crop productivity of saline soils. Frontiers in Microbiology. 2019;**10**:2791. DOI: 10.3389/fmicb.2019.02791

[52] Ansari O, Choghazardi H, Sharif Zadeh F, Nazarli H. Seed reserve utilization and seedling growth of treated seeds of mountain rye (*Secale montanum* L.) as affected by drought stress. Cercetări Agronomice în Moldova. 2012;**45**(2):43-48

[53] Rossi G, Guzzon F, Bickler C, Cauzzi P, Di Martino L, Di Cecco V, et al. Seed germination under osmotic stress across different wild populations of mountain rye (*Secale strictum* (C. Presl) C. Presl). Plant Bio Systems an International Journal Dealing with all Aspects of Plant Biology. 2020;**156**:1-8. DOI: 10.1080/11263504.2020.1857869

[54] Kopittke PM, Lombi E, Wang P, Schjoerring JK, Husted S. Nanomaterials as fertilizers for improving plant mineral nutrition and environmental outcomes. Environmental Science: Nano. 2019;**6**(12):3513-3524. DOI: 10.1039/C9EN00971J

[55] Buet A, Galatro A, Ramos-Artuso F, Simontacchi M. Nitric oxide and plant mineral nutrition: Current knowledge. Journal of Experimental Botany. 2019;**70**(17):4461-4476. DOI: 10.1093/jxb/erz129

[56] Khamsuk O, Sonjaroon W, Suwanwong S, Jutamanee K, Suksamrarn A. Effects of 24-epibrassinolide and the synthetic brassinosteroid mimic on chili pepper under drought. Acta Physiologiae Plantarum. 2018;**40**(6):1-12. DOI: 10.1007/s11738-018-2682-z

[57] Aminifard MH, Bayat H. Evaluation of seed germination and seedling growth characteristics of sweet pepper (*Capsicum annuum*) under salinity and drought treatment. Iranian Journal of Seed Research. 2020;**6**(2):137-149. DOI: 10.29252/yujs.6.2.137

[58] Fazeli-Nasab B, Sayyed RZ. Plant growth-promoting rhizobacteria and salinity stress: A journey into the soil. In: Sayyed RZ, Arora NK, Reddy MS, editors. Plant Growth Promoting Rhizobacteria for Sustainable Stress Management: Volume 1: Rhizobacteria in Abiotic Stress Management. Singapore: Springer Singapore; 2019. pp. 21-34. DOI: 10.1007/978-981-13-6536-2_2

[59] Rasouli H, Popović-Djordjević J, Sayyed RZ, Zarayneh S, Jafari M, Fazeli-Nasab B. Nanoparticles: A new threat to crop plants and soil rhizobia? In: Hayat S, Pichtel J, Faizan M, Fariduddin Q, editors. Sustainable Agriculture Reviews 41: Nanotechnology for Plant Growth and Development. Cham: Springer International Publishing; 2020. pp. 201-214. DOI: 10.1007/978-3-030-33996-8_11

[60] Hegedüs A, Erdei S, Horváth G. Comparative studies of H_2O_2 detoxifying enzymes in green and greening barley seedlings under cadmium stress.

Plant Science. 2001;**160**(6):1085-1093. DOI: 10.1016/S0168-9452(01)00330-2

[61] Pourtabrizi S, Pourseyedi S, Abdeshahi R, Naderneghad N. Effect of cadmium stress on gene expression and enzyme activity of glutathione reductase of milk thistle (*Silybum marianum*). Agricultural Biotechnology Journal. 2018;**9**(4):39-50. DOI: 10.22103/jab.2018.2013

[62] Omer AM, Osman MS, Badawy AA. Inoculation with *Azospirillum* brasilense and Pseudomonas geniculata reinforces flax (*Linum usitatissimum*) growth by improving physiological activities under saline soil conditions. Botanical Studies. 2022;**63**(1):15. DOI: 10.1186%2Fs40529-022-00345-w

[63] Bashan Y, De-Bashan LE. How the plant growth-promoting bacterium *Azospirillum promotes* plant growth—A critical assessment. Advances in Agronomy. 2010;**108**:77-136. DOI: 10.1016/S0065-2113(10)08002-8

[64] Kloepper J, Gutierrez-Estrada A, McInroy J. Photoperiod regulates elicitation of growth promotion but not induced resistance by plant growth-promoting rhizobacteria. Canadian Journal of Microbiology. 2007;**53**(2):159-167. DOI: 10.1139/w06-114

[65] Han J, Sun L, Dong X, Cai Z, Sun X, Yang H, et al. Characterization of a novel plant growth-promoting bacteria strain Delftia tsuruhatensis HR4 both as a diazotroph and a potential biocontrol agent against various plant pathogens. Systematic and Applied Microbiology. 2005;**28**(1):66-76. DOI: 10.1016/j.syapm.2004.09.003

[66] Roesti D, Gaur R, Johri B, Imfeld G, Sharma S, Kawaljeet K, et al. Plant growth stage, fertilizer management and bio-inoculation of arbuscular mycorrhizal fungi and plant growth promoting rhizobacteria affect the rhizobacterial community structure in rain-fed wheat fields. Soil Biology and Biochemistry. 2006;**38**(5):1111-1120. DOI: 10.1016/j.soilbio.2005.09.010

[67] Okçu G, Kaya MD, Atak M. Effects of salt and drought stresses on germination and seedling growth of pea (*Pisum sativum* L.). Turkish Journal of Agriculture and Forestry. 2005;**29**(4):237-242

[68] Lone AB, Unemoto LK, Ferrari EAP, Takahashi LSA, de Faria RT. The effects of light wavelength and intensity on the germination of pitaya seed genotypes. Australian Journal of Crop Science. 2014;**8**(11):1475-1480

[69] Vaisi G, Mohtadi A, Moradi A. The effect of different treatments on seed germination and dormancy breaking in seeds of Gundelia tournefortii. Nova Biologica Reperta. 2018;**5**(1):26-37. DOI: 10.29252/nbr.5.1.26

[70] Hamidi A, Asgharzadeh A, Choukan R. Study on the effect of plant growth promoting Rhizobacteria (PGPR) application on seed germination ability, seedling vigor and some related characters of late maturity maize (*Zea mays* L.) hybrids. Iranian Journal of Seed Sciences and Research. 2018;**5**(1):123-136. DOI: 10.22124/jms.2018.2905

[71] Mrkovacki N, Milic V. Use of Azotobacter chroococcum as potentially valuable for agricultural application. Annals of Microbiology. 2001;**51**(2):145-158

Chapter 10

Mechanisms and Use of Plant Growth-Promoting Bacteria to Improve Seed Germination in Adverse Environments

Gipumi Torres-Abe, Karina Alejandra Balderas-Ruíz, Gabriel Guillén Solís and Ángel Arturo Guevara-García

Abstract

Seed dormancy and germination are highly regulated processes under the control of various factors, among which stand out the internal balance of abscisic acid (ABA)/gibberellin (GA) and environmental conditions that impact this ratio. Germination determines when plants enter to natural or agricultural ecosystems. It represents the basis of agricultural production, and several agricultural strategies have been implemented to improve it. Plant Growth Promoting Rhizobacteria (PGPR) are ubiquitous soil microorganisms that favorably impact plant performance through pathogen control, nutrient uptake, tolerance to stress conditions, and even the production plant growth regulators, which impact both seed germination and subsequent plant development. Seeds germinate as soon as they are embedded and sown, so even when most studies have focused on the mechanisms that depend on plants themselves, it is evident that a better knowledge of plant-microorganism interactions could be useful to improve agricultural production and achieve sustainable and ecologically friendly agriculture. Here, after describing the endogenous mechanisms controlling germination and dormancy, we will update the information about the potential of PGPR to improve germination, and we will give a general overview of their use in agriculture systems to guarantee the food security in the years to come.

Keywords: seed dormancy, seed germination, seed biopriming, plant growth-promoting rhizobacteria (PGPR), sustainable agriculture

1. Introduction

The constant increase in the world population, which leads to a growing demand for food, coupled with climate change and increasing environmental pollution, aggravated by the indiscriminate use of chemical fertilizers and pesticides that increasingly compromise agricultural production, has motivated the search for cultivation methods that, while increasing yields, are more environmentally friendly. With this idea in mind, many investigations have focused on the study of the rhizosphere, considered the largest and most diverse ecosystem on the Earth, to identify those microorganisms, mainly bacteria, that in favor of their own survival positively impact the growth and health of plants whose roots are an important biotic component of their ecological niche. The study of the generically called Plant Growth Promoting Rhizobacteria dates back many years, but the strategies and approaches with which they are currently studied are novel. Among these strategies, molecular biology, omics technologies, and even bioinformatics approaches based on mathematical predictions and the use of big data stand out. An exhaustive review of each of these is beyond the scope of this review, instead, after giving an overview of the germination and dormancy processes, which represent the determining factors for the successful establishment of a crop, we focus on describing the composition of the rhizosphere in terms of PGPR, the mechanisms they use to interact with plants, the main strategy for their incorporation into agricultural programs, and the novel strategies that have already begun to be put into practice, to achieve sustainable agriculture that guarantees the efficient supply of food in the medium and long term.

2. Seed dormancy and germination

2.1 Seeds

The seed is the structural element that contains the plant embryo, allowing its dispersal in the environment, its maturation, and establishment of the seedling. Therefore, it is a critical stage to ensure the survival of the next generation of the plant. Its structure is composed of the embryo, the result of fertilization of the nucleus of the ovule by the nucleus of the male pollen tube, which will form the future seedling; the embryo is enveloped by two structures: the endosperm, a storage tissue for energy reserves, available for early growth of the seedling, and the seed coat, a substantial constituent that protects the embryo from the external environment [1].

2.1.1 Seed dormancy concept

Dormancy is the inability of a viable seed to germinate under favorable conditions [2]. This state is particularly important because it allows seeds to germinate in time and order, reducing the risk of premature death in conditions that are not favorable for plant growth, such as heat, cold, and drought [1].

There are five types of dormancy/latency described: physiological, morphological, morphophysiological, physical latency, and combined (physical latency + physiological latency). Physiological latency is the most common of all. Gymnosperm plants and all major clades of angiosperms present it. Within physiological dormancy, three levels are recognized: deep, intermediate, and non-deep; most seeds have the last class. Non-deep dormancy is characterized by producing normal seedlings; germination can be promoted by the phytohormone gibberellin; depending on the species, cold (0-10°C) or warm (≥15°C) stratification can break dormancy; seeds can subsequently mature in dry storage, and scarification can promote germination [3]. In addition, another generally accepted distinction of the concept is primary and secondary dormancy. Primary dormancy refers to seeds that are in the final stage of maturation. Then, the primary dormancy can be released post-ripening or through the stratification treatments described above. However, if the conditions required to end dormancy and induce germination are absent, the seed can re-enter a state of dormancy called secondary dormancy until conditions are again acceptable [4].

2.1.2 Germination concept

Germination is a process that begins when the seed absorbs water (imbibition) and ends with the emergence of the embryonic axis, generally the radicle, through the structures that surround it (seed coat and endosperm); the last event is known as visible germination [1]. With seed imbibition, various optimal conditions are needed to initiate the various mechanisms to promote germination such as the presence of optimal temperature, oxygen, and humidity. It may also need other factors such as light and/or nitrate, in addition to the absence of stress factors such as salinity, drought, cold, or heat [4, 5].

Germination is divided into three phases. In Phase I, the dry seed rapidly absorbs water which leads to a change in the temporal structure of the membrane upon hydration. During this phase, various solutes and low molecular weight metabolites are released into the imbibition solution that surrounds the seed. Then, water absorption stabilizes, and then quiescent seed resumes its metabolic activity as Phase II begins. This phase is the most important since various cellular and biochemical events are triggered that prepare the emergence of the radicle, such as the resumption of oxygen consumption, DNA repair, and reform of the cellular cytoskeleton. Because dormant seeds cannot complete germination, they do not enter the next phase. Finally, during Phase III, plant cells expand by absorbing water and loosening their cell walls, allowing the initiation of growth from the embryo to the formation of the seedling [1, 2].

2.2 Hormonal control during seed dormancy and germination

There is experimental evidence that dormancy and germination, like any other biological process, are regulated by different hormones working independently (co-regulation) or in coordination (crosstalk). At least seven different phytohormones (abscisic acid, gibberellins, ethylene, auxins, cytokinins, jasmonic acid, and brassinosteroids) have been involved in the regulation of both processes; however, it has been clearly demonstrated that abscisic acid (ABA) and gibberellins (GA) are the main hormones that antagonistically regulate dormancy and germination. The role of endogenous ABA in seeds is necessary, as in addition to inducing and maintaining dormancy, it prevents early germination, allowing seed maturation. On the other hand, increased GA levels facilitate the release of dormancy and the initiation of germination, when endogenous ABA levels decrease [6, 7]. The contribution of each of them is described in the following sections.

2.2.1 The role of ABA in the seed

Genetic studies demonstrate through reciprocal crosses of wild-type plants and ABA-deficient mutants of *Arabidopsis thaliana* L. that ABA is synthesized during the maturation stage, mostly by maternal tissues and then in lower concentrations in the embryo and endosperm. However, ABA produced by maternal tissues or supplied exogenously are not sufficient to induce dormancy, implying that it is a form of embryonic-controlled dormancy that depends on ABA synthesis in the embryo and/or tissue [8, 7]. However, overexpression of genes responsible for ABA biosynthesis can increase dormancy or delay germination [9, 10]. Furthermore, in the *A. thaliana* Cvi (Cape Verde Islands) ecotype, a decrease in ABA levels is described in the early stages of imbibition but in late stages, they increase and remain at high concentrations only in dormant seeds [11].

2.2.1.1 ABA biosynthesis

Active ABA is synthesized through indirect xanthophyll pathways (zeaxanthin, violaxanthin, and neoxanthin). Three types of genes responsible for the successive steps of ABA biosynthesis have been identified, such as zeaxanthin epoxidation (ZEP), oxidative cleavage of 9-cis-epoxycarotenoids (NCED), and abscisic aldehyde oxidation (AAO) [11]. Zeaxanthin is converted to violaxanthin by zeaxanthin epoxidase (ZEP) via the anteroxanthin intermediary pathway. Although there is a biosynthetic pathway of abscisic acid independent of zeaxanthin epoxidase, mutants defective in zeaxanthin epoxidation result in ABA deficiency and therefore reduced seed dormancy [8, 12]. The next critical gene in the next stage of ABA biosynthesis is NCED9, which cleaves the cis-isomers of violaxanthin and neoxanthin. The first NCED gene, VIVIPAROUS14, was cloned from the maize plant (*Zea mays* L.) after isolating the viviparous vp14 mutant deficient in the oxidation of the 9-cis epoxycarotenoid during ABA biosynthesis and consequently a decrease in ABA levels in the dry seed [13]. The last step for ABA biosynthesis is the oxidation of abscisic aldehyde by the abscisic aldehyde oxidase. Arabidopsis contains four aldehyde oxidases (AAO), of which AAO3 has been reported to have activity on the abscisic aldehyde. The AAO3 mutant alleles exhibit a decrease in ABA content and seed dormancy [14].

2.2.1.2 ABA signaling

ABA is detected by pyrabactin resistance1/pyrabactin-like/regulatory components of ABA receptors (PYR/PYL/RCAR) [15]. Currently, PYLs are the largest family of phytohormone receptors. In Arabidopsis, 14 PYL members with redundant functions have been reported, which mediate the ABA response by interacting with type 2C protein phosphatases (PP2C; negative regulators) and antagonizing their action. In the absence of ABA, PP2C proteins (ABA-insensitive 1/2 (ABI1/2) and ABA-hypersensitive germination3 (AHG3)) inhibit the activity of the ABA signaling proteins, sucrose nonfermenting 1-related protein kinase 2 s (SnRK2s; positive regulators) by dephosphorylation of its kinase activation loop. While in the presence of ABA, PYR/PYL/RCAR proteins form a complex with PP2C and inhibit the phosphatase activity of PP2C, allowing the function of SnRK2 to be activated [11].

Arabidopsis consists of three SnRK2s (SnRK2.2, SnRK2.3, and SnRK2.6) that act redundantly in the transmission of ABA signaling in seed maturation and dormancy induction. The main targets of SnRKs are the abscisic acid responsive element (ABRE) binding factors (ABF). The ABF family consists of nine members ABF1, ABF2/ABA-responsive element binding protein1 (AREB1), ABF3, ABF4/AREB3, AREB3, ABI5, bZIP15, bZIP67, and EEL of the bZIP subfamily, which are predominantly involved in the regulation of ABA-mediated transcription [11]. Furthermore, through genetic analysis, it has been revealed that the key transcription factors in the ABA response in the seed are ABI3 (B3 type), ABI4 (AP2 type), and ABI5 (bZIP type) [16]. The seed phenotypes of the ABA-insensitive response (abi) mutants of *A. thaliana* ABI1 to ABI5 demonstrate that ABI1 to ABI5 are involved in dormancy and/or germination [15].

The ABI3 gene together with two members of the leafy cotyledon class of regulators, FUSCA3 (FUS3) and leafy cotyledons 1 and 2 (LEC1/2) form a regulatory network involved in seed maturation and the embryo-to-embryo transition phase to the seedling. All *ABI3, LEC1*, *LEC2,* and *FUS3* mutants are affected in seed maturation and share common phenotypes, such as decreased maturation dormancy [17].

Another positive regulator of seed dormancy is the ABI4 gene. It has been reported that MYB96, the R2R3-type MYB transcription factor, positively regulates dormancy and negatively regulates germination through regulating the expression of ABI4 and biogenesis genes, such as NCED2 and NCED6 [6]. While ABI5 participates in the regulation of seed germination and early seedling growth in the presence of ABA and under abiotic stress [18].

2.2.1.3 ABA catabolism

In addition to ABA biosynthesis, catabolism is an important mechanism for regulating ABA levels. There are two main pathways of catabolism in Arabidopsis: the hydroxylation of ABA in the 8′ position by P-450 type monooxygenases to obtain an unstable intermediate (8'-OH-ABA) that is then isomerized to phasic acid (PA) and the esterification of ABA to ABA-glucose ester (ABA-GE) [19]. The enzymes responsible for inactivating ABA to phasic acid and therefore reducing ABA levels are ABA-8'hydroxylases, which are cytochromes P459 that code for the CYP707A family. These cyp707a mutants exhibit delayed germination [20]. In Arabidopsis, four members of the CYP707A family have been identified, demonstrating a different role during seed development and post germination growth. The CYP CYP7071 is mainly expressed during mid-maturation, while CYP707A2 is responsible for the regulation of ABA levels in the late maturation of germination [21].

2.2.2 The role of giberellins in the seed

The phytohormone gibberellin (GA) is essential to promote germination because GA-deficient mutants such as *GA1/GA2/GA3* are not able to germinate without exogenous GAs [22]. Furthermore, it was observed that the use of GA biosynthesis inhibitors, such as paclobutrazol and tetcyclacis, prevents germination, concluding that *de novo* biosynthesis of GAs is required during seed imbibition [23]. Likewise, GAs can promote germination due to their ability to overcome the limitations required for post-ripening (after-ripening) such as light and cold, as possibly the biosynthesis of GAs is regulated by the phytochrome pathway [24]. Also, induction of secondary dormancy due to sub/supraoptimal imbibition temperatures has been shown to be correlated with suppression of bioactive GA level via the expression of specific GA20ox and GA3ox genes [25].

The promotion of seed germination by high levels of GA or by GA signaling is possible by overcoming the mechanical barriers imposed by the layers that surround the embryo such as the seed coat [26] and by endosperm weakening through enzymatic degradation of mannan-rich cell walls [27].

2.2.2.1 GA biosynthesis

The biosynthesis of GAs is mainly regulated by the catalytic action of the enzyme GA 20-oxidase (GA20ox) and GA 3-oxidase (GA3ox), while their inactivation is controlled by GA 2-oxidase (GA2ox) [28]. Bioactive GAs accumulate before radicle emergence GA biosynthesis occurs at two separate locations within the embryo: (1) the early biosynthetic pathway, includes steps catalyzed by the ent-copayl diphosphate synthase (CPS, the gene GA1 of Arabidopsis) and the ent-kaurene oxidase (KO, the GA3 gene of Arabidopsis), in the provascular tissue where the activity of the promoter of the AtCPS gene and (2) the late biosynthetic pathway that includes the bioactive formation of GA is located by GA 3-oxidase, in the cortex and endodermis of the root where AtGA3ox2 transcripts and AtGA3ox2 gene promoter activity accumulate [29].

2.2.2.2 GA signaling

The signal transduction cascade of gibberellin pathway is triggered by the perception of bioactive GAs by its receptor gibberellin insensitive dwarf1 (GID1) [30]. The GA response in germination requires the induced degradation of DELLA (aspartic acid–glutamic acid–leucine–leucine–alanine domain) proteins, which are negative regulators of GA signaling. Binding of GA-GID1 promotes the formation of the GA–GID1–DELLA complex, which triggers the recognition of DELLA by the F-box protein, the central component of the SCFSLY1/GID2 ubiquitin E3 ligase complex for its poly-ubiquitination and subsequent degradation by the 26S proteasome [31].

The endosperm layer represents a physical restriction for the seedling to emerge; therefore, the endosperm weakening layer is vital to complete germination. GAs are involved in this process, as several genes regulated by GAs are known to encode

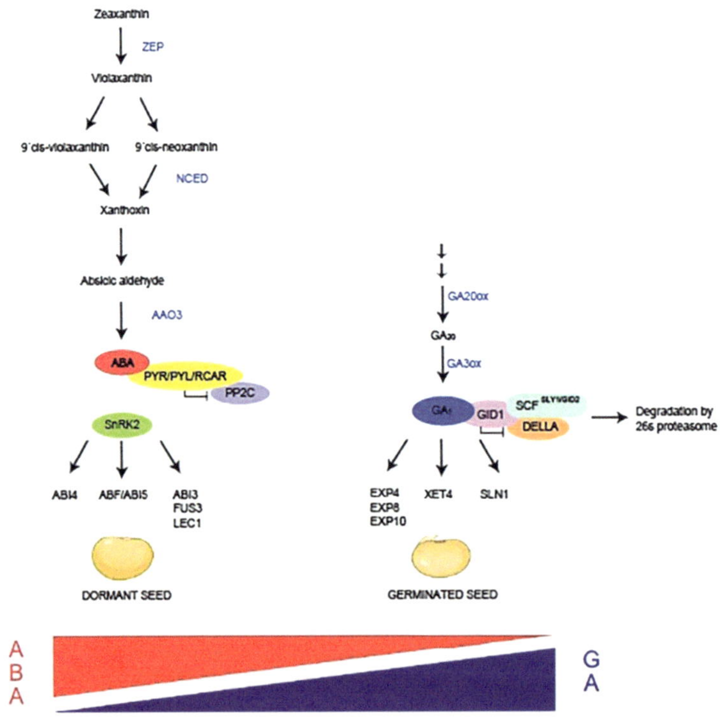

Figure 1.
Seed Dormancy and Germination Graphical Abstract. The balance of Abscisic Acid (ABA) and Gibberellic Acid (GA) levels during dormancy and germination processes is shown at the bottom. The main components of the ABA (left) and GA (right) biosynthesis and signaling pathways involved in both processes are shown at the top. ZEP (zeaxanthin epoxidase); NCED (cis-epoxycarotenoid oxidase); AAO3 (aldehyde oxidase 3); PYR/PYL/RCAR (pyrabactin resistance1/pyrabactin-like/regulatory components of ABA receptors); PP2C (type 2C protein phosphatases); SnRK2s (sucrose nonfermeting 1-related protein kinase 2); ABI (aba insensitive 3, 4 & 5); ABF (abscisic acid binding factors); FUS3 (Fusca 3); LEC1 (leafy cotyledon1). GAox (GAoxidase); GID1 (gibberellin insensitive dwarf1); SCFSLY1/GID2 (ubiquitin ligase); DELLA (GA repressors); EXP (expansins 4, 8 &10); XET4 (xyloglucan endotransglycosylases) SLN1 (DELLA preotein).

enzymes that modify the cell wall. For example, endo-β-mannases, expansins (EXP4, EXP8, and EXP10), and xyloglucan endotransglycosylases (XET4), which are expressed in different locations, endo-β-mannases, XET4, and EXP4 are expressed in the endosperm layer, EXP8 in the radicle cortex, and EXP10 in the embryo [32]. On the other hand, it has also been reported that GA, through aleurone cells, stimulates the synthesis and secretion of α-amylases that allow the mobilization of stored endosperm reserves necessary for seedling growth through target of rapamycin (TOR) signaling-dependent activation in wheat [33].

A graphical abstract of the components of the ABA and AG pathways involved in the signal transduction cascade controlling germination and dormancy processes is presented in **Figure 1**.

Much knowledge has been accumulated about the functions of the ABA and GA signaling pathways, but much remains to be discovered about the specificity and redundancy of the signaling of these and other hormones whose effects, in addition to being spatiotemporal and consequently, depending on the stage of development, are most likely also species-specific.

In the following sections, the participation of rhizosphere bacteria that impact seed dormancy/germination processes will be described, as well as their use in the biopriming strategy, which will most likely be the predominant alternative for the manipulation of both processes that have a great impact on agricultural production.

3. Growth promoting bacteria/rhizobacteria and their impact on germination

3.1 Plant-microorganisms interactions

Plant-microorganism interactions can be beneficial (mutualism), neutral (commensalism), or negative (pathogenic). The microorganisms that establish beneficial associations with the plant became plant's microbiota, plant's microbiota form a holobiont, maintaining a benefic or neutral interaction within the ecosystem [34, 35]. The plant has the capacity to select beneficial microorganisms by secreting compounds which function as substrates or signals to favor the microbial colonization, as seed germination stage to the root establishment. The interactions plant-microorganism become beneficial because some microorganisms stimulate plant growth through improving plant nutrition or due to hormone production, and on the other hand, the microorganisms can also protect or stimulate the plant's defense system to resist biotic and/or abiotic stress condi¬tions. The adverse abiotic conditions abate the plant health and additionally make it prone to diseases caused by biotic factors (phytopathogens or pests). Together, the stress factors affect not only the productivity of the crop but also the vigor of the seed and therefore the beginning of the next life cycle of the plant. Consequently, the establishment of beneficial interactions with microorganisms is a plant's strategy that helps to overcome different stress conditions caused by the environment.

3.2 Seed microbiome

Seeds are considered the end and beginning of plant's life cycle. Seed germination is the most important process in the plant development, and the seed microbiota is determinant in promoting the nutrition and health of seedling, which is highly desired from the beginning of the plant life cycle, especially in agricultural crops.

DNA metabarcoding analysis has been used to know the abundance, structure, and possible function of endophytic (microorganism that inhabits internal tissues of plants) bacteria in non-germinated and germinated seeds of economically important vegetable families as Apiaceae (parsley and carrot), Asteraceae (lettuce), Brassicaceae (cauliflower and broccoli), and Solanaceae (tomato) [36]. It was found that the microbiome abundance (quantification of the endophytic bacterial community) in these vegetables was in the range of 1.6×10^8 to 7.4×10^4 copies of 16S rRNA gene g–1 seed, which was associated with 23 different phyla of endophytic bacteria, and around 50 species were found in non-germinated seeds and 93 species in germinated seeds. The phyllum of *Pseudomonadota*, *Actinomycetota*, and *Bacillota* were the most dominant commu-endophytic bacteria, and the relative abundance of each phyllum was a function of a variety plant analyzed and the seed germination state. There were no statistically significant differences in alpha-diversity (measurement of the abundance and diversity of species within a community) between germinated and non-germinated seeds of vegetables in the families Apiaceae, Asteraceae, and Brassicaceae, and in contrast, differences were found within the Solanaceae family. In terms of beta-diversity (number of different species with respect to the set of species of various communities), it was found that diversity overlapped only between germinated and non-germinated seeds in the Solanaceae family. In the four plant families analyzed, the relative abundances of bacteria were mainly associated with the microbial function of chemoheterotrophy and fermentation in non-germinated and germinated seeds, followed by functions such as nitrate reduction, inhibition of phytopathogens, and manganese oxidation, nitrogen, and nitrate respiration. The growth-promoting activities of bacteria present in seeds were motility and nitrogen fixation and to a lesser extent in germinated seeds 1-aminocyclopropane-1-carboxylate (ACC) deaminase activity, auxin production, and phosphate solubilization. It was found that, in non-germinated seeds and in germinated seeds, respectively, 70 and 60% of the endobacteria had biological control activity against some phytopathogen such as *Xanthomonas* sp. RGM 2955, *Pseudomonas syringae* pv. *syringae* RGM 3354 or *Pseudomonas viridiflava* RGM 3342 [36].

Seed vigor is a quality that encompasses characteristics such as seed longevity, post-storage seed germination, seedling growth, and initial stress resistance; together, they determine the time and successful plant establishment. Germination rate is the most correlated characteristic with seed vigor [37]. In this sense, those bacteria that achieve the establishment in the spermosphere (soil around the seed affected by carbon deposition) are contributing to different biological processes with the vigor of the seed.

For example, *Ormosia henryi* seeds (species with high economic and ecological value in China) is known to have low germination rate due to the hardness of its protective layer, but a seed treatment with *Bacillus* or *Paraburkholderia* a 76% and 73% germination rate, respectively, was obtained [38]. In perspective, it was interesting to know how the microbiota of the spermosphere of *O. henryi* contributed to seed germination under natural conditions. The germination stage of *O. henryi* has five stages: (I) clarification, (II) imbibition, (III) radicle protrusion, (IV) radicle expansion, and (V) cotyledon elongation. Using metagenomic analysis, it was found that community abundance and richness of bacteria in the spermosphere of *O. henryi* changed depending on the germination stage, and there was a positive correlation between bacteria abundance and seed germination rate. In each germination stage, 50 genera of bacteria belonging to eight different phyla were detected in the spermosphere of *O. henryi*, with the highest abundance of Proteobacteria (32-36%) and Actinobacteriota (16-19%) phylum and the lower abundance of *Choroflexi*,

Bacteroidota, *Acidobacteroidota*, *Myxococcota*, *Verrucomicrobiota* and *Firmicutes*. The biological functions of bacteria varied depending on *O. henryi* seed germination stage. In germination stages IV and V, the highest abundance of bacteria was detected, and the dominant biological activities of bacteria were degradation of polysaccharides and fats as well as biological control.

Using microbial culture techniques and metagenomic analysis, the effect of temperature on microbiome dynamics of rice seeds was analyzed [39]. It was found that the amount of culturable endophytic bacteria in rice seeds (Shindongjin and Sukwang varieties) increased during seed maturation stage but decreased due to the effect of temperature (15 or 4°C) and storage time (2, 4 or 6 months). After 6 months of storage, the abundance of culturable bacteria and microbiome recovered from seeds was significantly higher with the condition of seed storage at 4°C compared to 15°C. In the seedlings, the total endobacteria from seeds with 6 months of storage were quantified in the range of 10^7 CFU g^{-1} for both rice varieties, and the storage temperatures were evaluated. The *Pantotea* genus was the most abundant in the seeds of both varieties and in the different conditions analyzed, followed by *Pseudomonas* and *Allorhizobium*. The storage temperature defined the migration (toward shoot or root), composition, and abundance of the microbiota in the seedling emerged from seeds with 6 months of storage. With storage at 4°C, 15 different endobacteria genera were preserved: the most abundant was *Pantotea*. In contrast, only nine genera were preserved at 15°C: the most abundant were *Herbaspirillum*, *Xhantomonas*, *Paenibacillus,* and *Pseudomonas* in residues of the germinated seed, *Pseudomonas* in root and *Pantotea* in shoot.

It is currently known that transference of plant microbiome can occur in three ways: (1) from plant to seed through sexual, asexual, or migration of endophytic microorganisms to the seed, (2) in seed dormancy by acquiring microorganisms during storage, or (3) from seed to seedling when endophytes migrate toward root or phyllosphere as well as other microorganisms acquired from the environment [40]. The microorganisms that remain stable in the seed make up the "core microbiota", those microorganisms of the "core microbiota" that remain inside the plant (endosphere), and that pass from generation to generation through the seed are known as "seed microbiota transferred", and it is recognized as a vertical transfer. However, the microbiota preserved in the external part of the seed (episphere) can also be transmitted from the seed to the plant and can maintain interaction throughout the whole plant's life cycle. On the other hand, it is known that the plant also acquires microbiota through horizontal transfer, being the soil, rhizosphere, stem, pollinators, flowers, and fruit sources of the microorganisms that will colonize the plant and will later be transferred through the seed [41].

3.3 Plant growth promoting bacteria

The use of beneficial microorganisms results in the improvement of structure and soil fertility, agricultural production, nutritional quality of the agricultural product, as well as reduction or elimination of use of chemical products for plant diseases and pest control. Plant Growth Promoting Bacteria (PGPB) are all those bacteria that through one or more biological functions are promoters of plant growth with mechanisms which are generally grouped as direct or indirect [42]. Direct mechanisms positively impact plant nutrition by improving soil nutrient availability and/or modifying root structure and architecture, which increases nutrient acquisition efficiency and therefore root functionality. The most frequent biological functions are nitrogen fixation, nutrients solubilization (i.e., phosphate and iron), and hormone production (i.e., auxins and cytokinins). In contrast, the indirect mechanisms of PGPB benefit to plant health having effect not only in the root but in different parts of the plant. The most frequent biological functions are space-nutrients competition, antimicrobial compound production, and the activation of induced systemic resistance (ISR) in the plant. In some cases, the bacteria have the capacity to form biofilms to attach it to plant tissues as part of its colonization process, and biofilms sometimes could be visible to our eyes. It must keep in mind that when a PGPB interacts with the plant, it exerts one or more biological functions simultaneously. This means that the effect of plant growth promotion is a result of direct and indirect mechanism interaction.

3.3.1 Direct mechanisms of PGPB

PGPB have ability to improve the availability of nutrients that under certain conditions the plant cannot acquire from the environment. One macroelement needed by plant and that significantly affects plant productivity is nitrogen. Some bacteria have the biological function of nitrogen fixation (they are diazotrophs). They take the atmospheric nitrogen (N_2) and convert it into ammonium ($NH4^+$) through an enzymatic reaction mediated by the "nitrogenase system" [43]. Bacteria such as those of the genus *Rhizobium* interact with roots of leguminous plants and form structures known as nodules; some others such as *Azotobacter*, *Azospirillum*, *Burkholderia*, *Paenibacillus*, *Bacillus*, *Pseudomonas* do not form nodules, but are found associated with plant or free-living [44]. Another element necessary for plant is phosphorus, which is frequently not available in the soil because the P ion is precipitated with ions such as Ca^{2+}, Fe^{3+}, or Al^{3+}. It has been found that certain strains such as those of genera *Acinetobacter*, *Burkholderia*, *Bacillus*, *Pseudomonas,* and *Brevibacillus* have the ability to make mineral (inorganic) and organic phosphorus available. Phosphate-solubilizing bacteria (PSB) that make mineral phosphate available use biological processes such as excretion of organic acids (i.e., gluconic), production of hydrogen sulfide, release of protons through nitrogen fixation and respiration, or also production of exopolysaccharides. However, that PSB make organic phosphate available using enzymes such as alkaline phosphatase, acid phosphatase, or phytase enzyme [44]. Iron is another nutrient that sometimes is not available to the plant because Fe is normally part of complexes such as hydroxides, hydroxy acids, and oxides, being the ferric (Fe^{3+}) and the ferrous (Fe^{2+}) state the most common in the soil and cannot be assimilated by the plant [45]. To acquire this nutrient, some PGPB produce siderophores (low molecular weight chelating compounds) in which the function is to sequester Fe from soil and making it available to the plant. Four types of siderophores produced by PGPB are known: catecholate, hydroxamate, carboxylates, and mixed [46]. These compounds are of particular interest because the siderophore-producing PGPB can inhibit the growth of phytopathogens by deprivation of Fe. Additionally, some siderophores have an affinity for other heavy metal ions such as Cr^{3+}, Al^{3+}, Cu^{2+}, and Pb^{2+} also contributing to soil bioremediation [47]. Some PGPB genera that produce siderophores are *Bacillus*, *Pseudomonas*, *Acinetobacter*, *Klebsiella*, and *Azotobacter* [46]. On the other hand, PGPB produce compounds that are detected by the plant trigger signaling cascades that generate changes in biosynthesis and homeostasis of phytohormones (auxins, salicylates, ethylene, cytokinins, gibberellins, brassinosteroids, jasmonates, abscisic acid, and strigolactones) affecting the plant growth, the activation of plant immune system, and tolerance to abiotic stress factors [48]. Several strains of PGPB in *Acetobacter*, *Acinetobacter*, *Aeromonas*, *Arthrobacter*, *Azospirillum*, *Azotobacter*, *Bacillus*, *Bradyrhizobium*, *Brevibacillus*, *Burkholderia*, *Enterobacter*, *Herbaspirillum*, *Mesorhizobium*, *Paenibacillus*, *Pantoea*, *Pseudomonas*, *Rhizobium*, *Rhodococcus*, *Serratia*, *Stenotrophomonas*, and *Streptomyces* are known to promote changes in the root due to auxin synthesis [49]. The effect of PGPB in root architecture is observed as follows: (1) inhibition of primary root growth, development of more lateral roots and root hairs, associated to an increase in root biomass or (2) increase in primary root length with an increase in shoot biomass [50]. With the increased root growth, the area of water and nutrient absorption increases, which is reflected in the increase in the total biomass of the plant. These changes will increase sites of the root with plant exudates from which PGPB will obtain substrates for their metabolism.

3.3.2 Indirect mechanisms of PGPB

The indirect mechanisms of PGPB benefit plant's health and have an effect not only in the root but also in different parts of the plant. An essential requirement for PGPB is to establish its interaction with the plant. The PGPB must have the ability of competition for space and nutrients to colonize the plant. Various biotic and abiotic stress factors such as salinity, heavy metals, drought, and phytopathogens affect the composition of root exudates and therefore the interaction with beneficial microorganisms. Some exudates serve as a substrate to support microbial growth and others as signals to activate clusters of genes that favor the colonization of beneficial microorganisms in the plant, for example, genes for motility, chemotaxis, and biofilm formation in plant tissues [51]. The genera *Bacillus* and *Pseudomonas* are examples of efficient biofilm-forming bacteria. In the case of *Bacillus*, it is known that cells within the biofilm are held together by an extracellular matrix, which is composed of exopolysaccharide (EPS), proteins (TasA and BslA), extracellular DNA (DNAe), and poly-γ-glutamic acid (γ-PGA) [52]. The biofilm allows the PGPB to remain attached to the plant tissues to ensure colonization and the acquisition of nutrients, and it also functions for the plant as a protective physical barrier against biotic (phytopathogens) and abiotic stress factors (drought, salinity, and heavy metals) [53, 54]. The production of compounds such as antibiotics, enzymes, and volatile compounds (inorganic and organic) is part of the biological strategy of the PGPB to colonize the plant and to compete for nutrients against other microorganisms. *Bacillus*, *Streptomyces*, *Pseudomonas*, and *Serratia* stand out for having species that in addition to being growth promoters, are highly effective as biological control agents because they produce antimicrobial compounds that inhibit phytopathogenic bacteria and fungi that cause diseases in a wide variety of plants. *Bacillus* antimicrobial compounds are lipopeptides, polyketides, volatile organic compounds (VOC), chitinase, protease, and siderophore that inhibit phytopathogens such as *Ralstonia*, *Penicillium*, and *Erwinia*. *Streptomyces* produce macrocyclic lactones, tripyrrole pigments, β-lactam antibiotics, biosurfactants, thiopeptide antibiotics, and polyketide-type compounds that inhibit phytopathogens such as *Fusarium*, *Rhizoctonia*, *Gaeumannomyces*, and *Magnaporthe*. *Pseudomonas* produces tricyclic compounds with nitrogen that inhibit phytopathogens such as *Pantoea*, *Thielaviopsis*, *Clavibacter*, and *Hyaloperonospora* [55]. On the other hand, the PGPB also produce compounds that, when are perceived by the plant, activate phytohormone signaling pathways that are involved in generating the stress response caused by phytopathogens. Some strains of *Bacillus*, *Pseudomonas*, *Enterobacter*, *Klebsiella*, *Azospirillum*, and *Paenibacillus* genus are recognized for activating ISR, which is known to be plant's mechanism to prepare and respond more strongly and quickly against the phytopathogen attack. VOC such as acetoin and 2,3-butanediol from *Bacillus* strains trigger signaling mediated by jasmonic acid and ethylene, causing ISR activation in the plant. Other bacterial compounds such as N-acyl-homoserine lactones (AHLs), siderophores, and cyclic lipopeptides could activate the ISR in some plants [56].

3.4 The PGPR in the rhizosphere

The rhizosphere is the closest part of soil that surrounds the plant root, and in this area, the greatest abundance of microorganisms has been detected. It is known that bacteria and fungi are the most abundant microorganisms forming part of plant microbiome; however, archeae, algae, nematodes and protists are also present, having different functions when interacting with the plant. These functions are determined by the diversity of environmental factors that the plant faces [57]. In particular, the term Plant Growth Promoting Rhizobacteria (PGPR) is used for bacterial strains that have been isolated from the rhizosphere. Currently, the use of PGPR has been proposed as a sustainable strategy to reduce the negative effects that cause biotic or abiotic stress on plants with the aim of increasing the productivity of agricultural crops [58]. There are several examples of cultivated plants where different PGPRs have been used to alleviate abiotic stress conditions to favor a more sustainable agriculture. Novel information on this topic can be found in the recent review by Aloo et al. [59]. In addition to being used as study models, some PGPR (i.e., *Bacillus velezensis* FZB42, *Pseudomonas putida* KT2440) are the active ingredient of products marketed in different parts of the world. Over the years, it has been observed that an ideal "PGPR" should have the following characteristics [42]:

- Highly competent to establish it in the rhizosphere
- Colonization of root plant in a significant way (in quantitative terms) after its inoculation
- Promoting plant growth
- Exhibition to a broad spectrum of action in terms of biological control
- Tolerance to physicochemical factors such as heat, desiccation, radiation, and antioxidant compounds

The PGPR promote the nutrition and health of plants through processes such as soil biofertilization, favoring plant nutrition, biostimulation due to the production of hormones, bioprotection against abiotic factors (tolerance to heat stress, drought, salinity, and heavy metals) and biotics (pests or phytopathogens) or bioremediation (degradation or mobilization of contaminating compounds, like heavy metals) [57]. The genera among which the largest number of PGPR have been identified are *Rhizobium*, *Frankia*, *Klebsiella*, *Clostridium*, *Nostoc*, *Anabaena*, *Azospirillum*, *Azotobacter*, *Burkholderia*, *Paenibacillus*, *Bacillus*, *Pseudomonas*, *Streptomyces*, and *Serratia* [60].

The communication between the plant and microorganisms has two basic requirements, the first is the generation of a plant signal and the second is the response of the microorganism to the signal [61]. Biotic and abiotic stress factors trigger internal plant signaling (orchestrated by phytohormones) resulting in the production of antimicrobial compounds, reinforcement of the cell wall, oxidative stress, senescence, among others, which are signals that determine the abundance and the biological functionality of beneficial microorganisms and the association with the plant [62]. Exudates are the communication bridge between the microorganism and the plant; the exudates contain some compounds that are chemical signals or substrates which stimulate microorganisms to colonize the plant. Plant exudates are classified as follows: (1) low molecular weight compounds (sugars, amino acids, phenols, volatile organic compounds, secondary metabolites, and organic acids) and (2) high molecular weight compounds (proteins and complex carbohydrates). The qualitative and quantitative composition of the exudates secreted by the root depends on the species and phenological stage of the plant, biotic and abiotic stress factors, the characteristics of the soil, and the diversity of microorganisms that surround it [61, 63].

In addition to the phenotypic characterization of a PGPR, genome analysis can be very useful to understand the potential of a microbial strain to establish beneficial interactions with plants. A clear example is the case of species of *Bacillus* genus. *Bacillus velezensis* is a bacterial species recognized as an efficient PGBR. The ability of this microorganism, as surely that of many others, to interact with plants is multifactorial, since it depends on: the production of hydrolytic enzymes (xylanase, cellulase and amylase) that participate in the degradation of plant cell walls allowing the assimilation of carbohydrates and giving it advantages to compete for nutrients present in plant exudates; synthesis of biofilm components to adhere to roots and efficiently compete for space with other microorganisms; synthesis of antimicrobial metabolites (difficidin, macrolactin, bacillaene, iturin, fengycin, surfactin and bacillibactin), which makes it an efficient biological control agent; synthesis of phytohormones, such as auxins and cytokinins, that stimulate growth and modify plant development; synthesis of VOC, such as acetoin and 2,3-butanediol, which trigger ISR in some plants, reducing the incidence and severity of diseases; as well as the synthesis of phytase, which helps to solubilize phosphate, making it available to the plant and promoting its nutrition [42, 63]. In contrast, in most cases, the other species of *Bacillus* (*B. subtilis*, *B. amyloliquefaciens*, *B. siamensis*, *B. licheniformis,* and *B. pumilus*) are mainly characterized by their performance as biological control agents due to the production of antimicrobial compounds. In the world, there are several bioproducts already marketed with strains of *Bacillus*, *Streptomyces*, *Pseudomonas*, *Serratia*, *Paenibacillus*, *Azospirillum*, *Pantotea*, *Rhizobium*, and *Azotobacter* as active ingredients [64, 65], and some bioproducts even contain several strains, taking advantage of the compatibility of biological functions between strains.

The use of metagenomic analysis has allowed us to know the diversity and abundance of microbial species in seeds, additionally to know the microbiome dynamics during storage, germination, and seedling development, even to know how the seed storage condition and germination stage affects the abundance and seed microbial diversity. Knowing the microbial genera and their biological functions at each plant stage is important to develop strategies for preserving the seed microbiota or for the inoculation of PGPR before sowing the seeds or other stages of plant development. Additionally, it is important to consider the effect of the environmental factors to which the plant-beneficial microorganism interaction is exposed.

Due to all potential biological functions of PGPB (or PGPR) that are currently known, their use is recommended as a sustainable agricultural practice to improve plant productivity, particularly for crops in soils with problems such as nutrient availability or presence of contaminating compounds, as well in regions of the world in which plants are exposed to stress conditions due to climate change (i.e., high temperatures, drought, and excess humidity) that reduce the health of plants and make them prone to diseases.

In the next section, we will further discuss PGPR with an emphasis on how they can be used to advance crop yield and sustainability.

4. Strategies to improve germination

The search and implementation of strategies that result in increases in crop production is a permanent concern of farmers and plant breeders. Thus, the selection and sowing of high-quality seeds with uniform germination that promotes the vigorous development of seedlings in early stages to guarantee the uniform establishment of crops in the field is a priority. With that idea in mind, seed priming has long been used as a strategy to enhance germination and consequently crop yields. More recently, based on the study of the rhizosphere and the interactions that plants establish with the microorganisms in their environment, biopriming has been considered one of the most promising and environmentally friendly strategies to achieve sustainable agriculture [66, 67].

Efficient and homogeneous seed germination is critical for proper crop establishment that is later reflected in greater production and sustainability. For this reason, farmers, growers and plant breeders are constantly searching for strategies that help improve seed quality and/or its performance at the time of germination. Within these strategies, the so-called biopriming is probably the most used and has the greatest potential to be used routinely to positively impact seed germination and consequently increase agricultural production [68].

4.1 Seed biopriming

In recent years, the so-called biopriming, based on the use of plant beneficial microorganisms, has received great attention as an effective alternative to improve seed germination with the idea of strengthening the development of sustainable agriculture by making more efficient use of natural resources. However, can be said that the concept of seed biopriming as such comes from when Callahan and collaborators reported the treatment of corn seeds with a strain of *Pseudomonas aureofaciens* for the biological control of *Pythium ultimum* [69]. From here on, the use of this strategy that combines the benefits of priming (i.e., homogeneous germination and increased seedling vigor) with those of the use of beneficial microorganisms (i.e. disease resistance and enhanced nutrient assimilation) has been the subject of numerous investigations focused on studying the potential of different microorganisms on different crops and application conditions, with the idea of optimizing protocols for their large-scale application [70]. Formulations for seed biopriming are already being marketed, and thanks to their effectiveness they are increasingly gaining popularity among farmers and plant breeders [71]. For the establishment of crops in stress conditions where germination is compromised, and the emerging seedlings vigor is essential, the use of these products is becoming the best alternative [72]. Ongoing research aims to unravel the biochemical and molecular mechanisms on which successful biopriming depends to extend its use to various crops and adverse environmental conditions [73]. Biotechnological approaches based on omics strategies will result in the identification of more effective microorganisms and the implementation of more efficient protocols for their incorporation into biopriming strategies [74–76]. In the next few years, seed biopriming will be recognized as a more ecologically friendly sustainable agricultural practice, which will necessarily reduce the indiscriminate use of pesticides and chemical fertilizers that characterizes current unsustainable farming practices [66, 68, 77, 78].

The bacterial genera best represented in the rhizosphere are *Beijerinckia*, *Arthrobacter*, *Burkholderia*, *Ochrobactrum*, *Derxia*, *Herbaspirillum*, *Klebsiella*, *Pantoae*, *Acinetobacter*, *Enterobacter*, *Rhodococcus*, *Gluconacetobacter*, *Bacillus*, *Alcaligenes*, *Acetobacter*, *Arthrobacter*, *Azospirillum*, *Pseudomonas*, *Lactobacillus*, *Azotobacter*, *Stenotrophomonas*, *Zoogloea*, *Peanibacillum*, *Azoarcus*, and *Serratia* [77, 79].

Some of them, such as *Pseudomonas* [80], *Bacillus* [81], *Azospirillum* [82], *Serratia* [83], and *Enterobacter* [84] due to their well-known properties as growth promoters and/or biological control agents, are already used for seed biopriming, but probably all/many of them, hand in hand with better knowledge of their properties as beneficial bacteria, could be incorporated into this type of strategies over the next few years. For some of those bacteria, such as *Pseudomonas* and *Bacillus*, it is already known that they produce phytohormones, like auxins and gibberellines, and solubilize Pi to decrease the level of ethylene in the root and as a result increase its growth and development [85, 86], but research about this is required to reveal the mechanisms that other PGPR use to improve plant performance. Research has shown that coating maize seeds with Bacillus promotes the growth of corn seedling, and this also could induce droughts and salt stress tolerance enhancing its germination performance [85]. In beans, cyanobacterial extracts seed biopriming control *Pythium ultimum* infection and promoted the growth parameters [87]. Even some other bacterial genera that are not very abundant in the rhizosphere, such as *Paenibacillus* and *Rhizobium*, have already demonstrated their efficiency in improving the performance of some plant species when used for seed biopriming [88]. Given that the estimated number of bacterial species that make up the rhizosphere reaches tens of thousands, our knowledge in this regard is clearly very limited. However, this same limitation allows us to see that the potential for using rhizospheric bacteria to improve plant performance through their incorporation into biopriming strategies is huge. Since the deficiency and/or availability of nitrogen is crucial for agricultural production, farmers around the world have historically used nitrogen fertilizers; unfortunately, the adverse effects of the indiscriminate and prolonged use of this type of agrochemicals are currently one of the most serious soil and water bodies contamination problems that affect animal and human health [89]. For this reason, of particular interest among PGPR with the potential to be incorporated in biopriming strategies are nitrogen-fixing rhizobacteria, either free-living or establishing symbiotic relationships with leguminous plants, like beans, with the ability to provide essential nitrogen to the plant [70, 82].

Some PGPR, known as endophytes, are capable of colonizing, establishing, and proliferating plant tissues from where they can even flow into the soil through root exudates [90]. An even greater number of rhizobacteria can form biofilms on roots to increase their survival [91]. In any case, plant-microbe interactions are mediated by diffusible and volatile chemical compounds that flow bidirectionally. The chemical nature of such compounds is diverse and depends on the species involved in the interaction. Regarding their impact on germination, it is intuitive to assume that diffusible compounds would be easier to apply for such purposes; however, both the use of encapsulation and sol-gel matrices represent viable alternatives for the application of volatile compounds in biopriming to large scale [92]. Some of the strategies for the encapsulation of bacterial compounds and/or complete bioactive microorganisms represent some of the most promising strategies for seed management that will allow to achieve sustainable agriculture.

4.2 Future prospects

4.2.1 Seed enhancement technologies (seed invigoration)

Soon, the integration of basic science with practical applications in agribusiness will make it possible the implementation of seed enhancement technologies (SETs). For instance, understanding the mechanisms of seed germination and the role of beneficial microorganisms in the rhizosphere will support innovative approaches such as seed biopriming [93]. By harnessing scientific knowledge on plant-microbe interactions, SETs will address critical issues in agribusiness, such as poor crop establishment and reduced yields, thereby contributing to sustainable agricultural practices. This synergy between basic research and applied science will not only advance our knowledge but also provide tangible solutions to real-world challenges faced by farmers and agribusinesses.

SETs include methods like biopriming, coating, and pelleting technologies that makes possible to treat seed surfaces with several materials such as nutrient components, plant growth regulators, beneficial microorganism, or even nutrient components and plant growth regulators [94]. Adopting seed biopriming with PGPR or coating technology that improve establishment rates are useful in early emergence, thereby promoting success in seed-based restoration efforts [95].

Currently, there are multiple studies on biofertilizers and PGPR that we can use with SETs and could be a hope for agriculture in marginal and polluted conditions. In this regard, there is a particular interest in identifying and using PGPR to promote the establishment of crops in saline, drought, and heavy metals contaminated environments, all of them increasing constantly and irremediably [96, 97]. These can include not only PGPR but also mycorrhizal fungi (another common inhabitants of the rhizosphere), which help plants by improving nutrient availability through N or P solubilizing effects, the release of growth regulators, and the reduction of diseases caused by soil-borne pathogens. A particular contribution of mycorrhizal fungi is to increase the surface coverage of roots to improve nutrient uptake [98].

4.2.2 Synthetic communities of microbes (SynCom)

Another revolutionary advance in agricultural biotechnology is the use of Synthetic Communities of microbes, or SynComs. These engineered microbial consortia exploit the characteristics of natural ecosystems to provide improved plant function, health, and productivity. By selectively assembling supportive microorganisms, SynComs can either mimic or enhance the types of rhizosphere interactions that already take place within ecosystems to make crops more resilient [99].

The merge of SynComs with SETs is a promising innovation for the agribusiness sector. The fusion of these technologies would further increase their efficiency, providing growers with a plant microbiome tailored to the microbes in which the crop germinates and develops [100].

For example, biopriming with SynComs placing PGPR directly to the seeds surface and creating symbiosis from the early days of plant development. This method not only should enhance the seed germination and vigor of seeding but also could provide the sustainability to cope up with various stresses and also improve plant resistance against different pathogens. Nitrogen-fixing bacteria, phosphate-solubilizing microbes, and growth-promoting hormones producing ones may further be selected as the microbial strains for SynComs which could provide all kind of supports indispensable to growing plant [101].

The incorporation of SynComs into SETs follows the guidelines behind sustainable agriculture, which uses fewer chemical fertilizers and pesticides to increase plant growth using natural processes. Not only is this good for the natural environment but it also makes agribusiness more economically sustainable due to lower input expensive and better crop yields [102].

In conclusion, SynComs offers an innovative perspective toward achievable food security right alongside environmental stewardship, whose use will surely soon be adopted to improve agricultural practices around the world. As they combine advanced biotechnological solutions with conventional agriculture, there is a strong possibility that the use of SynComs and PGPR-based SETs will indicate the future direction for improving productivity hand in hand with sustainability. Crops using these novel technologies will be robust, with high yields, and acceptance of these innovations could revolutionize conventional farming, ensuring food security and economic prosperity for farmers worldwide.

5. Conclusions

The main aspects that stand out from this review are that the search for alternatives that help increase agricultural production in increasingly limiting conditions is urgent to achieve the production levels required to satisfy the food demand of a constantly growing world population. A high rate of homogeneous seed germination has a positive effect over the efficiency of agricultural production, so any strategy that improves it will be useful for such purposes. Seed germination is mainly controlled by the endogenous balance of abscisic acid and gibberellins; however, the rhizosphere inhabits numerous bacteria, generically known as "Plant Growth Promoting Rhizobacteria (PGPR)", which favorably impact seed germination and plant development. The study of these bacteria, mainly focused on a better understanding of the PGPR-plant interaction, with different approaches that should include cutting-edge technologies and strategies, represents an excellent alternative that sooner or later will have an impact on the already unavoidable establishment of sustainable and ecologically friendly agriculture.

Acknowledgements

Group research is funded by DGAPA-UNAM, PAPIIT (IN213723) and CONAHCyT, MÉXICO (CF-2023-I-1605). GTA and KABR thank CONAHCyT for their support with the master's scholarship (CVU 1315223) and the posdoctoral scholarship (EPM- Application 2023-1, agreement I1200/311/2023), respectively.

Author details

Gipumi Torres-Abe, Karina Alejandra Balderas-Ruíz, Gabriel Guillén Solís
and Ángel Arturo Guevara-García*
Plant Signaling and Development Laboratory, Institute of Biotechnology-UNAM,
Cuernavaca, Morelos, México

*Address all correspondence to: arturo.guevara@ibt.unam.mx

References

[1] Klupczyńska EA, Pawłowski TA. Regulation of seed dormancy and germination mechanisms in a changing environment. International Journal of Molecular Sciences. 2021;**22**:1357. DOI: 10.3390/ijms22031357

[2] Shu K, Meng YJ, Shuai HW, Liu WG, Du JB, Liu J, et al. Dormancy and germination: How does the crop seed decide? Plant Biology (Stuttgart, Germany). 2015;**17**:1104-1112. DOI: 10.1111/plb.12356

[3] Wang JY, Bu ZJ, Poschlod P, Yusup S, Zhang JQ, Zhang ZX. Seed dormancy types and germination response of 15 plant species in temperate montane peatlands. Ecology and Evolution. 2024;**14**:e11671. DOI: 10.1002/ece3.11671

[4] Buijs G. A perspective on secondary seed dormancy in Arabidopsis thaliana. Plants (Basel). 2020;**9**:749. DOI: 10.3390/plants9060749

[5] Sajeev N, Koornneef M, Bentsink L. A commitment for life: Decades of unraveling the molecular mechanisms behind seed dormancy and germination. The Plant Cell. 2024;**36**:1358-1376. DOI: 10.1093/plcell/koad328

[6] Shu K, Liu XD, Xie Q, He ZH. Two faces of one seed: Hormonal regulation of dormancy and germination. Molecular Plant. 2016;**9**:34-45. DOI: 10.1016/j.molp.2015.08.010

[7] Verma V, Ravindran P, Kumar PP. Plant hormone-mediated regulation of stress responses. BMC Plant Biology. 2016;**16**:86. DOI: 10.1186/s12870-016-0771-y

[8] Sano N, Marion-Poll A. ABA metabolism and homeostasis in seed dormancy and germination. International Journal of Molecular Sciences. 2021;**22**:5069. DOI: 10.3390/ijms22105069

[9] Huang Y, Guo Y, Liu Y, Zhang F, Wang Z, Wang H, et al. 9-cis-Epoxycarotenoid dioxygenase 3 regulates plant growth and enhances multi-abiotic stress tolerance in Rice. Frontiers in Plant Science. 2018;**9**:162. DOI: 10.3389/fpls.2018.00162

[10] Huang Y, Jiao Y, Yang S, Mao D, Wang F, Chen L, et al. SiNCED1, a 9-cis-epoxycarotenoid dioxygenase gene in Setaria italica, is involved in drought tolerance and seed germination in transgenic Arabidopsis. Frontiers in Plant Science. 2023;**14**:1121809. DOI: 10.3389/fpls.2023.1121809

[11] Ali F, Qanmber G, Li F, Wang Z. Updated role of ABA in seed maturation, dormancy, and germination. Journal of Advanced Research. 2022;**35**:199-214. DOI: 10.1016/j.jare.2021.03.011

[12] Jia KP, Mi J, Ali S, Ohyanagi H, Moreno JC, Ablazov A, et al. An alternative, zeaxanthin epoxidase-independent abscisic acid biosynthetic pathway in plants. Molecular Plant. 2022;**15**:151-166. DOI: 10.1016/j.molp.2021.09.008

[13] Schwartz SH, Tan BC, Gage DA, Zeevaart JAD, McCarty DR. Specific oxidative cleavage of carotenoids by VP14 of maize. Science. 1997;**276**:1872-1874. DOI: 10.1126/science.276.5320.1872

[14] González-Guzmán M, Abia D, Salinas J, Serrano R, Rodríguez PL. Two new alleles of the abscisic aldehyde oxidase 3 gene reveal its role in Abscisic acid biosynthesis in seeds.

Plant Physiology. 2004;**135**:325-333. DOI: 10.1104/pp.103.036590

[15] Lin JH, Yu LH, Xiang CB. Arabidopsis nitrate regulated 1 acts as a negative modulator of seed germination by activating 3 expression. The New Phytologist. 2020;**225**:835-847. DOI: 10.1111/nph.16172

[16] Nakashima K, Yamaguchi-Shinozaki K. ABA signaling in stress-response and seed development. Plant Cell Reports. 2013;**32**:959-970. DOI: 10.1007/s00299-013-1418-1

[17] Yamamoto A, Yoshii M, Murase S, Fujita M, Kurata N, Hobo T, et al. Cell-by-cell developmental transition from embryo to post-germination phase revealed by heterochronic gene expression and ER-body formation in Arabidopsis leafy cotyledon mutants. Plant & Cell Physiology. 2014;**55**:2112-2125. DOI: 10.1093/pcp/pcu139

[18] Chen X, Han C, Yang R, Wang X, Ma J, Wang Y. Influence of the transcription factor ABI5 on growth and development in Arabidopsis. Journal of Plant Physiology. 2024;**302**:154316. DOI: 10.1016/j.jplph.2024.154316

[19] Finkelstein R. Abscisic acid synthesis and response. Arab Book. 2013;**11**:e0166. DOI: 10.1199/tab. 0166

[20] Okamoto M, Kushiro T, Jikumaru Y, Abrams SR, Kamiya Y, Seki M, et al. ABA 9′-hydroxylation is catalyzed by CYP707A in Arabidopsis. Phytochemistry. 2011;**72**:717-722. DOI: 10.1016/j.phytochem.2011.02.004

[21] Okamoto M, Kuwahara A, Seo M, Kushiro T, Asami T, Hirai N, et al. CYP707A1 and CYP707A2, which encode Abscisic acid 8′-hydroxylases, are indispensable for proper control of seed dormancy and germination in Arabidopsis. Plant Physiology. 2006;**141**:97-107. DOI: 10.1104/pp.106.079475

[22] Shani E, Hedden P, Sun TP. Highlights in gibberellin research: A tale of the dwarf and the slender. Plant Physiology. 2024;**195**:111-134. DOI: 10.1093/plphys/kiae044

[23] Otani M, Zheng L, Kawakami N. Genetic, epigenetic, and environmental control of seed dormancy and germination. Methods in Molecular Biology. 2024;**2830**:3-12. DOI: 10.1007/978-1-0716-3965-8_1

[24] Stawska M, Oracz K. phyB and HY5 are involved in the blue light-mediated alleviation of dormancy of Arabidopsis seeds possibly via the modulation of expression of genes related to light, GA, and ABA. International Journal of Molecular Sciences. 2019;**20**:5882. DOI: 10.3390/ijms20235882

[25] Tuan PA, Nguyen TN, Jordan MC, Ayele BT. A shift in abscisic acid/gibberellin balance underlies retention of dormancy induced by seed development temperature. Plant, Cell and Environment [Internet]. 2021;**44**:2230-2244. DOI: 10.1111/pce.13963

[26] Debeaujon I, Koornneef M. Gibberellin requirement for Arabidopsis seed germination is determined both by Testa characteristics and embryonic Abscisic Acid1. Plant Physiology. 2000;**122**:415-424. DOI: 10.1104/pp.122.2.415

[27] Farooq MA, Ma W, Shen S, Gu A. Underlying biochemical and molecular mechanisms for seed germination. International Journal of Molecular Sciences. 2022;**23**(15):8502

[28] Hedden P. The current status of research on gibberellin biosynthesis.

Plant & Cell Physiology. 2020;**61**:1832-1849. DOI: 10.1093/pcp/pcaa092

[29] Urbanova T, Leubner-Metzger G. Gibberellins and seed germination. In: Roberts JA, editor. Annual Plant Reviews Online. 2018. pp. 253-284. DOI: 10.1002/9781119312994.apr0538

[30] Yamaguchi I, Nakajima M, Park SH. Trails to the gibberellin receptor, gibberellin insensitive dwarf1. Bioscience, Biotechnology, and Biochemistry. 2016;**80**:1029-1036. DOI: 10.1080/09168451.2016

[31] Bouré N, Arnaud N. Molecular GA pathways as conserved integrators for adaptive responses. Plant Biology (Stuttgart, Germany). 2023;**25**:649-660. DOI: 10.1111/plb.13549

[32] Xu H, Lantzouni O, Bruggink T, Benjamins R, Lanfermeijer F, Denby K, et al. A molecular signal integration network underpinning Arabidopsis seed germination. Current Biology. 2020;**30**:3703-3712. DOI: 10.1016/j.cub.2020.07.012

[33] Alybayev S, Smekenov I, Kuanbay A, Sarbassov D, Bissenbaev A. Gibberellic-acid-dependent expression of α-amylase in wheat aleurone cells is mediated by target of rapamycin (TOR) signaling. Current Plant Biology. 2024;**37**:100312. DOI: 10.1016/j.cpb.2023.100312

[34] Bai B, Liu W, Qiu X, Zhang J, Zhang J, Bai Y. The root microbiome: Community assembly and its contributions to plant fitness. Journal of Integrative Plant Biology. 2022;**64**:230-243. DOI: 10.1111/jipb.13226

[35] Liu S, He F, Kuzyakov Y, Xiao H, Hoang DTT, Pu S, et al. Nutrients in the rhizosphere: A meta-analysis of content availability and influencing factors. Science Total Environment. 2022;**826**:153908. DOI: 10.1016/j.scitotenv.2022.153908

[36] Acuña JJ, Hu J, Inostroza NG, Valenzuela T, Perez P, Epstein S, et al. Endophytic bacterial communities in ungerminated and germinated seeds of commercial vegetables. Scientific Reports. 2023;**13**:19829. DOI: 10.1038/s41598-023-47099-4

[37] Reed RC, Bradford KJ, Khanday I. Seed germination and vigor: Ensuring crop sustainability in a changing climate. Heredity. 2022;**128**:450-459. DOI: 10.1038/s41437-022-00497-2

[38] Ge M, Wei X. Spermosphere bacteria promote Ormosia henryi seed germination by activating metabolic pathways. Forests. 2023;**14**:1136. DOI: 10.3390/f14061136

[39] Dutta S, Choi SY, Lee YH. Temporal dynamics of endogenous bacterial composition in rice seeds during maturation and storage and spatial dynamics of the bacteria during seedling growth. Frontiers in Microbiology. 2022;**13**:877781. DOI: 10.3389/fmicb.2022.877781

[40] Abdelfattah A, Tack AJM, Lobato C, Wassermann B, Berg G. From seed to seed: The role of microbial inheritance in the assembly of the plant microbiome. Trends Microbiology. 2023;**31**:346-355. DOI: 10.1016/j.tim.2022.10.009

[41] Zeng Q, Ding X, Wang J, Han X, Iqbal HMN, Bilal M. Insight into soil nitrogen and phosphorus availability and agricultural sustainability by plant growth-promoting rhizobacteria. Environmental Science and Pollution Research. 2022;**29**:45089-45106. DOI: 10.1007/s11356-022-20399-4

[42] Basu A, Prasad P, Das SN, Kalam S, Sayyed RZ, Reddy MS, et al.

Plant growth promoting Rhizobacteria (PGPR) as green bioinoculants: Recent developments constraints and prospects. Sustainability. 2021;**13**:1140. DOI: 10.3390/su13031140

[43] Chakraborty S, Venkataraman M, Infante V, Pfleger BF, Ané JM. Scripting a new dialogue between diazotrophs and crops. Trends Microbiology. 2024;**32**:577-589. DOI: 10.1016/j.tim.2023.08.007

[44] Zeng Q, Zhao Y, Han SW, D, Yang M. Seed-to-seed: Plant core vertically transmitted microbiota. Journal of Agricultural and Food Chemistry. 2023;**71**:19255-19264. DOI: 10.1021/acs.jafc.3c07092

[45] Das PP, Singh KR, Nagpure G, Mansoori A, Singh RP, Ghazi IA, et al. Plant-soil-microbes: A tripartite interaction for nutrient acquisition and better plant growth for sustainable agricultural practices. Environmental Research. 2022;**214**:113821. DOI: 10.1016/j.envres.2022.113821

[46] Deb CR, Tatung M. Siderophore producing bacteria as biocontrol agent against phytopathogens for a better environment: A review. South African Journal of Botany. 2024;**165**:153-162. DOI: 10.1016/j.sajb.2023.12.031

[47] Garg G, Kumar S, Bhati S. Siderophore in plant nutritional management: Role of endophytic bacteria. In: Maheshwari DK, Dheeman S, editors. Endophytes: Mineral Nutrient Management, Vol. 3; Sustainable Development and Biodiversity. Vol. 26. Berlin, Germany: Springer, Champ; 2021. DOI: 10.1007/978-3-030-65447-4_14

[48] Jalmi SK, Sinha AK. Ambiguities of PGPR-induced plant signaling and stress management. Frontiers in Microbiology. 2022;**13**:899563. DOI: 10.3389/fmicb.2022.899563

[49] Keswani C, Singh SP, Cueto L, García-Estrada C, Mezaache-Aichour S, Glare TR, et al. Auxins of microbial origin and their use in agriculture. Applied Microbiology and Biotechnology. 2020;**104**:8549-8565. DOI: 10.1007/s00253-020-10890-8

[50] Verbon EH, Liberman LM. Beneficial microbes affect endogenous mechanisms controlling root development. Trends in Plant Science. 2016;**21**:218-229. DOI: 10.1016/j.tplants.2016.01.013

[51] Chen L, Liu Y. The function of root exudates in the root colonization by beneficial soil rhizobacteria. Biology. 2024;**13**:95. DOI: 10.3390/biology13020095

[52] Arnaouteli S, Bamford NC, Stanley-Wall N, Kovács ÁT. Bacillus subtilis biofilm formation and social interactions. Nature Reviews Microbiology. 2021;**19**:600-614. DOI: 10.1038/s41579-021-00540-9

[53] Ajijah N, Fiodor A, Pandey AK, Rana A, Pranaw K. Plant growth-promoting bacteria (PGPB) with biofilm-forming ability: A multifaceted agent for sustainable agriculture. Diversity. 2023;**15**:112. DOI: 10.3390/d15010112

[54] Li Y, Narayanan M, Shi X, Chen X, Li Z, Ma Y. Biofilms formation in plant growth-promoting bacteria for alleviating agro-environmental stress. Science Total Environment. 2024;**907**:167774. DOI: 10.1016/j.scitotenv.2023.167774

[55] Lee J, Kim S, Jung H, Koo B-K, Han JA, Lee H-S. Exploiting bacterial genera as biocontrol agents: Mechanisms interactions and applications in sustainable agriculture. Journal of Plant Biology. 2023;**66**:485-498. DOI: 10.1007/s12374-023-09404-6

[56] Zhu L, Huang J, Lu X, Zhou C. Development of plant systemic resistance by beneficial rhizobacteria: Recognition initiation elicitation and regulation. Frontiers Plant Science. 2022;**13**:952397. DOI: 10.3389/fpls.2022.952397

[57] Trivedi P, Leach JE, Tringe SG, Sa T, Singh BK. Plant-microbiome interactions: From community assembly to plant health. Nature Reviews Microbiology. 2020;**18**:607-621. DOI: 10.1038/s41579-020-0412-1

[58] Poveda J, González-Andrés F. Bacillus as a source of phytohormones for use in agriculture. Applied Microbiology Biotechnology. 2021;**105**:8629-8645. DOI: 10.1007/s00253-021-11492-8

[59] Aloo BN, Dessureault-Rompré J, Tripathi V, Nyongesa BO, Were BA. Signaling and crosstalk of rhizobacterial and plant hormones that mediate abiotic stress tolerance in plants. Frontiers in Microbiology. 2023;**14**:1171104. DOI: 10.3389/fmicb.2023.1171104

[60] Khatoon Z, Huang S, Rafique M, Fakhar A, Kamran MA, Santoyo G. Unlocking the potential of plant growth-promoting rhizobacteria on soil health and the sustainability of agricultural systems. Journal of Environmental Management. 2020;**273**:111118. DOI: 10.1016/j.jenvman.2020.111118

[61] Upadhyay SK, Srivastava AK, Rajput VD, Chauhan PK, Bhojiya AA, Jain D, et al. Root exudates: Mechanistic insight of plant growth promoting rhizobacteria for sustainable crop production. Frontiers in Microbiology. 2022;**13**:916488. DOI: 10.3389/fmicb.2022.916488

[62] Bastías DA, Balestrini R, Pollmann S, Gundel PE. Environmental interference of plant-microbe interactions. Plant Cell Environment. 2022;**45**:3387-3398. DOI: 10.1111/pce.14455

[63] Rabbee MF, Ali MS, Choi J, Hwang BS, Jeong SC, Baek KH. Bacillus velezensis: A valuable member of bioactive molecules within plant microbiomes. Molecules (Basel). 2019;**24**:1046. DOI: 10.3390/molecules24061046

[64] Lahlali R, Ezrari S, Radouane N, Kenfaoui J, Esmaeel Q, El Hamss H, et al. Biological control of plant pathogens: A global perspective. Microorganisms. 2022;**10**:596. DOI: 10.3390/microorganisms10030596

[65] Nwachukwu BC, Babalola OO, Hassen AI. Rhizosphere competence and applications of plant growth-promoting rhizobacteria in food production-a review. Scientific African. 2024;**23**:e02081. DOI: 10.1016/j.sciaf.2024.e02081

[66] Srivastava S, Tyagi R, Sharma S. Seed biopriming as a promising approach for stress tolerance and enhancement of crop productivity: A review. Journal of the Science of Food and Agriculture. 2024;**104**:1244-1257. DOI: 10.1002/jsfa.13048

[67] Sher A, Sarwar T, Nawaz A, Ijaz M, Sattar A, Ahmad S. Methods of seed priming. In: Hasanuzzaman M, Fotopoulos V, editors. Priming and Pretreatment of Seeds and Seedlings. Singapore: Springer; 2019. DOI: 10.1007/978-981-13-8625-1_1

[68] Bayanati M, Lajayer BA, Al-Tawaha AR, Astatkie T. Seed bio-priming: An emerging tool towards improved germination and agricultural sustainability. In: Aftab T, editor. New Front Plant-Environment Interactions. Environmental Science Engineering. Berlin, Germany: Springer; 2023. DOI: 10.1007/978-3-031-43729-8_9

[69] Callan NW, Mathre D, Miller JB. Bio-priming seed treatment for biological control of Pythium ultimum pre-emergence damping-off in sh-2 sweet corn. Plant Disease. 1990;**74**:368-372. DOI: 10.1094/PD-74-0368

[70] Chaurasia J, Poudel B, Mandal T, Acharya N, Ghimirey V. Effect of micronutrients, rhizobium, salicylic acid, and effective microorganisms in plant growth and yield characteristics of green gram [*Vigna radiata* (L.) Wilczek] in Rupandehi, Nepal. Heliyon. 2024;**10**:e26821. DOI: 10.1016/j.heliyon.2024.e26821

[71] Das S, Kundu S, Meena K, Jha RK, Varma A, Bahuguna RN, et al. Seed biopriming with potential bioagents influences physiological processes and plant defense enzymes to ameliorate sheath blight induced yield loss in rice (*Oryza sativa* L.). World Journal of Microbiology and Biotechnology. 2023;**39**:136. DOI: 10.1007/s11274-023-03576-6

[72] Singh A, Patani A, Patel M, Vyas S, Verma RK, Amari A, et al. Tomato seed bio-priming with Pseudomonas aeruginosa strain PAR: A study on plant growth parameters under sodium fluoride stress. Frontiers in Microbiology. 2024;**14**:1330071. DOI: 10.3389/fmicb.2023.1330071

[73] Hussain S, Hussain S, Khaliq A, Ali S, Khan I. Physiological, biochemical, and molecular aspects of seed priming. In: Hasanuzzaman M, Fotopoulos V, editors. Priming and Pretreatment of Seeds and Seedlings. Singapore: Springer; 2019. DOI: 10.1007/978-981-13-8625-1_3

[74] Muñoz-Ramírez ZY, González-Escobedo R, Avila-Quezada GD, Ramírez-Sánchez O, Higareda-Alvear VM, Zapata-Chávez E, et al. Exploring microbial rhizosphere communities in asymptomatic and symptomatic apple trees using amplicon sequencing and shotgun metagenomics. Agronomy. 2024;**14**:357. DOI: 10.3390/agronomy14020357

[75] Nam NN, Do HDK, Loan Trinh KT, Lee NY. Metagenomics: An effective approach for exploring microbial diversity and functions. Food. 2023;**12**:2140. DOI: 10.3390/foods12112140

[76] Nwachukwu BC, Babalola OO. Metagenomics: A tool for exploring key microbiome with the potentials for improving sustainable agriculture. Front Sustainable Food Systems. 2022;**6**:886987. DOI: 10.3389/fsufs.2022.886987

[77] Singh P, Vaishnav A, Liu H, Xiong C, Singh HB, Singh BK. Seed biopriming for sustainable agriculture and ecosystem restoration. Microbial Biotechnology. 2023;**16**:2212-2222. DOI: 10.1111/1751-7915.14322

[78] de Andrade LA, Santos CHB, Frezarin ET, Sales LR, Rigobelo EC. Plant growth-promoting Rhizobacteria for sustainable agricultural production. Microorganisms. 2023;**11**:1088. DOI: 10.3390/microorganisms11041088

[79] Chepsergon J, Moleleki LN. Rhizosphere bacterial interactions and impact on plant health. Current Opinion in Microbiology. 2023;**73**:102297. DOI: 10.1016/j.mib.2023.102297

[80] Papadopoulou A, Matsi T, Kamou N, Avdouli D, Mellidou I, Karamanoli K. Decoding the potential of a new pseudomonas putida strain for inducing drought tolerance of tomato (Solanum lycopersicum) plants through seed biopriming. Journal of Plant Physiology. 2022;**271**:153658. DOI: 10.1016/j.jplph.2022.153658

[81] Miljaković D, Marinković J, Tamindžić G, Milošević D, Ignjatov M, Karačić V, et al. Bio-priming with bacillus isolates suppresses seed infection and improves the germination of garden peas in the presence of Fusarium strains. Journal of Fungi (Basel, Switzerland). 2024;**10**:358. DOI: 10.3390/jof10050358

[82] Rozier C, Gerin F, Czarnes S, Legendre L. Biopriming of maize germination by the plant growth-promoting rhizobacterium Azospirillum lipoferum CRT1. Journal of Plant Physiology. 2019;**237**:111-119. DOI: 10.1016/j.jplph.2019.04.011

[83] Fiodor A, Ajijah N, Dziewit L, Pranaw K. Biopriming of seed with plant growth-promoting bacteria for improved germination and seedling growth. Frontiers in Microbiology. 2023;**14**:1142966. DOI: 10.3389/fmicb.2023.1142966

[84] Roslan MAM, Zulkifli NN, Sobri ZM, Zuan ATK, Cheak SC, Abdul Rahman NA. Seed biopriming with P- and K-solubilizing *Enterobacter hormaechei* sp. improves the early vegetative growth and the P and K uptake of okra (*Abelmoschus esculentus*) seedling. PLoS One. 2020;**15**:e0232860. DOI: 10.1371/journal.pone.0232860

[85] Chaudhary P, Khati P, Chaudhary A, Gangola S, Kumar R, Sharma A. Bioinoculation using indigenous *Bacillus* spp. improves growth and yield of Zea mays under the influence of nanozeolite. 3 Biotech. 2021;**11**:1-12. DOI: 10.1007/s13205-020-02561-2

[86] Rojas-Sánchez B, Orozco-Mosqueda MDC, Santoyo G. Field assessment of a plant growth-promoting pseudomonas on phytometric, nutrient, and yield components of maize in a Milpa agrosystem. Agricultural Research. 2024;**10**:1-16. DOI: 10.1007/s40003-024-00756-0

[87] Toribio AJ, Jurado MM, Suárez-Estrella F, López MJ, López-González JA, Moreno J. Seed biopriming with cyanobacterial extracts as an eco-friendly strategy to control damping off caused by Pythium ultimum in seedbeds. Microbiological Research. 2021;**248**:126766. DOI: 10.1016/j.micres.2021.126766

[88] Mashabela MD, Tugizimana F, Steenkamp PA, Piater LA, Dubery IA, Mhlongo MI. Untargeted metabolite profiling to elucidate rhizosphere and leaf metabolome changes of wheat cultivars (*Triticum aestivum* L.) treated with the plant growth-promoting rhizobacteria *Paenibacillus alvei* (T22) and *Bacillus subtilis*. Frontiers in Microbiology. 2022;**13**:971836. DOI: 10.3389/fmicb.2022.971836

[89] Zayed O, Hewedy OA, Abdelmoteleb A, Ali M, Youssef MS, Roumia AF, et al. Nitrogen journey in plants: From uptake to metabolism, stress response, and microbe interaction. Biomolecules. 2023;**13**:1443. DOI: 10.3390/biom13101443

[90] Shaffique S, Khan MA, Wani SH, Pande A, Imran M, Kang SM, et al. A review on the role of endophytes and plant growth promoting rhizobacteria in mitigating heat stress in plants. Microorganisms. 2022;**10**:1286. DOI: 10.3390/microorganisms10071286

[91] Karimi E, Aliasgharzad N, Esfandiari E, Karimi E, Aliasgharzad N, Esfandiari E, et al. Biofilm forming rhizobacteria affect the physiological and biochemical responses of wheat to drought. AMB Express. 2022;**12**:93. DOI: 10.1186/s13568-022-01432-8

[92] Sharifi R, Ryu CM. Biogenic volatile compounds for plant disease diagnosis and health improvement. Plant

Pathology Journal. 2018;**34**:459-469. DOI: 10.5423/PPJ.RW.06.2018.0118

[93] Deshmukh AJ, Jaiman RS, Bambharolia RP, Patil VA. Seed biopriming–a review. International Journal of Economic Plants. 2020;7:038-043. DOI: 10.23910/2/2020.0359

[94] Panda D, Mondal S. Seed enhancement for sustainable agriculture: An overview of recent trends. Plant Archives. 2020;**20**:2320-2332. DOI: 10.51470/PLANT ARCHIVES.2019.v20

[95] Beveridge FC, Williams A, Adkins SW. Seed enhancement technologies to improve germination and emergence of Australian native Poaceae. Seed Science Research. 2020;**30**:293-303. DOI: 10.1017/S0960258520000276

[96] Lorenz C, Vitale E, Hay-Mele B, Arena C. Plant growth promoting rhizobacteria (PGPR) application for coping with salinity and drought: A bibliometric network multi-analysis. Plant Biology (Stuttgart, Germany). 2024;**26**:777-788. DOI: 10.1111/plb.13661

[97] Gupta R, Khan F, Alqahtani FM, Hashem M, Ahmad F. Plant growth-promoting rhizobacteria (PGPR) assisted bioremediation of heavy metal toxicity. Applied Biochemistry and Biotechnology. 2024;**196**:2928-2956. DOI: 10.1007/s12010-023-04545-3

[98] Han J, Zhang Y, Xi H, Zeng J, Peng Z, Ali G, et al. Maize, wheat, and soybean root traits depend upon soil phosphorus fertility and mycorrhizal status. Mycorrhiza. 2023;**33**:359-368. DOI: 10.1007/s00572-023-01126-4

[99] Shayanthan A, Ordoñez PAC, Oresnik IJ. The role of synthetic microbial communities (SynCom) in sustainable agriculture. Front Agronomy. 2022;**4**:896307. DOI: 10.3389/fagro.2022.896307

[100] Yin C, Hagerty CH, Paulitz TC. Synthetic microbial consortia derived from rhizosphere soil protect wheat against a soilborne fungal pathogen. Frontiers in Microbiology. 2022;**13**:908981. DOI: 10.3389/fmicb.2022.908981

[101] Pradhan S, Tyagi R, Sharma S. Combating biotic stresses in plants by synthetic microbial communities: Principles, applications and challenges. Journal of Applied Microbiology. 2022;**133**:2742-2759. DOI: 10.1111/jam.15799

[102] Parnell JJ, Vintila S, Tang C, Wagner MR, Kleiner M. Evaluation of ready-to-use freezer stocks of a synthetic microbial community for maize root colonization. Microbiology Spectrum. 2024;**12**:e0240123. DOI: 10.1128/spectrum.02401-23

Chapter 11

Unveiling the Role of Arbuscular Mycorrhiza in Seed Germination

Suman Dutta, Debashis Paul, Sourav Garai, Anirban Roy and Sudhir Kumar

Abstract

Seed germination is a pivotal stage in the life cycle of plants, marking the transition from dormancy to active growth. Arbuscular Mycorrhizal (AM) fungi form symbiotic associations with the roots of most terrestrial plants, profoundly influencing various aspects of plant growth and development. This chapter provides a comprehensive exploration of the intricate relationship between AM fungi and seed germination, spanning genetic, molecular, ecological, and practical dimensions. Moreover, insights into hormonal regulation, mycorrhizal networks, soil conditions, and ecological implications mediated by AM fungi in modulating seed germination processes are also discussed. The chapter also addresses the role of AM symbiosis in conferring abiotic stress tolerance to seeds, particularly under drought, salinity, and temperature stress. Practical applications and management strategies involving the utilization of AM fungi in agriculture and restoration ecology are outlined, emphasizing their potential to improve seed germination rates and facilitate ecosystem restoration efforts. Challenges and future directions are discussed, underscoring the need for continued research to fully harness the benefits of AM symbiosis for sustainable plant growth and ecosystem management. In conclusion, this chapter offers a comprehensive synthesis of the role of AM fungi in seed germination, emphasizing its multidimensional impacts and potential applications across various fields.

Keywords: arbuscular mycorrhiza, seed germination, nutrient availability, hormonal regulation, abiotic stress tolerance

1. Introduction

A specific group of fungi known as arbuscular mycorrhiza (AM) is essential to the complex system of interactions between soil and plants that keep ecosystems alive [1]. Plant growth and development depend on these symbiotic partnerships, which are defined by the connection between AM fungus and plant roots [2]. This study explores the many facets of AM fungus, including their extensive distribution in many environments and their structural and functional characteristics. Another crucial stage in the life cycle of plants is seed germination, which signifies the change from a dormant seed to an active, developing plant. Seed germination is a complicated process that is crucial for plant establishment and ecosystem dynamics because it is impacted by a wide range of physiological processes and environmental stimuli [3]. This chapter investigates the relationship between AM fungus and seed germination at the interface of these two fields. AM fungus modulates plant hormones and improves nutrient absorption, which can have a major effect on seedling establishment and development. Comprehending the communication channels and mycorrhizal networks between AM fungus and seeds/plants offers a valuable understanding of the manner in which microbial populations and soil conditions influence these interactions. This article also investigates the molecular and genetic bases of plant responses to AM fungus during seed germination. The investigation also looks at how abiotic stressors like salt, dryness, and severe temperatures affect how well AM symbiosis promotes seed germination. Lastly, the wider ecological effects of AM-mediated seed germination are discussed, with a focus on succession patterns and community dynamics.

2. Understanding arbuscular mycorrhiza and seed germination

AM (Arbuscular Mycorrhiza) is an obligate eukaryote in nature involved in a unique mutualistic association with eukaryotic plants, which not only colonize root tissue but shoot, rhizomes, etc. [4]. AM fungi form mycorrhizal networks by connecting the roots of multiple plants with multiple species, causing positive fitness in both organisms [5]. Coevolution and reciprocal genetic changes of ancestral free-living fungi and ancestral terrestrial plants may have played a major role in this association [6]. AM are endomycorrhiza group that forms tree-like branches inside plants upon symbiosis and are obligate. AM are members of the ancient fungus group, which enhances nutrient acquisition among symbionts through signal transduction [7]. They are a member of the monophyletic group *Glomeromycota*. Pre-penetration appendages make epidermal bridges for plant and AM fungi communication through hyphae [8]. A specialized hyphal network for nutrient and water uptake in AM fungi helps in plant association, which completes its lifecycle using plant carbon as a nutrient source [9]. Besides, the plant supplies lipids to AM fungi, which enhance the fungi life cycle [10].

The transition of an inactive embryo into an active growth phase occurs through seed imbibition and cell elongation. This process enhances cell numbers and initiates new growth in plant seeds through germination, which also signals metabolic activation upon hydration. The emergence of seedlings from embryos, with the potential to develop into healthy plants, completes the seed germination process [11]. Seed germination is characterized by the activation of mitochondrial enzymes, increased gibberellic acid (GA) biosynthesis, and a heightened requirement for oxygen. This leads to the hydrolysis of starch, proteins, lipids, and phytic acids, followed by the flow of sucrose to the embryonic axis [12]. Various phytohormones play crucial roles in seed germination and dormancy, with the abscisic acid (ABA) ratio being particularly important due to their antagonistic relationship [13]. Additionally, strigolactone enhances seed germination by increasing α-amylase activity [14].

3. AM fungi influence on seed germination

Phytohormones and enzyme releases from AM fungi enhance seed germination and help convert nutrients into bioavailable forms, promoting plant tolerance under nutrient-deficient conditions [15]. The inactivation of GA (gibberellic acid), which leads to mycorrhizal symbiosis and enhanced germination, has also been observed [16]. Orchid seeds, which lack endosperm, require external nutrient sources for germination. The initial absence of photosynthesis reduces the carbon supply to AM fungi, which is supplemented until the first green leaf emerges [17]. The potential of AM fungi to enhance seed germination is significant across various orchid species, improving the quality and quantity of secondary metabolites in some plants [18]. Mycorrhizal associations are crucial for the successful *in vitro* germination of several endangered orchid species [3, 19]. AM fungi are particularly beneficial for terrestrial orchids, supplying nitrogen and phosphorus in exchange for carbon, thus enhancing germination even under low light conditions [20]. AM fungi have been reported as germination enhancers for *Cinchona officinalis* [21]. They also improve nitrogen nutrition from organic matter through symbiosis, converting nutrients into bioavailable forms [22]. Acting as biofertilizers, AM fungi promote growth and protection from multiple abiotic stresses [23]. Under heavy metal contamination, antioxidant activity reduces. Since antioxidants are crucial for stress tolerance and germination, AM fungi help phytostabilize heavy metals [24, 25]. The light requirement varies among terrestrial orchids, but AM fungi associations mitigate the effects of limited light, enhancing germination [20, 26]. The presence of trehalase in the orchid genome underscores the importance of AM fungi in orchid germination [27]. Seed coating with fungal mycelia reduces costs and enhances germination efficiency and growth [28]. The universal effect of AM fungi on orchid seed germination supports conservation efforts for critically endangered orchids by providing essential initial support (**Table 1**). In conclusion, AM fungi play a crucial role in enhancing seed germination, nutrient uptake, and plant tolerance to various stresses, making them invaluable for agricultural practices, ecological

Role of AM fungi	Effect	Reference
Protection from High Light	Enhances Germination in Orchid	Alghamdi et al. [20]
Protection from Harsh environments, Sustainable Crop production	Acts as Biofertilizer, Low chemical input	Begum et al. [23]; Lanfranco et al. [29]
Nitrogen and Phosphate Acquisition and Nutrient Cycling	Growth Promotion	Artursson et al. [30]; Bonfante and Genre [31]
Plant Growth	Synergistic effect for growth	Nanjundappa et al. [32]
GA inactivation	Seed germination	Miura et al. [16]
Protection of Endangered Orchids	Seed germination and seedling growth of *Paphiopedilumhirsutissimun*	Tian et al. [3]
Disease resistance in Nonhost species	Multifunctional Systemic resistance	Fernández et al. [33]
Ecosystem Function and Global Nitrogen Cycle	Enhanced Nutrient availability	Hestrin et al. [22]
Nutrient Exchange	Mutualism of Survivability	Parniske et al. [7]
Secondary Metabolite Secretion	Disease and Stress resistance	Smith and Read [34]
Phytostabilization	Heavy Metal Toxicity reduction	Bano and Ashfaq [24]

Table 1.
Effect of beneficial role of AM fungi in plants physiology and seed germination.

restoration, and the conservation of endangered species. Strigolactone secretion in the rhizosphere induces AM fungi association in plant roots by attracting the fungi and promoting hyphal branching. A complex lipopolysaccharide '*Myc*' factor initiates AM fungal association with plants through a Common Signaling Symbiosis Pathway (CSSP), facilitated by N-acetylglucosamine-based molecules via a tri-level structured hierarchy of perception, transmission, and transcription of lipopolysaccharide modulation [35]. Continuous oscillation and spiking of Ca^{2+} in the root hair nuclear membrane enhance root colonization and symbiosis [31]. While strigolactone enhances the germination of parasitic plants, its role in enhancing AM fungi interaction also inhibits the germination of these parasitic plants [36]. The association between fungi and plants helps transfer multiple responses, including defense signals and allelochemicals, between plants [5].

4. Genetic and molecular insights

Most terrestrial plants have symbiotic relationships with AM fungi in their roots, which are essential for nutrient absorption, stress tolerance, and overall plant health. Plant-AM fungal interactions during seed germination influence numerous physiological processes that affect seedling establishment and future development. Understanding the genetic basis of these interactions is crucial for deciphering the underlying processes and harnessing their potential for agricultural benefits [37]. Researchers are investigating the genetic basis of plant responses to AM fungi during seed germination to better understand the complex mechanisms regulating these symbiotic relationships (**Table 2**). This work focuses on examining various genetic factors that influence the initiation and development of these symbiotic interactions, which in turn affect seed germination and subsequent plant growth. By analyzing these genetic components, researchers aim to uncover the molecular pathways and regulatory systems that govern plant-fungus interactions. This research not only addresses fundamental questions in plant biology but also has the potential to lead to improved agricultural practices [44]. The

S.N.	AM (type/Strain)	Gene/QTL/Transcription factor involved	Response	References
1	*Glomus intraradices*	*NPR1* (Non-expressor of Pathogenesis-Related Genes 1)	Reduced reactive oxygen species (ROS) activity	Zhang et al. [38]
2	*Funneliform ismosseae*	*LjERF1* (Ethylene Responsive Factor 1)	Inhibit seedling growth	Berrocal-Lobo [39]
3	*Rhizophagus irregularis*	*PT4* (Phosphate Transporter 4)	Increase in enzyme activity	Zhang et al. [40]
4	*Gigaspora margarita*	*MYB72* (Transcription Factor *MYB72*)	Increased Root Exudation	Massalha et al. [41]
5	*Glomus mosseae*	*PHT1* (Phosphate Transporter 1)	Increase (P) phosphate uptake	Victor Roch et al. [42]
6	*Rhizophagus clarus*	*RAM1* (Required for ArbuscularMycorrhization 1)	Increase (Colonization Efficiency)	Oliveira et al. [43]

Table 2.
Plant responses to major AM strain and gene/factors implicated.

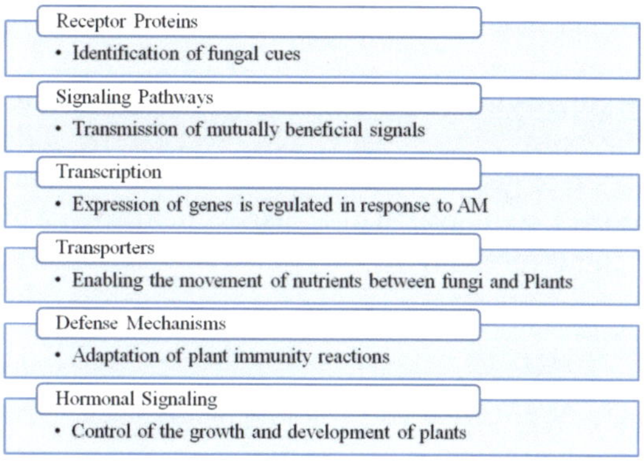

Figure 1.
Essential genetic components and roles of AM fungus.

main genetic variables linked to plant responses to AM fungi during seed germination are illustrated in **Figure 1**.

4.1 Genetic variables affecting plant responses to AM fungi

Understanding the molecular mechanisms behind plant-AM fungus symbiosis is essential for enhancing agricultural practices. Key signaling pathways, including those mediated by strigolactones, phosphate availability, and plant hormones, along with transcriptional regulation and complex regulatory networks, play crucial roles in establishing and maintaining these symbiotic relationships.

4.1.1 Signaling pathways

Important signaling molecules that facilitate the identification and formation of symbiotic connections between plants and AM fungus include strigolactones and jasmonic acid. Genetic research has revealed the receptors and downstream signaling elements involved in these pathways, providing insight into the molecular processes controlling the start of symbiosis [45]. The three main signaling pathways in plant and fungi interactions are described with the following:

a. *Strigolactone Signaling*: The significance of strigolactone receptors in plant-fungus interactions has been highlighted by recent research that has clarified their role in mediating the formation of symbiosis and emphasizing downstream signaling components [46].

b. *Phosphate Signaling*: The symbiotic connection between plants and AM fungus is regulated by the availability of phosphate. In the symbiotic interface, phosphate transporters and regulatory proteins control phosphate signaling pathways, which affects fungal colonization and nutrition exchange [47].

c. *Plant Hormones*: Plant-fungi interactions involve a variety of hormonal signaling pathways, such as those involving ethylene, cytokinins, and jasmonic acid. Many features of symbiosis, including colonization, nutrition transport, and defensive responses, are regulated by crosstalk between hormone pathways and symbiosis-related genes [48].

4.1.2 Transcriptional regulation

In response to AM fungus invasion, transcription factors are essential in coordinating changes in gene expression. Research has revealed transcriptional regulators that affect the expression of genes related to defense mechanisms, nutrition transport, and the formation of symbiosis. These discoveries provide light on the genetic regulation of plant-fungus interactions [40].

4.1.3 Regulatory networks

Complex regulatory circuits that control how plants react to AM fungus are created by the integration of many signaling pathways and transcriptional networks. These regulatory networks are starting to be untangled by genetic studies, which emphasize the linkages between various pathways and the hierarchical structure of gene regulation in symbiotic relationships [49].

4.2 Molecular signaling between AM fungus and seeds/plants

Molecular signaling is a key component in understanding the complex interactions that result in symbiotic relationships that are vital for nutrient exchange and plant health between seeds and AM fungus [37]. Several studies clarify the signaling processes controlling the establishment and maintenance of symbiosis through an examination of current developments in this field, providing insights into the molecular underpinnings of plant-fungus interactions and their consequences for agriculture. Agriculture will be greatly impacted by our understanding of the molecular signaling pathways and genes associated with symbiosis that is engaged in plant-fungus interactions. By using this information, methods for increasing crop resistance to environmental stressors, increasing nutrient absorption efficiency, and encouraging sustainable agriculture practices may be developed [50]. There are symbiosis-related genes for the connections that play important roles.

a. *Mycorrhiza-Induced Genes*: When AM fungi colonize plants, they cause the expression of genes linked to symbiosis, such as those that code for phosphate transporters (PTs) specialized to symbiosis and proteins called Arbuscular Mycorrhiza-induced proteins (MIPs). These genes create an environment that is favorable for symbiotic contact and allows the flow of nutrients [2].

b. *Transcription Factors*: Transcription factors, particularly those from the myeloblastosis (MYB) and gibberellin-acid insensitive (GAI)-repressor of GA1 (RGA)- and -scarecrow (SCR) (GRAS) transcription factors families, play a pivotal role in orchestrating plant-fungal interactions. These proteins regulate the activation of genes involved in symbiosis by responding to fungal presence. Their function is essential in the genetic changes that occur when plants and fungi establish a symbiotic relationship [37].

5. Abiotic stress tolerance

Exciting new insights into the relationships between plants and fungi have been revealed by studies on how AM fungi promote seed germination under severe salt

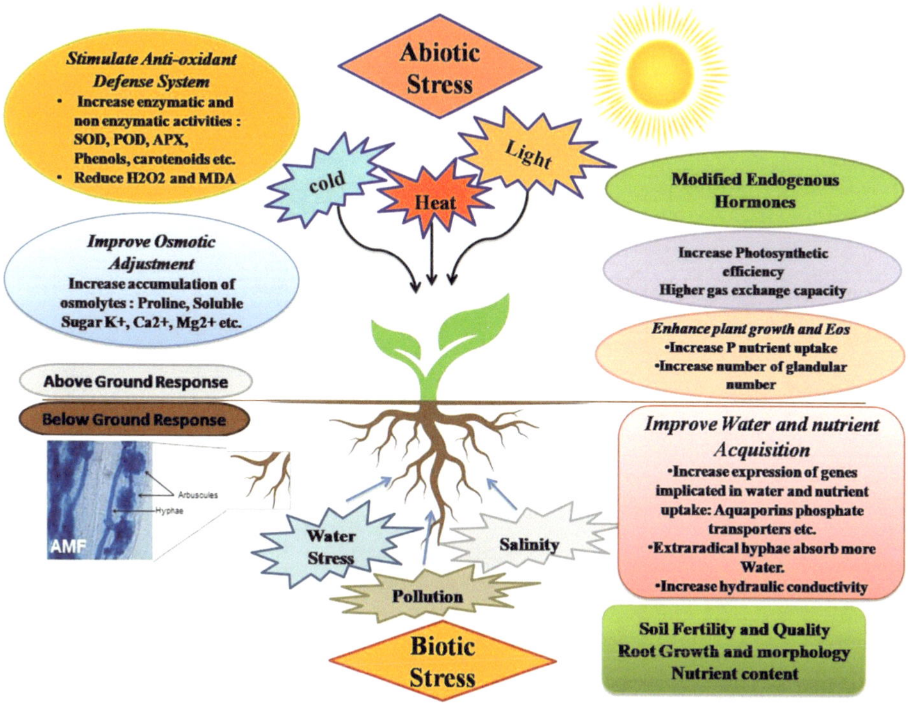

Figure 2.
Response of AM fungi symbiosis with host under stress.

and water stress conditions (**Figure 2**). AM fungi are crucial mediators that help plants adapt and become resilient to drought and salinity (**Table 3**). By enhancing water intake and increasing water use efficiency through the expansion of the plant root system, AM fungi improve seed germination under water stress [75]. Additionally, these fungi aid in the production and accumulation of osmoprotectants like proline and trehalose, which mitigate the negative effects of water deficit on seed germination. Similarly, by regulating ion homeostasis, reducing sodium absorption, and increasing potassium accumulation in plant tissues, AM fungi play a vital role in alleviating salt stress in high salinity conditions. Furthermore, AM fungi stimulate the production of antioxidant enzymes and compatible solutes, which reduces oxidative stress and enhances seed germination in saline environments. Overall, understanding the mechanisms behind the symbiotic relationships between plants and AM fungi holds potential for developing strategies to increase crop resistance to salinity and drought, thereby advancing sustainable agriculture in affected areas [76].

6. Community dynamics and succession

AM fungi play a pivotal role in enhancing plant growth and survival rates, facilitating phosphorus uptake, and shaping the dynamics and succession of plant communities. Smith and Read [34] assert that the mutualistic relationship between AM fungi and plants, initiated during seed germination, results in stronger seedling development and higher germination success rates. Compared to non-inoculated controls,

S.N.	Abiotic stress	Crop	Arbuscular mycorrhiza (Type/Strain)	AMF inoculation-responses/Enhancement in germination % and other traits	References
1.	Salinity	*Acacia nilotica*	*Glomus fasciculate*	Shoot and root dry weight are improved along with Zn, Cu, and P contents.	Giri et al. [51]
2.	Drought	Tomato	*Glomus aggregatum*	Alleviated oxidative stress and increased proline accumulation.	Chandrasekaran et al. [52]
3.	Flooding tolerance	*Lonicera japonica*	*G. versiforme*	Cd content is reduced in roots and shoots.	Jiang et al. [53]
4.	Salinity	Maize	*Rhizophagus irregularis*	Enhanced seedling length and biomass.	Chen et al. [1]
5.	Flood tolerance	*Panicum hemitomon Leersia hexandra*	*Acaulospora trappei, Glomus leptotichum, Scutellosporaheterogama, Glomus leptotichum, G. etunicatum and G. gerdemannii*	Enhanced phosphorus content in plants.	Miller et al. [54]
6.	Salinity	Maize	*Glomus intraradices*	Enhanced expression of stress-responsive genes and osmolyte accumulation.	Jahromi et al. [55]
7.	Drought	Tomato	*Rhizophagus irregularis*	Improved water use efficiency and reduced transpiration rate.	Leventis et al. [56]
8.	Flood tolerance	*Trifoliumpratense*	*Glomus mosseae*	Root and Shoot concentrations are decreased along with reduction in Zn uptake.	Li et al. [57]
9.	Drought	Barley	*Rhizophagus irregularis*	Improved germination rate and seedling establishment.	Khalvati et al. [58]
10.	Salinity	*Cucumis sativus*	*Glomus mosseae, Glomus etunicatum, Glomus intraradices*	Biomass is increased, Biosynthesis of antioxidant enzymes and photosynthetic pigments.	Hashem et al. [59]
11.	Flood tolerance	*Zea mays*	*Glomus isolates*	Enhanced production of Mg, P and K are increased, improved dry weight and distinction in the distribution of essential and heavy metals in cells.	Kaldorf et al. [60]
12.	Flood tolerance	*Trigonellafoenum-graceum*	*F. mosseae*	Improves the plant growth and yield.	Kelkar et al. [61]
13.	Salinity	Wheat	*Rhizophagusintraradices*	Enhanced root morphology and sodium exclusion.	Elgharably et al. [62]
14.	Flood tolerance	*Populus alba Populusnigra*	*R. irregularis or F. mosseae*	Alleviation in the Zn and Cu phytotoxicity.	Lingua et al. [63]

S.N.	Abiotic stress	Crop	Arbuscular mycorrhiza (Type/Strain)	AMF inoculation-responses/Enhancement in germination % and other traits	References
15.	Salinity	*Aeluropus littoralis*	*Claroideoglomus etunicatum*	Stomatal conductance, root and shoot dry biomass, free alpha amino acids, soluble sugars and Na and Cl uptake is increased.	Hajiboland et al. [64]
16.	Flood tolerance	*Pterocarpus officinalis*	*Glomus intraradices*	Higher plant growth and phosphorus content in leaves.	Fougnies et al. [65]
17.	Salinity	Soybean	*Funneliform ismosseae*	Enhanced nodulation and nitrogen fixation capacity.	Amani Machiani et al. [66]
18.	Flood tolerance	*Aster tripolium*	*Glomus geosporum*	Higher sugars and proline content.	Solís-Rodríguez et al. [67]
19.	Salinity	Tomato	*Rhizophagusirregularis*	Growth hormone promotes plant health, increasing root and shoot weight and improving leaf structure.	Hashem et al. [68]
20.	Salinity	*Leymus chinensis*	*Glomus mosseae*	Positive outcomes include elevated phosphorus and nitrogen levels, enhanced seedling weight, and increased plant water content.	Liang et al. [69]
21.	Salinity	*Cucumis sativus* L.	*Claroideoglomus etunicatum, Rhizophagus intraradices, Funneliformis mosseae*	Enhanced growth, higher antioxidant enzyme activity, elevated proline and phenolic content, and improved uptake of vital mineral elements were observed. Furthermore, the absorption of sodium ions was reduced.	Shahvali et al. [70]
22.	Salinity	*Medicago sativa*	*Glomus mosseae*	Among the crucial nutrients for plants, phosphorus (P), nitrogen (N), and potassium (K) play pivotal role.	Parihar et al. [71]
23.	Cold stress	*Cucumis sativus L.*	*Rhizophagusirregularis*	Photosynthetic efficiency and carbon sink both showed improvement.	Mirshad et al. [72]
24.	High temperature stress	Grapevine *(Vitis vinifera L.)*	*Rhizoglomus irregulare, Funneliformis mosseae*	The growth rate increased alongside enhanced substrate carbon conversion efficiency and stomatal conductance.	Rani et al. [73]
25	Drought	*Ipomoea batata*	*Glomus spp.*	Osmo-protectants played a role in adjusting osmotic potential.	Pal et al. [74]

Table 3.
AM interactions with various crops under abiotic stress conditions.

Bever [77] found that seedlings inoculated with AM fungi exhibit higher germination success, providing these plants with a competitive advantage for resources such as light, water, and nutrients, potentially altering competitive hierarchies. Furthermore, AM fungi significantly influence plant succession, aiding pioneer species in establishing themselves in primary succession by enhancing soil nutrient availability [78]. In secondary succession, AM fungi promote ecosystem recovery and stability by facilitating plant reestablishment following disturbances like fire or human activity [79]. The abundance and diversity of plant species within ecosystems are profoundly affected by AM fungi. While mycorrhizal interactions can sustain biodiversity by promoting cohabitation, they may also confer competitive advantages to certain species, potentially reducing overall biodiversity [80]. Johnson et al. [81], who note that different species offer distinct advantages to host plants, influencing interspecies interactions and community composition, highlight the importance of functional diversity in AM fungi. Mycorrhizal networks, according to Van der Heijden et al. [82], enhance community resilience and stability by facilitating nutrient flow across plants, particularly in stressful conditions. Environmental factors and the functional diversity of AM fungi significantly influence these interactions [81, 83]. Overall, AM fungi enhance nutrient uptake and competitiveness, thereby influencing community structure and dynamics, while also fostering cohabitation and biodiversity within ecosystems.

7. Practical applications

AM symbiosis is a critical component for sustainable agricultural systems, offering numerous benefits such as enhanced seed germination, plant growth, host plant nutrition, and mineral cycling [84]. AM fungi absorb nutrients and water through two primary modes. Initially, in the asymbiotic phase, hyphal morphogenesis occurs when root exudates from host plants become available [85]. The germling hyphae, which elongate and branch, physically contact host roots, penetrate root cells using appressoria, and form arbuscules within the root cortex [86, 87]. This extensive hyphal network enhances absorption capacity and surface area, thereby improving the uptake and translocation of essential nutrients like phosphorus and nitrogen [88]. These extraradical hyphae can access nutrients and water from fine soil crevices beyond the reach of roots [89]. The expansive mycelial network significantly increases the absorptive surface area and aids in soil moisture retention [90]. Beyond nutrient and water uptake, AM fungi play a vital role in improving soil structure and quality. Their external hyphal networks promote soil aggregation by creating a skeletal structure in the mycorrhizosphere. AM fungi enhance soil structure by releasing various organic compounds, including the highly effective glomalin protein, which binds soil particles, stabilizing these aggregates for up to 6 months even after the hyphal network has disappeared. Arbuscular mycorrhization boosts soil organic matter content and water-holding capacity [75, 91], contributing to soil ecosystem conservation. Additionally, the extended hyphae help mitigate water deficits in dry soils and reduce evaporation [92]. AM fungi play a significant role in nutrient solubilization by secreting acid and alkaline phosphatases and organic acids, which help mineralize and release nutrients [93]. The uptake of nutrients, particularly phosphorus (P), is influenced by the physiology of both the plant and the fungus. The translocation of nutrients occurs through the processes of loading and unloading via extracellular hyphae and cortical arbuscules, and this is affected by the plant's nutrient requirements, arbuscular mycorrhizal fungi (AMF) species, strains, and environmental conditions [94]. AM fungi are particularly crucial for phosphate uptake in P-deficient soils, as they mobilize P from rock phosphate [95]. Additionally, AM hyphae can decompose larger organic molecules [23], facilitating nitrogen transfer from organic matter to plant tissues and thus increasing plant biomass [96]. The AM hyphal network also aids in the uptake of potassium [97] and essential micronutrients such as magnesium, zinc, copper, calcium, sulfur, sodium, manganese, boron, molybdenum, and iron, all of which are vital for plant growth [64]. By participating in nutrient cycling, AM fungi ensure adequate nutrient availability even in infertile or less fertile soils [98]. Additionally, AM symbiosis significantly boosts the levels of chlorophyll, carotenoids, and phenolics in plants [99]. This early enhancement in germination and growth improves plant vigor and reproductive health, ultimately increasing yield. The positive effects of AM inoculation on plant growth and productivity have been well-documented [100]. Recent studies on crops such as tomato, rice, wheat, maize, yam, and potato have demonstrated these benefits [101]. Furthermore, AM colonization has been shown to enhance food quality by increasing levels of antioxidants, flavonoids, and vitamin C [102]. AM can serve as a valuable amendment to improve soil fertility, crop productivity, yield quality, and to revitalize agro-ecosystems [103]. Incorporating the benefits of AM inoculation into farming practices can promote sustainable agricultural systems by enhancing plant growth and product quality [104]. Additionally, AM increases nodule formation in leguminous plants and enhances free nitrogen fixation [105]. By acting as a biofertilizer, AM can reduce the need for chemical fertilizers in crop production. Consequently, this reduces reliance on chemical fertilizers. AM plants also produce beneficial phytochemicals like carotenoids and flavonoids, which help reduce oxidative damage and benefit human health [106].

8. Management strategies

8.1 Fertilizer management

The application of slow-releasing fertilizers, such as rock phosphate, can enhance AM colonization and activity. Rock phosphate, as an unconventional yet effective alternative to inorganic phosphate fertilizers, promotes mycorrhizal growth and function [84]. Plants colonized by AM and supplemented with rock phosphate exhibit greater growth compared to non-AM plants, given double the amount of superphosphate. Additionally, low to moderate nitrogen levels boost AM colonization, sporulation, plant growth, and root development. Initial nitrogen supply can sometimes aid the establishment of mycorrhizae [107]. AM fungi perform better even with reduced fertilizer doses [108]. Low to moderate doses of fertilizer, especially organic ones, enhance AM-mediated plant growth and biocontrol capabilities. Without mycorrhizal application, increased phosphorus supply can decrease biomass, while low phosphorus levels promote root branching for better phosphorus acquisition [109]. Mycorrhizae increase biomass at low fertilization levels, but high fertilization levels reduce biomass. Mycorrhizae also improve resistance to herbivores at medium fertilization levels, with the strongest resistance observed when plants are fertilized with phosphorus-rich organic fertilizers [110]. Modifying conventional agricultural practices toward an integrated eco-friendly approach, such as avoiding over-fertilization and applying beneficial soil microorganisms and mycorrhizae-helper bacteria, can improve phosphorus use efficiency, reduce phosphorus toxicity in soil, and minimize phosphorus leaching into groundwater [111].

8.2 Tillage improvement

Conservation agriculture practices, which include minimal soil disturbance and retaining more than 30% of crop residue, are increasingly adopted globally. These practices are known to enhance soil health by optimizing key soil attributes. Conservation tillage combined with the application of organic manure can support the survival and inoculation of beneficial organisms, improve soil aggregation, and enhance phosphorus uptake [112].

8.3 Choice of crops, cultivars, and management

Choosing highly mycotrophic crops and cultivars with root architectures that efficiently access phosphorus and form active symbioses with arbuscular mycorrhizal fungi (AMF) is essential [86]. It is beneficial to avoid crop rotations involving non-mycorrhizal families like Brassicaceae and Amaranthaceae, as they release fungi-static compounds into the soil. Instead, growing mycorrhizal cover crops after these non-mycorrhizal crops can prepare the soil for the next planting [113]. Domestication may have reduced the ability of plants to respond positively to AMF in high-phosphorus soils. Mathur et al. [114] found that both wild and domesticated species of 27 crops benefited similarly from AMF under low phosphorus conditions. However, under high phosphorus conditions, the response of wild varieties to AMF did not differ significantly, whereas growth was strongly reduced in domesticated species. This suggests that domesticated crops only benefit from mycorrhizal associations at low soil phosphorus concentrations. Although AM fungi do not have strict host specificity, they do exhibit host preferences, which can vary with geographical distribution and land use [115]. The AMF community in different hosts may respond similarly to soil phosphorus gradients. Crop rotation with highly mycotrophic crops from families such as Fabaceae or Poaceae can positively influence AM function. While AM species composition may vary with plant species and take time to adjust, the subsequent mycotrophic crop will benefit from reduced phosphorus requirements compared to fallow land without vegetation [116].

8.4 Soil-host compatibility

The native arbuscular mycorrhizal fungi (AMF) flora in upland ecosystems has shown high efficiency and responsiveness to upland rice. Crop rotation with rice has been found to enhance soil phosphorus content and boost native AM inoculum [117]. Although AMF generally does not thrive in wetland habitats, and their application is minimal in lowland rice cultivation, some species like *Glomus etunicatum*, *G. mosseae*, and *G. intraradices* have performed well. These species increase phosphorus uptake and root colonization under flooded conditions, in both high- and low-fertility soils, thereby promoting nutrient acquisition and yield in rice [118]. Additionally, *Glomus intraradices* has been found to enhance growth response, photosynthetic efficiency, and antioxidative responses in rice plants under drought stress [119]. Host-fungi compatibility is a significant factor, as intraspecific strain diversity impacts the efficiency of AMF. Host cultivars exhibit differential responses to AMF, with various genotypes showing different behaviors in distinct environments [120]. The molecular mechanisms determining fungal performance in plants are largely unknown and may relate to the amounts of carbohydrates and lipids shared. Studies have shown that the expression pattern of monosaccharide transporter genes in *Rhizophagus irregularis* varies between intraradical and extraradical hyphae depending on the host plant and phosphorus uptake [121]. Biomass gain correlates with extraradical mycelial mass, indicating complex genotype-environment interactions [122]. Research on molecular diversity in roots has revealed differences in AM community composition associated with wheat and nitrogen-fixing crops [115]. The AM community composition associated with wheat varies with the growing season, phosphorus fluxes, and fertilization levels [123]. Dominant AM species differ between conventional (*Funneliformis* spp.) and organic systems (*Claroideoglomus* spp.) [124], highlighting the variation in AM efficiency with different fertilizers, especially phosphorus [125]. Soil-AM strain compatibility is also crucial for effective AM application, as soil conditions determine the most effective AM species for a given crop. Studies by Aguilera et al. [126] found *Acaulospora* and *Scutellospora* to be dominant in acidic soils under continuous wheat cropping, while Castillo et al. [127] reported a prevalence of *Acaulospora* and *Claroideoglomus* in similar conditions. Thus, applying a few AM inocula indiscriminately in any soil for any crop may not be effective. The growth response of a single host crop species can vary with different AM fungal species, and similarly, the same AM fungal isolate can lead to different growth responses with various plant species, cultivars, or genotypes [128]. Soil conditions have been found to control native AM dominance and effectiveness in the rhizosphere, as observed in rice fields in Ghana with *Rhizophagus*, *Glomus*, *Scutellospora*, and *Acaulospora* [129].

9. Conclusion future and directions

The practical applications of AM symbiosis offer promising solutions for sustainable agricultural systems, contributing to enhanced seed germination, plant growth, host plant nutrition, and mineral cycling. AM fungi play crucial roles in nutrient absorption and water uptake, influencing soil structure and quality and promoting ecosystem stability. However, despite these benefits, several challenges and future directions need to be addressed to fully leverage the potential of AM symbiosis in agricultural management. One major challenge lies in optimizing fertilizer management strategies to enhance AM colonization and activity. While slow-releasing fertilizers like rock phosphate can promote mycorrhizal growth and function, balancing fertilizer doses, especially nitrogen levels, is crucial for maximizing AM-mediated benefits without compromising plant growth. Integrating organic fertilizers and minimizing over-fertilization can further enhance phosphorus use efficiency and reduce phosphorus leaching, contributing to sustainable soil fertility management. Improving tillage practices, such as adopting conservation agriculture techniques and retaining crop residues, can support the survival and inoculation of beneficial microorganisms, including AM fungi. Conservation tillage combined with organic manure application can enhance soil aggregation and phosphorus uptake, thereby improving soil health and crop productivity. The choice of crops, cultivars, and crop management practices also plays a significant role in optimizing AM symbiosis. Selecting highly mycotrophic crops and cultivars with efficient root architectures can enhance phosphorus acquisition and symbiotic interactions with AM fungi. Avoiding crop rotations involving non-mycorrhizal families and promoting mycorrhizal cover crops can prepare the soil for optimal mycorrhizal colonization and subsequent plant growth. Ensuring soil-host compatibility is essential for effective AM application, particularly in diverse agricultural ecosystems. Understanding the native AM fungal flora and its responsiveness to different host plants and soil conditions can guide the selection of suitable AM inocula for specific crops and soil types. Molecular diversity studies can further elucidate the genotype-environment interactions influencing AM fungal performance and host plant responses. In conclusion, addressing these challenges and implementing targeted management strategies can enhance the practical applications of AM symbiosis in agriculture, leading to improved soil fertility, crop productivity, and sustainability. By integrating AM symbiosis into agricultural practices, we can foster resilient and environmentally friendly farming systems that promote long-term soil health and food security.

Author details

Suman Dutta[1], Debashis Paul[2], Sourav Garai[1], Anirban Roy[1] and Sudhir Kumar[3]*

1 Ramakrishna Mission Vivekananda Educational and Research Institute, West Bengal, India

2 ICAR-Central Institute for Cotton Research, Regional Station, Sirsa, Haryana, India

3 ICAR-Indian Agricultural Research Institute, New Delhi, India

*Address all correspondence to: sudhirnpf@gmail.com

References

[1] Chen Q, Deng X, Elzenga JTM, Van Elsas JD. Effect of soil bacteriomes on mycorrhizal colonization by *Rhizophagus irregularis*—Interactive effects on maize (Zea mays L.) growth under salt stress. Biology and Fertility of Soils. 2022;**58**(5):515-525

[2] Bücking H, Liepold E, Ambilwade P. The role of the mycorrhizal symbiosis in nutrient uptake of plants and the regulatory mechanisms underlying these transport processes. Plant Science. 2012;**4**:108-132

[3] Tian F, Wang JC, Bai XX, Yang YB, Huang L, Liao XF. Symbiotic seed germination and seedling growth of mycorrhizal fungi in paphiopedilum hirsutissimun (Lindl.Ex hook.) stein from China. Plant Signaling & Behavior. 2023;**18**(1):2293405. DOI: 10.1080/15592324.2023.2293405

[4] Duckett JC, Ligrone R. A comparative cytological analysis of fungal endophytes in the sporophyte rhizomes and vascularised gametophytes of tmesipteris and psilotum. Canadian Journal of Botany. 2005;**83**:1443-1456

[5] Gorzelak MA, Asay AK, Pickles BJ, Simard SW. Inter-plant communication through mycorrhizal networks mediates complex adaptive behaviour in plant communities. AoB Plants. 2015;**7**. DOI: 10.1093/aobpla/plv050

[6] Cairney JW. Evolution of mycorrhiza systems. Die Naturwissenschaften. 2000;**87**(11):467-475. DOI: 10.1007/s001140050762

[7] Parniske M. Arbuscular mycorrhiza: The mother of plant root endosymbioses. Nature Reviews. Microbiology. 2008;**6**:763-775. DOI: 10.1038/nrmicro1987

[8] Genre A, Chabaud M, Timmers T, Bonfante P, Barker DG. Arbuscular mycorrhizal fungi elicit a novel intracellular apparatus in medicago truncatula root epidermal cells before infection. The Plant Cell. 2005;**17**:3489-3499

[9] Becard G, Kosuta S, Tamasloukht M, Séjalon-Delmas N, Roux C. Partner communication in the arbuscular mycorrhizal interaction. Canadian Journal of Botany. 2004;**82**:1186-1197

[10] Keymer A, Pimprikar P, Wewer V, Huber C, Brands M, Bucerius SL, et al. Lipid transfer from plants to arbuscular mycorrhiza fungi. eLife. 2017:6. DOI: 10.7554/eLife.29107. Published on: Jul 20, 2017

[11] Parihar SS, Dadlani M, Lal SK, Tonapi VA, Nautiyal PC, Basu S. Effect of seed moisture content and storage temperature on seed longevity of hemp (*Cannabis sativa* L.). Indian Journal of Agricultural Sciences. 2014;**84**(11):1303-1309

[12] Nautiyal PC, Sivasubramaniam K, Dadlani M. Seed dormancy and regulation of germination. In: Dadlani M, Yadava DK, editors. Seed Science and Technology. Singapore: Springer; 2023. DOI: 10.1007/978-981-19-5888-5_3

[13] MacGregor DR, Kendall SL, Florance H, Fedi F, Moore K, Paszkiewicz K, et al. Seed production temperature regulation of primary dormancy occurs through control of seed coat phenylpropanoid metabolism. The New Phytologist. 2015;**205**:642-652

[14] Al-Amri AA, Alsubaie QD, Alamri SA, Siddiqui MH. Strigolactone analog GR24 induces seed germination and improves growth performance of

different genotypes of tomato. Journal of Plant Growth Regulation. 2023;**42**:5653-5666. DOI: 10.1007/s00344-023-10947-8

[15] Saboor A, Ali MA, Danish S, Ahmed N, Fahad S, Datta R, et al. Effect of arbuscular mycorrhizal fungi on the physiological functioning of maize under zinc-deficient soils. Scientific Reports. 2021;**11**:18468. DOI: 10.1038/s41598-021-97742-1

[16] Miura C, Furui Y, Yamamoto T, Kanno Y, Honjo M, Yamaguchi K, et al. Autoactivation of mycorrhizal symbiosis signaling through gibberellin deactivation in orchid seed germination. Plant Physiology. 2024;**194**(1):546-563. DOI: 10.1093/plphys/kiad517

[17] Xing X, Ma X, Men J, Chen Y, Guo S. Phylogenetic constrains on mycorrhizal specificity in eight dendrobium (orchidaceae) species. Science China. Life Sciences. 2017;**60**(5):536-544

[18] Zhao Y, Cartabia A, Lalaymia I, Declerck S. Arbuscular mycorrhizal fungi and production of secondary metabolites in medicinal plants. Mycorrhiza. 2022;**32**:221-256. DOI: 10.1007/s00572-022-01079-0

[19] Jiang X, Zhao Z, Jacquemyn H, Ding G, Ding W, Xing X. Addition of fungal inoculum increases seed germination and protocorm formation in a terrestrial orchid. Global Ecology Conservation. 2022;**38**. DOI: 10.1016/j.gecco.2022.e02235

[20] Alghamdi SA. Influence of mycorrhizal fungi on seed germination and growth in terrestrial and epiphytic orchids. Saudi Journal of Biological Sciences. 2019;**26**(3):495-502. DOI: 10.1016/j.sjbs.2017.10.021

[21] Fernandez-Zarate FH, Huaccha-Castillo AE, Quiñones-Huatangari L, Vaca-Marquina SP, Sanchez-Santillan T, Morales-Rojas E, et al. Effect of arbuscular mycorrhiza on germination and initial growth of *Cinchona officinalis* L. (Rubiaceae). Forest Science and Technology. 2022;**18**(4):182-189. DOI: 10.1080/21580103.2022.2124318

[22] Hestrin R, Hammer EC, Mueller CW, Lehmann J. Synergies between mycorrhizal fungi and soil microbial communities increase plant nitrogen acquisition. Communications Biology. 2019;**2**:233. DOI: 10.1038/s42003-019-0481-8

[23] Begum N, Qin C, Ahanger MA, Raza S, Khan MI, Ashraf M, et al. Role of arbuscular mycorrhizal fungi in plant growth regulation: Implications in abiotic stress tolerance. Frontiers in Plant Science. 2019;**10**:1068. DOI: 10.3389/fpls.2019.01068

[24] Bano S, Ashfaq D. Role of mycorrhiza to reduce heavy metal stress. Natural Science. 2013;**5**:16-20. DOI: 10.4236/ns.2013.512A003

[25] Li W, Chen K, Li Q, Tang Y, Jiang Y, Su Y. Effects of arbuscular mycorrhizal fungi on alleviating cadmium stress in *Medicago truncatula* gaertn. Plants (Basel). 2023;**12**(3):547. DOI: 10.3390/plants12030547

[26] Tsulsiyah B, Farida T, Lembayung C, Semiarti E. Important role of mycorrhiza for seed germination and growth of dendrobium orchids. Research Journal of Botany. 2021;**6**:60805. DOI: 10.22146/jtbb.60805

[27] Li MH, Liu KW, Li Z, Lu HC, Ye QL, Zhang D, et al. Genomes of leafy and leafless platanthera orchids illuminate the evolution of mycoheterotrophy. Nature Plants. 2022;**8**:373-388

[28] Oliveira RS, Rocha I, Ma Y, Vosátka M, Freitas H. Seed coating with

arbuscular mycorrhizal fungi as an ecotechnological approach for sustainable agricultural production of common wheat (*Triticum aestivum* L.). Journal of Toxicology and Environmental Health. Part A. 2016;**79**(7):329-337. DOI: 10.1080/15287394.2016.1153448

[29] Lanfranco L, Bonfante P, Genre A. The mutualistic interaction between plants and arbuscular mycorrhizal fungi. Microbiology Spectrum. 2016;**4**(6): 349-366. DOI: 10.1128/microbiolspec. FUNK-0012-2016

[30] Artursson V, Finlay RD, Jansson JK. Interactions between arbuscular mycorrhizal fungi and bacteria and their potential for stimulating plant growth. Environmental Microbiology. 2006;**8**(1):1-10. DOI: 10.1111/j.1462-2920.2005.00942.x

[31] Bonfante P, Genre A. Mechanisms underlying beneficial plant–fungus interactions in mycorrhizal symbiosis. Nature Communications. 2010;**1**:48. DOI: 10.1038/ncomms1046

[32] Nanjundappa A, Bagyaraj DJ, Saxena AK, Kumar M, Chakdar H. Interaction between arbuscular mycorrhizal fungi and bacillus spp. in soil enhancing growth of crop plants. Fungal Biology and Biotechnology. 2019;**6**:23. DOI: 10.1186/ s40694-019-0086-5

[33] Fernández I, Cosme M, Stringlis IA, Yu K, De Jonge R, Van Wees SM, et al. Molecular dialogue between arbuscular mycorrhizal fungi and the nonhost plant Arabidopsis thaliana switches from initial detection to antagonism. New Phytologist. 2019;**223**(2):867-881. DOI: 10.1111/nph.15798

[34] Smith SE, Read DJ. Mycorrhizal Symbiosis. San Diego, CA: Academic Press; 2008

[35] MacLean AM, Bravo A, Harrison MJ. Plant signaling and metabolic pathways enabling arbuscular mycorrhizal Symbiosis. The Plant Cell. 2017;**29**(10):2 319-2335. DOI: 10.1105/tpc.17.00555

[36] Bouwmeester HJ, Roux C, Lopez-Raez JA, Bécard G. Rhizosphere communication of plants, parasitic plants and AM fungi. Trends in Plant Science. 2007;**12**(5):224-230. DOI: 10.1016/j. tplants.2007.03.009

[37] Ho-Plágaro T, García-Garrido JM. Molecular regulation of arbuscular mycorrhizal symbiosis. International Journal of Molecular Sciences. 2022;**23**(11):5960. DOI: 10.3390/ ijms23115960

[38] Zhang HZ, Cai XZ. Nonexpressor of pathogenesis-related genes 1 (NPR1): A key node of plant disease resistance signaling network. Sheng Wu Gong Cheng Xue Bao. 2005;**21**(4):511-515

[39] Berrocal-Lobo M, Molina A. Ethylene response factor 1 mediates *Arabidopsis* resistance to the soilborne fungus *Fusarium oxysporum*. Molecular Plant-Microbe Interactions. 2004;**17**(7):763-770. DOI: 10.1094/MPMI.2004.17.7.763

[40] Zhang Q, Wang S, Xie Q, Xia Y, Lu L, Wang M, et al. Control of arbuscule development by a transcriptional negative feedback loop in medicago. Nature Communications. 2023;**14**(1):5743

[41] Massalha H, Korenblum E, Tholl D, Aharoni A. Small molecules below-ground: The role of specialized metabolites in the rhizosphere. The Plant Journal. 2017;**90**(4):788-807

[42] Victor Roch GG, Maharajan T, Ceasar SA, Ignacimuthu S. The role of PHT1 family transporters in the acquisition and redistribution of

phosphorus in plants. Critical Reviews in Plant Sciences. 2019;**38**(3):171-198

[43] Oliveira TC, Cabral JS, Santana LR, Tavares GG, Santos LD, Paim TP, et al. The arbuscular mycorrhizal fungus rhizophagus clarus improves physiological tolerance to drought stress in soybean plants. Scientific Reports. 2022;**12**(1):9044. DOI: 10.1038/s41598-022-13059-7

[44] Kobayashi Y, Maeda T, Yamaguchi K, Kameoka H, Tanaka S, Ezawa T, et al. The genome of rhizophagus clarus HR1 reveals a common genetic basis for auxotrophy among arbuscular mycorrhizal fungi. BMC Genomics. 2018;**19**:465. DOI: 10.1186/s12864-018-4853-0

[45] Berger F, Gutjahr C. Factors affecting plant responsiveness to arbuscular mycorrhiza. Current Opinion in Plant Biology. 2021;**59**:101994

[46] Soliman S, Wang Y, Han Z, Pervaiz T, El-Kereamy A. Strigolactones in plants and their interaction with the ecological microbiome in response to abiotic stress. Plants. 2022;**11**(24):3499

[47] Zhang Y, Feng H, Druzhinina IS, Xie X, Wang E, Martin F, et al. Phosphorus/nitrogen sensing and signaling in diverse root–fungus symbioses. Trends in Microbiology. 2024;**32**(2):200-215

[48] Nambara E, Van Wees SC. Plant hormone functions and interactions in biological systems. The Plant Journal. 2021;**105**(2):287-289

[49] Gupta P, Singh SK. Gene regulatory networks: Current updates and applications in plant biology. Molecular Appl Plant Biology Environment Challenge. 2019:395-417. DOI: 10.1007/978-981-15-0690-1_18

[50] Shah NH, Kuriyan J. Understanding molecular mechanisms in cell signaling through natural and artificial sequence variation. Nature Structural & Molecular Biology. 2019;**26**(1):25-34

[51] Giri B, Kapoor R, Mukerji KG. Improved tolerance of Acacia nilotica to salt stress by arbuscular mycorrhiza, glomus fasciculatum, may be partly related to elevated K/Na ratios in root and shoot tissues. Microbial Ecology. 2007;**54**:753-760

[52] Chandrasekaran M, Boopathi T, Manivannan P. Comprehensive assessment of ameliorative effects of AMF in alleviating abiotic stress in tomato plants. Journal of Fungi. 2021;**7**(4):303

[53] Jiang QY, Zhuo F, Long SH, Zhao HD, Yang DJ, Ye ZH, et al. Can arbuscular mycorrhizal fungi reduce Cd uptake and alleviate Cd toxicity of lonicera japonica grown in Cd-added soils? Scientific Reports. 2016;**6**:21805

[54] Miller SP, Sharitz RR. Manipulation of flooding and arbuscular mycorrhiza formation influences growth and nutrition of two semiaquatic grass species. Functional Ecology. 2000;**14**:738-748

[55] Jahromi F, Aroca R, Porcel R, Ruiz-Lozano JM. Influence of salinity on the in vitro development of glomus intraradices and on the in vivo physiological and molecular responses of mycorrhizal lettuce plants. Microbial Ecology. 2008;**55**:45-53

[56] Leventis G, Tsiknia M, Feka M, Ladikou EV, Papadakis IE, Chatzipavlidis I, et al. Arbuscular mycorrhizal fungi enhance growth of tomato under normal and drought conditions, via different water regulation mechanisms. Rhizosphere. 2021;**19**:100394

[57] Li XL, Christie P. Changes in soil solution Zn and pH and uptake of Zn by arbuscular mycorrhizal red clover in Zn-contaminated soil. Chemosphere. 2001;**42**:201-207

[58] Khalvati M, Bartha B, Dupigny A, Schröder P. Arbuscular mycorrhizal association is beneficial for growth and detoxification of xenobiotics of barley under drought stress. Journal of Soils and Sediments. 2010;**10**:54-64

[59] Hashem A, Alqarawi AA, Radhakrishnan R, Al-Arjani AF, Aldehaish HA, Egamberdieva D, et al. Arbuscular mycorrhizal fungi regulate the oxidative system, hormones and ionic equilibrium to trigger salt stress tolerance in *Cucumis sativus* L. Saudi Journal of Biological Sciences. 2018;**25**:1102-1114

[60] Kaldorf M, Kuhn AJ, Schröder WH, Hildebrandt U, Bothe H. Selective element deposits in maize colonized by a heavy metal tolerance conferring arbuscular mycorrhizal fungus. Journal of Plant Physiology. 1999;**154**:718-728

[61] Kelkar TS, Bhalerao SA. Beneficiary effect of arbuscular mycorrhiza to Trigonella foenum-graceum in contaminated soil by heavy metal. Research Journal of Recent Sciences. 2013;**2**:29-32

[62] Elgharably A, Nafady NA. Inoculation with arbuscular mycorrhizae, Penicillium funiculosum and Fusarium oxysporum enhanced wheat growth and nutrient uptake in the saline soil. Rhizosphere. 2021;**18**:100345

[63] Lingua G, Franchin C, Todeschini V, Castiglione S, Biondi S, Burlando B, et al. Arbuscular mycorrhizal fungi differentially affect the response to high zinc concentrations of two registered poplar clones. Environmental Pollution. 2008;**153**:137-147

[64] Hajiboland R, Dashtebani F, Aliasgharzad N. Physiological responses of halophytic C4 grass, aeluropus littoralis to salinity and arbuscular mycorrhizal fungi colonization. Photosynthetica. 2015;**53**(4):572-584. DOI: 10.1007/s11099-015-0131-4

[65] Fougnies L, Renciot S, Müller F, Plenchette C, Prin Y, De Faria SM, et al. Arbuscular mycorrhizal colonization and nodulation improve flooding tolerance in pterocarpus officinalis Jacq. Seedlings. Mycorrhiza. 2007;**17**:159-166

[66] Amani Machiani M, Javanmard A, Morshedloo MR, Janmohammadi M, Maggi F. Funneliformis mosseae application improves the oil quantity and quality and eco-physiological characteristics of soybean (Glycine max L.) under water stress conditions. Journal of Soil Science and Plant Nutrition. 2021;**21**:3076-3090

[67] Solís-Rodríguez UR, Ramos-Zapata JA, Hernández-Cuevas L, Salinas-Peba L, Guadarrama P. Arbuscular mycorrhizal fungi diversity and distribution in tropical low flooding forest in Mexico. Mycological Progress. 2020;**19**:195-204

[68] Hashem A, Abd Allah E, Alqarawi A, Al Huqail AA, Egamberdieva D, Wirth S. Alleviation of cadmium stress in Solanum lycopersicum L. by arbuscular mycorrhizal fungi via induction of acquired systemic tolerance. Saudi Journal of Biological Sciences. 2016;**23**:272-281

[69] Liang SC, Jiang Y, Li MB, Zhu WX, Xu N, Zhang HH. Improving plant growth and alleviating photosynthetic inhibition from salt stress using AMF in alfalfa seedlings. Journal of Plant Interactions. 2019;**14**:482-491

[70] Shahvali R, Shiran B, Ravash R, Fallahi H, Đeri BB. Effect of symbiosis

with arbuscular mycorrhizal fungi on salt stress tolerance in GF677 (peach-almond) rootstock. Scientia Horticulturae. 2020;**272**:109535

[71] Parihar M, Rakshit A, Rana K, Tiwari G, Jatav SS. The effect of arbuscular mycorrhizal fungi inoculation in mitigating salt stress of pea (*Pisum sativum* L.). Communications in Soil Science and Plant Analysis. 2020;**51**:1545-1559

[72] Mirshad P, Puthur JT. Arbuscular mycorrhizal association enhances drought tolerance potential of promising bioenergy grass (Saccharum arundinaceum retz.). Environmental Monitoring and Assessment. 2016;**188**:425

[73] Rani B. Effect of Arbuscular Mycorrhiza Fungi on Biochemical Parameters in Wheat (*Triticum aestivum* L.) under Drought Conditions [Doctoral Dissertation]. Hisar, India: CCSHAU; 2016

[74] Pal A, Pandey S. Role of arbuscular mycorrhizal fungi on plant growth and reclamation of barren soil with wheat (*Triticum aestivum* L.) crop. International. Journal of Soil Science. 2016;**12**:25-31

[75] Bitterlich M, Franken P, Graefe J. Arbuscular mycorrhiza improves substrate hydraulic conductivity in the plant available moisture range under root growth exclusion. Frontiers in Plant Science. 2018;**9**:301. DOI: 10.3389/fpls.2018.00301

[76] Singh M, Bisht S, Singh S, Sharma JG. Implications of abiotic stress tolerance in arbuscular mycorrhiza colonized plants: Importance in plant growth and regulation. Journal of Applied Biology & Biotechnology. 2022;**10**(6):1-11

[77] Bever JD. Negative feedback within a mutualism: Host-specific growth of mycorrhizal fungi reduces plant benefit. Proceedings of the Biological Sciences. 2002;**269**(1509):2595-2601

[78] van der Heijden MGA, Bardgett RD, van Straalen NM. The unseen majority: Soil microbes as drivers of plant diversity and productivity in terrestrial ecosystems. Ecology Letters. 2008;**11**(3):296-310

[79] Brundrett MC. Mycorrhizal associations and other means of nutrition of vascular plants: Understanding the global diversity of host plants by resolving conflicting information and developing reliable means of diagnosis. Plant and Soil. 2009;**320**:37-77

[80] Smith FA, Smith SE. Roles of arbuscular mycorrhizas in plant nutrition and growth: New paradigms from cellular to ecosystem scales. Annual Review of Plant Biology. 2011;**62**:227-250

[81] Johnson NC, Graham JH, Smith FA. Functioning of mycorrhizal associations along the mutualism–parasitism continuum. The New Phytologist. 1997;**135**(4):575-586

[82] Van der Heijden MGA, Horton TR. Socialism in soil? The importance of mycorrhizal fungal networks for facilitation in natural ecosystems. Journal of Ecology. 2009;**97**(6):1139-1150

[83] Hart MM, Reader RJ. Host plant benefit from association with arbuscular mycorrhizal fungi: Variation due to differences in experimental conditions. Mycorrhiza. 2002;**12**:273-279

[84] Thirkell TJ, Charters MD, Elliott AJ, Sait SM, Field KJ. Ecological solutions to global food security: Are mycorrhizal fungi our sustainable saviours? Considerations for achieving food security. Journal of Ecology. 2017;**105**:921-929. DOI: 10.1111/1365-2745.12788

[85] Coelho LCS, Mignoni DSB, Silva FSB, Braga MR. Seed exudates of Sesbania virgata (Cav.) Pers. stimulate the asymbiotic phase of the arbuscular mycorrhizal fungus Gigaspora albida Becker & Hall. Hoehnea. 2019;**46**:e272018. DOI: 10.1590/2236-8906-27/2018

[86] Berruti A, Lumini E, Balestrini R, Bianciotto V. Arbuscular mycorrhizal fungi as natural biofertilizers: let's benefit from past successes. Frontiers in Microbiology. 2016;**6**:1559. DOI: 10.3389/fmicb.2015.01559

[87] Giovannini L, Palla M, Agnolucci M, Avio L, Sbrana C, Turrini A, et al. Arbuscular mycorrhizal fungi and associated microbiota as plant biostimulants: Research strategies for the selection of the best performing inocula. Agronomy. 2020;**10**(1):106. DOI: 10.3390/agronomy10010106

[88] Baum C, El-Tohamy W, Gruda N. Increasing the productivity and product quality of vegetable crops using arbuscular mycorrhizal fungi: A review. Scientia Horticulturae. 2015;**187**:131-141

[89] Püschel D, Bitterlich M, Rydlová J. Facilitation of plant water uptake by an arbuscular mycorrhizal fungus: A Gordian knot of roots and hyphae. Mycorrhiza. 2020;**30**:299-313. DOI: 10.1007/s00572-020-00949-9

[90] Augé RM, Toler HD, Saxton AM. Arbuscular mycorrhizal symbiosis alters stomatal conductance of host plants more under drought than under amply watered conditions: A meta-analysis. Mycorrhiza. 2015;**25**:13-24. DOI: 10.1007/s00572-014-0585-4

[91] Zhang Q, Zhu D, Ding J, Zheng F, Zhou S, Lu T, et al. The fungicide azoxystrobin perturbs the gut microbiota community and enriches antibiotic resistance genes in enchytraeus crypticus. Environment International. 2019;**131**:104965. DOI: 10.1016/j.envint.2019.104965

[92] Jayne B, Quigley M. Influence of arbuscular mycorrhiza on growth and reproductive response of plants under water deficit: A meta-analysis. Mycorrhiza. 2014;**24**:109-119

[93] Sato T, Ezawa T, Cheng WG, Tawaraya K. Release of acid phosphatase from extraradical hyphae of arbuscular mycorrhizal fungus rhizophagus clarus. Soil Science & Plant Nutrition. 2015;**61**(2):269-274. DOI: 10.1080/00380768.2014.993298

[94] Cao Y, Wu X, Zhukova A, Tang Z, Weng Y, Li Z, et al. Arbuscular mycorrhizal fungi (AMF) species and abundance exhibit different effects on saline-alkaline tolerance in Leymus chinensis. Journal of Plant Interactions. 2020;**15**(1):266-279. DOI: 10.1080/17429145.2020.1802524

[95] Etesami H, Jeong BR, Glick BR. Contribution of arbuscular mycorrhizal fungi, phosphate-solubilizing bacteria, and silicon to P uptake by plant. Frontiers in Plant Science. 2021;**12**:699618. DOI: 10.3389/fpls.2021.699618

[96] Thirkell TJ, Cameron DD, Hodge A. Resolving the 'nitrogen paradox' of arbuscular mycorrhizas: Fertilization with organic matter brings considerable benefits for plant nutrition and growth. Plant, Cell & Environment. 2016;**39**:1683-1690

[97] Zhao R, Guo W, Bi N, Guo J, Wang L, Zhao J, et al. Arbuscular mycorrhizal fungi affect the growth, nutrient uptake and water status of maize (Zea mays L.) grown in two types of coal mine spoils under drought stress. Applied Soil Ecology. 2015;**88**:41-49. DOI: 10.1016/j.apsoil.2014.11.016

[98] Johnson NC, Wilson GWT, Wilson JA, Miller RM, Bowker MA. Mycorrhizal phenotypes and the law of the minimum. The New Phytologist. 2015;**205**:1473-1484

[99] Baslam M, Esteban R, García-Plazaola JI, Goicoechea N. Effectiveness of arbuscular mycorrhizal fungi (AMF) for inducing the accumulation of major carotenoids, chlorophylls and tocopherol in green and red leaf lettuces. Applied Microbiology and Biotechnology. 2013;**97**(7):3119-3128. DOI: 10.1007/s00253-012-4526-x

[100] Elbon A, Whalen JK. Phosphorus supply to vegetable crops from arbuscular mycorrhizal fungi: A review. Biological Agriculture and Horticulture. 2015;**31**:73-90

[101] Hijri M. Analysis of a large dataset form field mycorrhizal inoculation trials on potato showed highly significant increase in yield. Mycorrhiza. 2016;**26**:209-214. DOI: 10.1007/s00572-015-0661-4

[102] Lu F, Lee C, Wang C. The influence of arbuscular mycorrhizal fungi inoculation on yam (Dioscorea spp.) tuber weights and secondary metabolite content. Peer J. 2015;**3**:e1266. DOI: 10.7717/peerj.1266

[103] Chen M, Arato M, Borghi L, Nouri E, Reinhardt D. Beneficial services of arbuscular mycorrhizal fungi—From ecology to application. Frontiers in Plant Science. 2018;**9**:1270. DOI: 10.3389/fpls.2018.01270

[104] Bardgett RD, van der Putten WH. Belowground biodiversity and ecosystem functioning. Nature. 2014;**515**:505-511

[105] Wang Y, Wang M, Li Y. Effects of arbuscular mycorrhizal fungi on growth and nitrogen uptake of chrysanthemum morifolium under salt stress. PLoS One. 2018;**13**:e0196408

[106] Sbrana C, Avio L, Giovannetti M. Beneficial mycorrhizal symbionts affecting the production of health promoting phytochemicals. Electrophoresis. 2014;**35**:1535-1546

[107] Getman-Pickering ZL, Stack GM, Thaler JS. Fertilizer quantity and type alter mycorrhizae-conferred growth and resistance to herbivores. Journal of Applied Ecology. 2021;**58**:931-940. DOI: 10.1111/1365-2664.13833

[108] Jarosz Z, Michałojć Z, Pitura K, Dzida K, Koter M. Influence of fertilization and mycorrhizae on the nutritional status of rhododendron (rhododendron hybridum) in a nursery. Agriculture. 2021;**11**:538. DOI: 10.3390/agriculture11060538

[109] García-Caparrós P, Lao MT, Preciado-Rangel P, Sanchez E. Phosphorus and carbohydrate metabolism in green bean plants subjected to increasing phosphorus concentration in the nutrient solution. Agronomy. 2021;**11**(2):245. DOI: 10.3390/agronomy11020245

[110] Zhu XC, Song FB, Liu FL, Liu S, Tian C. Carbon and nitrogen metabolism in arbuscular mycorrhizal maize plants under low-temperature stress. Crop & Pasture Science. 2015;**66**(1):62-70. DOI: 10.1071/CP14159

[111] Battini F, Grønlund M, Agnolucci M, Jakobsen I. Facilitation of phosphorus uptake in maize plants by mycorrhizosphere bacteria. Scientific Reports. 2017;**7**:4686. DOI: 10.1038/s41598-017-04959-0

[112] Wilkes TI, Warner DJ, Edmonds-Brown V, Davies KG, Denholm I. Zero tillage systems conserve arbuscular

mycorrhizal fungi, enhancing soil glomalin and water stable aggregates with implications for soil stability. Soil System. 2021;**5**(1):4. DOI: 10.3390/soilsystems5010004

[113] Njeru EM, Avio L, Sbrana C, Turrini A, Bocci G, Barberi P, et al. First evidence for a major cover crop effect on arbuscular mycorrhizal fungi and organic maize growth. Agronomy for Sustainable Development. 2014;**34**:841-848. DOI: 10.1007/s13593-013-0197-y

[114] Mathur S, Sharma MP, Jajoo A. Improved photosynthetic efficacy of maize (Zea mays) plants with arbuscular mycorrhizal fungi (AMF) under high temperature stress. Journal of Photochemistry and Photobiology. 2018;**180**:149-154

[115] Higo M, Isobe K, Miyazawa Y, Matsuda Y, Drijber RA, Torigoe Y. Molecular diversity and distribution of indigenous arbuscular mycorrhizal communities colonizing roots of two different winter cover crops in response to their root proliferation. Journal of Microbiolial Ecology. 2016;**54**:86-97. DOI: 10.1007/s12275-016-5379-2

[116] Higo M, Isobe K, Kondo T, et al. Temporal variation of the molecular diversity of arbuscular mycorrhizal communities in three different winter cover crop rotational systems. Biology and Fertility of Soils. 2015;**51**:21-32. DOI: 10.1007/s00374-014-0945-4

[117] Maiti D, Singh RK, Variar M. Rice-based crop rotation for enhancing native arbuscular mycorrhizal (AM) activity to improve phosphorus nutrition of upland rice (*Oryza sativa* L.). Biology and Fertility of Soil. 2012;**48**(1):67-73. DOI: 10.1007/s00374-011-0609-6

[118] Watanarojanaporn N, Boonkerd N, Tittabutr P, Longtonglang A, Young JP, Teaumroong N. Effect of rice cultivation systems on indigenous arbuscular mycorrhizal fungal community structure. Microbes and Environments. 2013;**28**(3):316-324. DOI: 10.1264/jsme2.ME13011

[119] Ruiz-Sánchez M, Aroca R, Muñoz Y, Polón R, Ruiz-Lozano JM. The arbuscular mycorrhizal symbiosis enhances the photosynthetic efficiency and the antioxidative response of rice plants subjected to drought stress. Journal of Plant Physiology. 2010;**167**(11):862-869. DOI: 10.1016/j.jplph.2010.01.018

[120] Chu Q, Wang X, Yang Y, Chen F, Zhang F, Feng G. Mycorrhizal responsiveness of maize (Zea mays L.) genotypes as related to releasing date and available P content in soil. Mycorrhiza. 2013;**23**:497-505. DOI: 10.1007/s00572-013-0492-0

[121] Ait Lahmidi N, Courty PE, Brulé D, Chatagnier O, Arnould C, Doidy J, et al. Sugar exchanges in arbuscular mycorrhiza: riMST5 and RiMST6, two novel rhizophagus irregularis monosaccharide transporters, are involved in both sugar uptake from the soil and from the plant partner. Plant Physiology and Biochemistry. 2016;**107**:354-363. DOI: 10.1016/j.plaphy.2016.06.023

[122] Sawers RJH, Svane SF, Quan C, Grønlund M, Wozniak B, Gebreselassie MN, et al. Phosphorus acquisition efficiency in arbuscular mycorrhizal maize is correlated with the abundance of root-external hyphae and the accumulation of transcripts encoding PHT1 phosphate transporters. The New Phytologist. 2017;**214**(2):632-643. DOI: 10.1111/nph.14403

[123] Qin H, Lu K, Strong PJ, Xu Q, Wu Q, Xu Z, et al. Long-term fertilizer application effects on the soil, root arbuscular mycorrhizal fungi and

community composition in rotation agriculture. Agriculture, Ecosystems & Environment Appl Soil Ecol. 2015;**89**:35-43. DOI: 10.1016/j.apsoil.2015.01.008

[124] Dai M, Hamel C, Bainard LD, Arnaud MS, Grant CA, Lupwayi NZ, et al. Negative and positive contributions of arbuscular mycorrhizal fungal taxa to wheat production and nutrient uptake efficiency in organic and conventional systems in the Canadian prairie. Soil Biology and Biochemistry. 2014;**74**:156-166. DOI: 10.1016/j.soilbio.2014.03.016

[125] Cruz-Paredes C, López-García Á, Rubæk GH, Hovmand MF, Sørensen P, Kjøller R. Risk assessment of replacing conventional P fertilizers with biomass ash: Residual effects on plant yield, nutrition, cadmium accumulation and mycorrhizal status. Science Total Environment. 2017;**575**:1168-1176. DOI: 10.1016/j.scitotenv.2016.09.194

[126] Aguilera P, Marín C, Oehl F, Godoy R, Borie F, Cornejo P. Selection of aluminum tolerant cereal genotypes strongly influences the arbuscular mycorrhizal fungal communities in an acidic andosol. Agriculture, Ecosystems and Environment. 2017;**246**:86-93. DOI: 10.1016/j.agee.2017.05.031

[127] Castillo CG, Borie F, Oehl F, Sieverding E. Arbuscular mycorrhizal fungi biodiversity: Prospecting in southern-central zone of Chile. A review. Journal of Soil Science and Plant Nutrition. 2016a;**16**:400-422. DOI: 10.4067/S0718-95162016005000036

[128] Castillo CG, Oehl F, Sieverding E. Arbuscular mycorrhizal fungal diversity in wheat agro-ecosystems in southern Chile and effects of seed treatment with natural products. Journal of Soil Science and Plant Nutrition. 2016b;**16**:967-978. DOI: 10.4067/S0718-95162016005000069

[129] Sarkodee-Addo E, Yasuda M, Lee CG, Kanasugi M, Fujii Y, Omari RA, et al. Arbuscular mycorrhizal fungi associated with rice (*Oryza sativa* L.) in Ghana: Effect of regional locations and soil factors on diversity and community assembly. Agronomy. 2020;**10**:559. DOI: 10.3390/agronomy10040559